S0-BWX-240

SYMPOSIA OF THE SOCIETY FOR EXPERIMENTAL BIOLOGY

NUMBER III

Other Publications of the Company of Biologists

THE JOURNAL OF EXPERIMENTAL BIOLOGY
THE QUARTERLY JOURNAL OF MICROSCOPIC SCIENCE

SYMPOSIUM NUMBER I: NUCLEIC ACID

SYMPOSIUM NUMBER II: GROWTH IN RELATION TO DIFFERENTIATION AND MORPHOGENESIS

CA(AESL)
3/5/63
48990

SYMPOSIA OF THE
SOCIETY FOR EXPERIMENTAL BIOLOGY

NUMBER III

SELECTIVE TOXICITY
AND
ANTIBIOTICS

Published for the Company of Biologists
on behalf of the Society for Experimental Biology

CAMBRIDGE: AT THE UNIVERSITY PRESS

1949

AGRI-
CULTURAL
LIBRARY

PUBLISHED BY
THE SYNDICS OF THE CAMBRIDGE UNIVERSITY PRESS

London Office: Bentley House, N.W. 1

Agents for U.S.A., Academic Press Inc.,
125 East 23rd Street, New York 10
Agents for Canada, India, and Pakistan: Macmillan

Printed in Great Britain at the University Press, Cambridge
(Brooke Crutchley, University Printer)

CONTENTS

Agriculture Blackwell 1-14-63

48490

PREFACE

This number contains most of the papers read at a Symposium of the Society for Experimental Biology, which was held at Edinburgh in July 1948. It is the third of an annual series of symposium reports.

The Symposium for 1949 will be on 'Physiological Mechanisms in Animal Behaviour'. It will be held at Cambridge.

The Society is deeply indebted to Imperial Chemical Industries Ltd., for a grant in aid of publication. We are also most appreciative of the facilities and help made available to us by members of Edinburgh University.

The editors wish to thank the Cambridge University Press for the kindness and courtesy with which they have assisted us in producing this report.

J. F. Danielli
R. Brown

Honorary Secretaries
Society for Experimental Biology

THE ORGANIC CHEMIST'S APPROACH TO CHEMOTHERAPY

By W. A. SEXTON

Imperial Chemical Industries Ltd., Research Laboratories, Blackley, Manchester

The invention of a new chemotherapeutic agent involves a search for a compound which combines in the best possible manner a number of essential properties. First, it must possess the requisite type of biological activity, that is to say, it must be capable of attacking the parasite or interfering with the function concerned. Secondly, it must have selective action, at least to the extent that it is not toxic to the treated subject. Thirdly, it must have properties which enable it to be administered in a reasonable dose, capable of being absorbed by the subject, transported to the site of action and able to persist for a sufficient time to produce its characteristic effect. All these properties may be related to the one fundamental subject which is of prime interest to the organic chemist, namely, the relationship between the chemical constitution and the behaviour of a molecule in a living system. It follows that knowledge gained in related fields such as the study of insecticides, or in applied botany, is fully relevant to the subject of chemotherapy.

In this paper, therefore, the interpretation of biological activity and the design of biologically active molecules will be considered as a branch of applied chemistry. Of all the branches of applied chemistry, the synthesis of biologically active substances is perhaps unique in the important respect that it is concerned with dynamic as opposed to static chemical systems. It is necessary at the outset to emphasize this point, for it colours the whole consideration of the subject. Recent metabolic studies using isotopic elements have emphasized the important fact that many of the chemical constituents of living matter are in a constant state of change, in that their synthesis and breakdown occur continually throughout life. It is not known whether this dynamic picture applies to certain structural elements upon which an organism depends for its mechanical strength, and of course materials such as food and vitamins which are provided from external sources are subject only to the processes associated with their utilization and in this sense are not classed as body constituents. With these exceptions, however, it seems that the dynamic conception is a general one. A natural product must be considered therefore not as having a static structure, but must be regarded in relation to its biogenetic chemical precursors, its chemical reactivity and its degradation products.

In attempting a speculative analysis of the effects of structural variation on the selective toxicity of organic compounds towards living organisms, two postulates will be made. The first postulate is that the biological activity of a substance is due to its combination, whether it is a natural constituent of the organism or whether it is a drug, with some substance or substances in the organism; and that this combination is either responsible for maintaining the normal balance of the dynamic chemical processes of the organism, or in the case of a drug, results in upsetting these processes. The term combination is to be interpreted widely as comprising unions of varying nature and force. At the one extreme there are the powerful unions formed through covalent links which unite the drug molecule with the cell constituent through reactions which are practically irreversible. As an example may be cited the union of a heavy metallic ion with a thiol group. At the other extreme there are the weaker forces which cause loose attachment such as van der Waals forces and those due to polar interaction and which, at least where small molecules containing few such linkages are concerned, are easily broken. In such cases, biological activity is often only of a temporary nature, being lost when the invading molecules are eliminated. This applies, for example, to the narcotic action of certain simple compounds such as ether and chloroform. Between these two extremes lies a gradation of types of combination including the relatively firm unions which result from the combination of macromolecular species in stereochemical juxtaposition, so that the molecules are held firmly together by a multiplicity of points of contact at each one of which the force is only small. It is this complex type of multipoint union which determines the specific properties of proteins and complex carbohydrates and which governs the phenomena of enzymology and immunochemistry. This will be referred to again later.

It is next postulated that the combination of biologically active molecule with cell constituent may be modified to a greater or lesser degree by variation of the physico-chemical properties of the active molecule. This variation may be deliberately brought about by structural alterations and can have two major effects. First, it can modify the chemical reactivity or the capacity for combination with cell constituents quantitatively, and secondly, it can for various reasons either prevent or facilitate access of the molecule to the cell constituent concerned or it may render the access feasible only in such organisms as are possessed of the appropriate chemico-morphological constitution. Thus the variation of physico-chemical properties through structural modifications becomes a basis for the selection of susceptible species, for specificity of action between different organs of the more complicated biological structures and for the selection of the most potent individual compound within a chemical class. One of the most familiar examples of the effects of physical properties in this respect is the variation of biological activity within members of homologous series, some typical examples of which are given in Table 1. The usual effect in an homologous series is a gradual rise to maximum activity at a certain chain length, followed by a fall as the series is ascended further (curve *A* in Fig. 1). This peak effect is often due to the fact that at a certain point, solubility in water becomes a limiting factor. Sometimes the 'peak' may be as low as one or two carbon atoms (curve *B* in Fig. 1), that is to say, the first part of the curve *A* is missing. A curve such as *A* is perhaps typical of the modification of biological activity through the interaction of two opposing factors, and a similarly shaped curve was obtained by Bell & Roblin (1942), who plotted the *in vitro* bactericidal activity of sulphonamide drugs against their acid dissociation constants.

The variation of activity as an homologous series is ascended may on occasion take on more complex patterns, due to the complicating effect of other factors. Thus Stiles & Rees (1935) observed that the toxicity of *n* fatty acids to potato tuber slices decreased gradually from C_1 to C_6 and then increased rapidly to C_9. This result was attributed to the additive contributions of two factors to the overall toxic effect, the toxicity of hydrogen ions and the toxicity of the undissociated molecule. In formic acid the effect was mainly due to hydrogen ions, and through the first half-dozen members of the series, this predominance of the hydrogen-ion contribution gradually lessened. After this, the main contribution was from the undissociated molecule, and activity rose with molecular weight to C_9. Higher acids were not examined, but doubtless a peak would have been reached at a certain chain length (see curve *C* in Fig. 1, the broken part of which is hypothetical).

Another effect in homologous series is what may be termed the odd-even effect, the alternation of activity between successive members of a series. Synerholm & Zimmerman (1947) observed that in the series

$$\text{Cl}\langle\text{C}_6\text{H}_3(\text{Cl})\rangle\text{O[CH}_2]_n\text{COOH},$$

auxin-like activity was found only in those compounds in which *n* was an odd number, and it was suggested that this might be explained by the fact that only with odd-numbered compounds could β-oxidation give rise to 2:4-dichlorophenoxyacetic acid, which was believed to be the true active agent in each case. A similar alternation in activity has been observed with certain antimalarial drugs.

Table 1. *Variation of biological activity in homologous series*

General formula ($R = n$-alkyl)	Biological response	Maximum activity (no. of C atoms)
$R.SCN$	Toxicity to aphids Toxicity to animals	10–14 1
2-SR-benzothiazole (benzothiazole ring with S, N and SR)	Toxicity to flies Fungistatic activity	2–3 1
$R.\overset{+}{N}Me_3$	Bacteriostasis Curare effect (frog)	10–12 2–3
$RO{-}C_6H_4{-}CH(CCl_3){-}C_6H_4{-}OR$	Contact insecticide	1
$R{-}C_6H_4{-}OH$	Bactericide ('Phenol coefficient')	5
$RO.CSSNa$	Fungicide	1
$C_6H_5{-}NH.COOR$	Inhibition of oat-seed germination	2
$C_{10}H_7{-}OCH_2COOR$	Inhibition of rape-seed germination	3
$Cl{-}C_6H_4{-}NH.C({=}NH).NH.C({=}NH).NHR$	Antimalarial	3
$HN{=}C(NH_2)(CH_2)_nC({=}NH)NH_2$	Trypanocide	$n = 10–14$
$(RO)_2P({=}O)F$	Inhibition of cholinesterase	Not less than 3

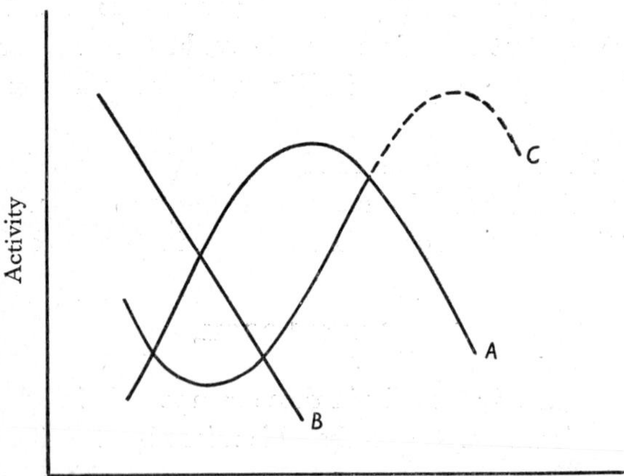

Fig. 1. Figure illustrating types of variation of activity as an homologous series is ascended.

The lesson of this for the organic chemist is that in synthesizing new molecular species for biological evaluation, it is usually not sufficient to examine a single member as representative of a chemical type, for a variation of physical properties might make all the difference between success and failure. This has been well illustrated in various fields, e.g. the thiocyanate insecticides, the phenoxyacetic acid herbicides and the diguanide antimalarials. The other point requiring emphasis is that variation of physical properties can become the basis for selective toxicity. To the examples given in the table may be added a few more in further illustration.

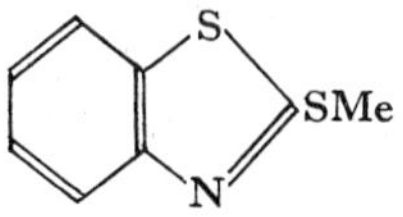

S SMe N

I

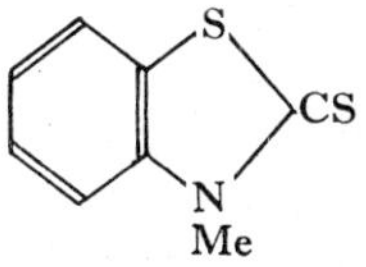

S CS N Me

II

The *S*-methyl derivative of 2-mercaptobenzthiazole (I) is toxic to certain insects but not to others. The isomeric *N*-methyl compound (II) differs from it in physical properties, but in their capacity for combining with cell constituents the two isomers would not be considered to be very different. The susceptibility of insect species is quite different for the two compounds (Davies & Sexton, 1948). The various penicillins differ not so much in their chemical reactivity as in their physical properties, and they are characterized biologically by different bacterial 'spectra'. At a concentration of 0·05%, emulsified hexyl thiocyanate caused 63% mortality of black chrysanthemum aphids, but at this concentration, the mortality of green chrysanthemum aphids or of red spider was very little more than that brought about by the emulsifier (potassium oleate) alone (Bousquet, Salzberg & Dietz, 1935).

We have considered above an analysis of the contributions towards biological activity of 'physical' and 'chemical' factors. This is of course an over-simplification, for the two groups of factors are by no means independent and there is no hard and fast line of demarcation between them. For this reason the terms 'physical' and 'chemical' are given in inverted commas. Nevertheless, it is a convenient conception in quite a large number of instances. Where this type of analysis is applicable, the overall biological effect can be seen to be due to an appropriate balance of 'physical' and 'chemical' effects. There are certain instances, however, in which either the 'chemical' or the 'physical' effect is dominant. If a markedly reactive grouping is contained in a small molecule which is at the same time devoid of other groupings likely to confer distinctive properties in relation to reactions with cell constituents, then biological activity may be expected to be manifested as a general toxic effect. In other words, dominance of the 'chemical' effect will result in general toxicity. Only when 'physical' effects become significant will marked specificity arise. Three simple examples are the lower alkylmercury salts, methyl thiocyanate and quinone. Methyl thiocyanate is generally toxic, but as the series is ascended one arrives at members in which the physical properties determine the specific toxicity to certain insects and mites. In the same way, quinone is generally toxic, but with the elaboration of the molecule by the introduction of appropriate substituent groups and rings one arrives at substances with characteristic activities such as the K vitamins, echinochrome, selective bactericides and certain anthelmintics. Similarly, there are cases in which 'physical' effects so dominate the situation that the reactivity of the substance with cell constituents is difficult to discern. Again, the 'physical' effects may have such overriding importance that two drugs of appropriate physical properties may produce similar biological effects, even though their chemically reactive groups are of a fundamentally different nature. This may be illustrated by examples. There are certain simple organic molecules, not highly reactive in the conventional chemical sense, but which nevertheless exhibit marked and characteristic biological activity. Such substances as the anaesthetics ethylene, *cyclo*propane, ether and chloroform come into this category, and it is necessary to consider how they may conform to the first postulate, namely, that they combine with cell constituents. It has been emphasized that combination with cell constituents can include unions of a loose nature, and herein must be found the reconciliation of the biological activity of these substances with this first postulate. The aliphatic hydrocarbon chains, for example, will have a capacity for attachment through van der Waals forces to lipophilic groups in macromolecular species, for example, to certain side-chains of amino-acid components of polypeptides. Similar attachments may be formed by such molecules as ether and chloroform. These attachments may well be sufficient to distort the delicate framework of macromolecular architecture and thus to interfere with dynamic function. In such cases the effects of variation in physical properties will be completely dominant, and it follows that groups of structurally unrelated chemical compounds, provided that they contain no grouping of outstanding chemical reactivity, may produce similar biological effects. Good examples of this phenomenon are provided by Ferguson (1939). Again, in their insecticidal action against blowflies, the *n*-alkyl thiocyanates *R*.SCN show maximum activity at about C_{10}–C_{14}. The 2-alkylthiobenzthiazoles have greatest activity at C_2–C_3. The characteristic chemical reactivity of the thiocyanates is due to the affinity of the carbon atom for electrons. On the other hand, the 2-alkylthiobenzthiazoles are electron-donating. Thus in the case of the two insecticides, dodecylthiocyanate and

2-ethylthiobenzthiazole, one has a reactive group which is fundamentally acidic in nature and the other has a basic group. What is important, however, is that, given the appropriate physical properties, e.g. fat solubility or fat/water partition, there must also apparently be a group capable of combining to an appropriate degree of firmness with some cell constituent (Davies & Sexton, 1948). If the functional group is not sufficiently reactive, or if it is too reactive, activity will be reduced, e.g. lauronitrile, lauric acid, dodecyl alcohol and dodecyl *iso*thiocyanate. Perhaps in the latter instance the reactive NCS group is fixed in a harmless manner before the molecule can reach a site where combination would result in a toxic effect.

Small molecules of low chemical reactivity, that is to say, small molecules which can become attached to cell constituents by forces less than those of covalence, may be expected to produce general effects, and whether or not activity is exhibited in any particular organism will be more dependent on the precise biochemical characteristics of the organism than is the case with the more reactive molecules. With a highly reactive molecule there are many points in cell constituents for chemical attack. With a less reactive one, the points of attack are less numerous, and significant combination with cell constituents will be to a large extent governed by the details of macromolecular architecture. The effects will be more easily abolished on removal of the drug, and there may be mentioned narcosis and bacteriostasis as easily reversible conditions brought about by simple molecules, such as, for example, ether, urethane and phenol. Specificity of action is here greatly dependent on the physical properties of the molecule, and this is a clear case of the dominance of the 'physical' factor. It might be argued that the characteristic effects of 'physical' dominance, e.g. the homologous series effect coupled with a generalized biological activity (toxicity) could be taken as diagnostic of an interference with macromolecular function.

Generally speaking, the larger or more complex the molecule, the greater is the likelihood of its biological activity being of a highly specific nature. This follows from the general argument that biological activity results from an appropriate combination of physical properties with chemical reactivity. In a complex molecule quite minor structural variations may affect both properties profoundly and thus upset the balance between them. This finds its maximum expression in the highly specific activities of enzymes and other proteins and the phenomena of immunochemistry, but it is also exhibited in molecules of much less than macromolecular complexity. There may be mentioned in this connexion the vitamins and coenzymes in which little structural variation is permissible, the synthetic oestrogens and the insecticide, γ-hexachloro*cyclo*hexane, where activity is associated with a precise stereochemical arrangement, and the highly specific antiprotozoal drugs such as the diguanide and heterocyclic types.

No account of the structural basis of selective toxicity would be complete without at least brief reference to the brilliant work of Landsteiner and of Pauling (see particularly Landsteiner (1936) and Pauling (1940)) on the significance of molecular structure in relation to serological reactions. This properly belongs to the subject of this paper, for the responses of an animal to foreign proteins introduced into the blood either by parenteral injection or formed there as a result of the interactions of a foreign substance of relatively low molecular weight with the body proteins comprise an exhibition of selective toxicity having the highest possible degree of specificity. Here there is a formidable body of evidence to show that the biological activity of a protein is associated primarily with its stereochemical structure, and experimental demonstrations of this have been achieved by modifying proteins through the attachment by means of firm chemical unions, e.g. the azo group and the peptide or ureido groups, of aromatic nuclei bearing chosen substituents. The specificity of action of such modified proteins has been shown to be associated with the nature of their molecular surfaces, and to be in accord with the view that they owe their behaviour to the precise arrangement of the multiple points of attachment to other large molecules. All this work is at present in the stage where a theoretical explanation of serological reactions, based upon considerations of the structure of atoms and molecules, may be advanced with considerable confidence. One cannot help feeling that the further advancement of this subject will eventually provide new principles and new techniques whereby medicinally useful substances can be manufactured in the laboratory without recourse to biological methods of production.

It is not always possible by examining the structure of a biologically active molecule to analyse the contributions made by the component groupings to the 'physical' and 'chemical' aspects of its properties, for it is perhaps the rule rather than the exception that molecular changes may affect both physical properties and the capacity for combining with cell constituents at the same time. Such dual alterations may be mutually complementary or mutually antagonistic in respect of the overall biological effect, and it cannot be too strongly emphasized that the decisive factor is the behaviour of the molecule as a whole. This is well exemplified in the rival views concerning the mode of action of the insecticide D.D.T. (III).

Cl–C_6H_4–CH(CCl_3)–C_6H_4–Cl

III

It is generally agreed that in this (and also in certain other contact insecticides) activity is associated with affinity for lipoids combined with a capacity for chemical reactivity. Molecular alteration affects both factors, and neither the highest reactivity nor the highest lipoid solubility is always to be associated with maximum insecticidal potency (see, for example, Kirkwood & Phillips (1946), who give references to earlier papers). Reconciliation of the apparently contradictory results can probably be achieved by considering the molecule as a whole, for what is required is probably an appropriate balance of these two properties.

The importance of the effect of structural modifications on physicochemical properties having been stressed, attention is now directed to other methods of achieving selective toxicity. It is obvious that if an organism is dependent for its functioning on a particular enzyme reaction to the extent that it is to some degree biochemically unique, then this particular enzyme reaction can be made the basis of a specific attack upon the organism or process concerned. The processes of photosynthesis are peculiar to green plants, and plants also have their own characteristic growth-regulatory mechanisms. This knowledge has assisted the synthesis of selective herbicides. Certain metabolite antagonists are only toxic to those bacteria which need to be supplied with the metabolite concerned in the nutrient medium or which can make it only in limiting amounts. This is a line of approach which is particularly attractive to the organic chemist, for it demands a knowledge of the structure of natural substances and of their dynamic function; and it has been the subject of intensive work since the concept of metabolite antagonism was first formally expressed, in particular relation to bacteria, by Fildes (1940). It is desirable to consider, therefore, the position of this concept in the light of the approach to the general question of correlating structures with activity here discussed. In its broadest and simplest form this hypothesis states that interference with the life of a micro-organism, or with a specific process in an organ of a more complicated biological entity, may result from a chemical attack upon an essential metabolite, thereby dislocating the balance of metabolic processes, often with far-reaching consequences. In practice, two general types of metabolite antagonism are recognized, antagonism by neutralization, and antagonism by competition. In the former, an essential metabolite is immobilized by chemical interaction with the drug, and in the latter the place of the metabolite in its next reaction in the sequence of biochemical events comprising the life process is taken by a substance of analogous structure which enters into the reaction concerned to a limited degree only, the subsequent reaction or reactions in the sequence being impossible under natural conditions. The distinction between inhibition by neutralization and competitive inhibition is perhaps artificial, or at least is only a matter of degree. The enzymatic conversion of metabolite A to metabolite B may be represented by the scheme:

$$A + \text{Enzyme} \xrightarrow[\text{I}]{\text{Combination}} \text{Complex} \xrightarrow[\text{II}]{\text{Reaction}} \text{Modified Complex} \xrightarrow[\text{III}]{\text{Dissociation}} B + \text{Enzyme}.$$

If now a competitive metabolite A' enters the system, A and A' will compete for the enzyme, the degree of competition being determined by the extent of the reversibility of reaction I. Reaction II is possible only with metabolite A and cannot proceed with A'. Hence, though A' is spoken of as competitive antagonist of A, it must also be viewed as a neutralizing antagonist for the enzyme; for the enzyme is just as much an essential metabolite as is A. Thus it is seen that the phenomenon of competitive metabolite antagonism accords with the first of the two postulates made above, and, indeed, is a special example of a more general phenomenon. The experimental observations available regarding the relationship between structure and the ability of a substance to function as a metabolite antagonist are also in accord with the second postulate, namely, that its efficacy may be very considerably modified by changes, through structural alterations, in its physical properties. An example of the homologous series effect in the field of metabolite antagonism is shown in certain biotin (IV) analogues. Compounds of formula V have been shown by Dittmer & du Vigneaud (1947) to be competitive antagonists of biotin in the metabolism of yeast and *Lactobacillus casei*, and in the series where $n = 3$–6, maximum activity was shown where $n = 4$.

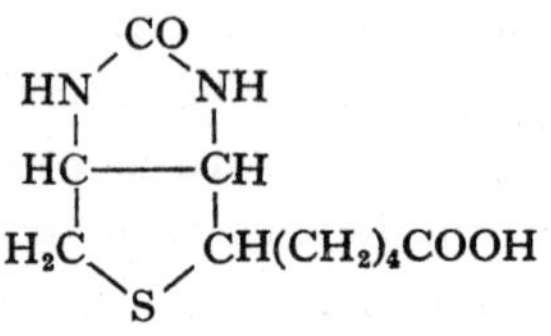

```
          CO                              CO
        /    \                          /    \
     HN        NH                    HN        NH
      |         |                     |         |
     HC ------- CH                 H2C ------- CH
      |         |                               |
    H2C        CH(CH2)4COOH                    CH2(CH2)nCOOH
       \      /
          S

          IV                                V
```

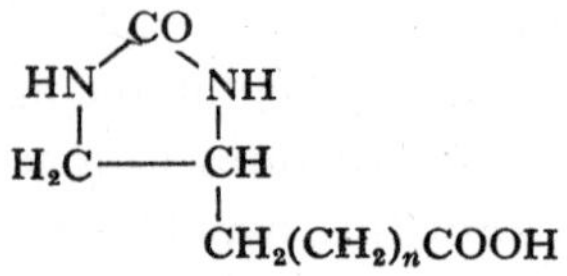

Similar effects are shown in certain analogues of pantothenic acid (VI):

$$\text{HO.CH}_2\text{CMe}_2\text{CHOH.CONH.CH}_2\text{CH}_2\text{COOH}$$

VI

Competitive inhibition usually demands some similarity of structure between the competing molecules. The resemblance may be in the shape of the molecules or, more precisely, in the electrochemical similarity and spatial disposition of these groups which are responsible for the attachment to the relevant cell constituents. The best known examples are the competition of malonic and succinic acids in the succinic dehydrogenase system and of *p*-aminobenzoic acid and sulphanilamide in the synthesis of pteroylglutamic acid. Examples have multiplied greatly in recent years, particularly amongst vitamins, coenzymes and amino-acids, and no doubt more will be heard of this from other speakers in this Symposium. It will suffice here to record a few general remarks upon the subject. First, in regard to molecular shape, the danger of considering two-dimensional formulae as drawn on paper cannot be too strongly emphasized. Molecules are rarely planar, though a good example of an essentially planar molecule of considerable complexity is provided in the porphin ring system. Secondly, the organic chemist in devising new structures, makes use of the fact that freedom of rotation can occur about a single bond. Thus, one may write phenoxyacetic acid as VII or VIII:

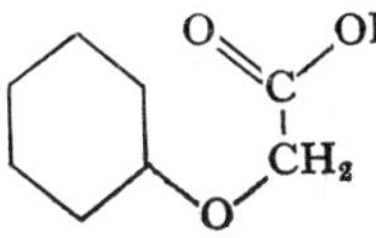

VII 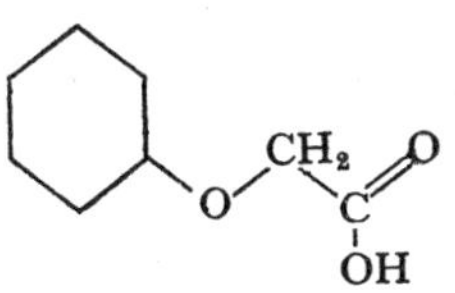 **VIII**

How many of the molecules at a given time are in form VII, form VIII or in other forms not depictable on paper is partly a matter for statistical computation, but in living surroundings is perhaps mainly determined by the molecular environment. Thirdly, it is necessary to emphasize the similarity of the valence angles of carbon, oxygen and nitrogen. In consequence of this, similarity of molecular shape is a common phenomenon, but this may have no biochemical significance except to point to the common origin of complex molecules in simple precursors such as the amino-acids and the intermediates of carbohydrate metabolism. There is thus great scope for irrelevant coincidence, and the chemist must be on guard against deceptive structural analogies. To exemplify, the arrangement IX is found in riboflavin, the pterins and the purines. This is doubtless one of the reasons

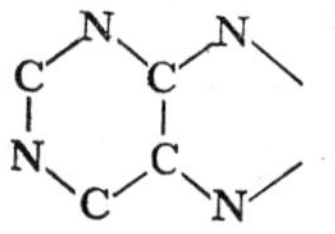

IX

why the synthesis of new biologically active molecules is often achieved as a result of faulty reasoning or why structurally plausible explanations so often follow quickly after empirical discovery.

Although facts are fast accumulating regarding the structural relationship between biologically functioning molecules and their competitive antagonists, in the present state of knowledge it would be dangerous to attempt generalizations. It cannot yet be predicted how far it is necessary to deviate from a given structure in order to produce an antagonistic effect, although certain well-tried procedures are open to the organic chemist. One such is the substitution of a —COOH group by $—SO_2NH_2$ or $—SO_3H$, which produces antagonistic substances in the classical cases of *p*-aminobenzoic acid and nicotinic acid. In both these cases the —COOH group in the natural product has a chemical functional significance, in the former by combination with glutamic acid in the folic acid molecule and in the latter by conversion to amide in coenzymes 1 and 2. On the other hand, changing α-naphthylacetic acid (synthetic plant growth regulator) to α-naphthylmethane sulphonic acid does not produce antagonism, for both substances have the same kind of activity. In this example, the —COOH group is believed to function by virtue of its polar character; that is to say, it functions as such without conversion to a derivative, and other acidic groups may therefore be expected to behave similarly. On this argument it might perhaps be suggested that pantothenic acid, which is antagonized by the corresponding sulphonic acid, pantoyltaurine, functions through conversion of the —COOH group to a derivative. An illustration of the delicacy of balance between the structures of naturally functioning substances and their competitive antagonists is provided by the varying response of different organisms to the same substance. Desthiobiotin, for example, has biotin-like action in promoting yeast growth, but antagonizes biotin in the metabolism of *L. casei*. Another type of illustration is provided in the fact that comparatively minor structural differences may occur between substances which simulate the action of a natural substance, which antagonize it or which are more or less without action. There are many examples of this in structural analogues of such substances as *p*-aminobenzoic acid, pantothenic acid, aneurin and biotin.

In conclusion, it is necessary to complete the general picture by some reference to the complications introduced by consideration of host-parasite relationships. This is of the greatest importance in chemotherapy, and one must consider not only the effect of the drug on the host and the parasite, but also possible effects of the host upon the drug. In the case of a parasitic organism, e.g. bacteria or Protozoa, the parasite must first be reached through the vehicle of the host. There will, therefore, be a set of factors determining absorption and distribution of the drug by the host, and those factors which dispose towards adequate distribution in the host are not necessarily the same as those which favour maximum activity against the parasite. This is why the majority of powerful bactericides are quite ineffective *in vivo*. The question of correlating antimalarial activity with structure is rendered very difficult by reason of the fact that in the present state of knowledge experiments *in vitro* are exceedingly difficult, and experiments conducted *in vivo* comprise a measurement of the combined effects of drug distribution and action upon the parasite. The animal body can use its chemical resources to bring about modification of drugs which are administered to it. Oxidation, reduction, hydrolytic and condensation reactions are performed with the greatest of ease, and these are usually referred to as detoxification mechanisms. This term is to some degree deceptive, for although such reactions do in fact very frequently result in the conversion of a drug to a less toxic compound or to one which is readily eliminated from the body, there is no reason to suppose that the body is equipped with a special chemical apparatus with the sole function of destroying miscellaneous poisons. The reactions are carried out by the normal functional enzymes of the body. It is possible therefore that a drug with little or no *in vitro* activity against a particular parasite may be converted by the host into a substance of marked activity. An example from recent work is provided by the hydrolysis of penicillin esters in the bodies of certain animals, and it has been recently suggested (Hawking, 1947) that this may apply also in the case of the antimalarial drug, 'Paludrine'.

REFERENCES

BELL, P. H. & ROBLIN, R. O. (1942). *J. Amer. Chem. Soc.* **64**, 2905.

BOUSQUET, E. W., SALZBERG, P. L. & DIETZ, H. F. (1935). *J. Indust. Engng Chem.* **27**, 1342.

DAVIES, W. H. & SEXTON, W. A. (1948). *Biochem. J.* (in the Press).

DITTMER, K. & DU VIGNEAUD, V. (1947). *J. Biol. Chem.* **169**, 63.

FERGUSON, J. (1939). *Proc. Roy. Soc.* B, **127**, 387.

FILDES, P. (1940). *Lancet*, **1**, 955.

HAWKING, F. (1947). *Nature, Lond.*, **159**, 409.

KIRKWOOD, S. & PHILLIPS, P. H. (1946). *J. Pharmacol.* **87**, 375.

LANDSTEINER, C. (1936). *The Specificity of Serological Reactions*. Springfield, Ill.

PAULING, L. (1940). *J. Amer. Chem. Soc.* **62**, 2643.

STILES, W. & REES, W. J. (1935). *Protoplasma*, **22**, 518.

SYNERHOLM, M. E. & ZIMMERMAN, P. W. (1947). *Contr. Boyce Thompson Inst.* **14**, 369.

THE DIFFERENTIAL ACTIVITY OF CERTAIN ANTI-MALARIAL AGENTS AGAINST PROTOZOA AND BACTERIA

By FRANCIS LESLIE ROSE
Imperial Chemical Industries, Ltd., Blackley, Manchester

I. INTRODUCTION

The present decade has witnessed a research in chemotherapy, hitherto unparalleled in its intensity, directed towards the discovery of new anti-malarial drugs. In America, the screening of many thousands of compounds disclosed the most promising drugs in the quinoline series. In the British Isles, a somewhat different approach to the problem had to be made. In these laboratories we selected another heterocyclic system, that of pyrimidine, on which to base new drug structures. A little later, the related acyclic diguanide system was studied, and culminated in the discovery of optimum antimalarial activity in the substance N_1-*p*-chlorophenyl-N_5-*iso*propyldiguanide (III; 'Paludrine').

These researches were initiated on the basis of a structure-activity concept which received modification and elaboration as the work progressed, but although the continuity of the chemical line of thought is readily apparent, it is now clear from a study of the records of the biological and biochemical effects produced by the many substances prepared, that their influence on micro-organismal life does not follow any discernible parallel course. Aided by a prolific 'working hypothesis' and by access to a biological assay technique so rapid and precise that the effect of structural modifications could be noted and incorporated almost immediately in new synthesis, the therapeutic effects achieved in this research, at any rate with respect to malaria, have so far outstripped the capacity of the workers concerned to offer adequate explanation.

These results are now in course of publication,* and it is ultimately proposed to tabulate and discuss the experimental findings in detail in a comprehensive series of papers. At the same time, many of the substances prepared in the course of this work have been submitted to test for other

* The chemical synthetic work, together with a rough indication of the activity in experimental malaria of some of the compounds, will be found in a series of papers appearing in the *Journal of the Chemical Society*. The experimental biological and biochemical researches, and the results of clinical investigations, are to be found in current and recent numbers of the *Annals of Tropical Medicine and Parasitology*, the *Transactions of the Royal Society of Tropical Medicine and Hygiene*, the *British Journal of Pharmacology and Chemotherapy*, and the *Biochemical Journal*.

purposes, activated in some instances by the prospect of eliciting some clue as to their possible mode of action against the malaria parasite from observation of their effect on other micro-organisms such as the bacteria, which lend themselves to more precise cultural control than the former. In addition, features in the stereochemistry of many of the drugs suggested some connexion with the prosthetic grouping associated with certain coenzymes and called for an investigation from the enzymological standpoint. Further researches have been concerned with the physical behaviour of selected compounds, both intrinsic and extrinsic, such, for example, as base strength and distribution in the tissues of experimental animals.

So far, attempts to rationalize the many results available have not been conspicuously successful, and the purpose of this contribution is to collect together and to display much of the relevant information so that it may receive wider consideration. For the most part, this has required nothing more than the restatement of published results and the inclusion, in some instances, of isolated and hitherto undisclosed observations made in these laboratories. In this connexion, the author would like, at the outset, to acknowledge his indebtedness to his colleagues of Imperial Chemical Industries Ltd., whose names appear throughout this communication, and whose experimental work and comments thereon provide the basis for this contribution to the Symposium.

As already implied, the antimalarial aspects of our researches called for the preparation of many hundreds of new substances. Complete experimental data of the type indicated is only available for a comparatively small number of these. It is proposed to select for detailed study, in addition to paludrine itself (III*a*) or (III*b*), those substances bearing in early publications the code numbers 2666(I) and 3349(II), as representative of the two main chemical types from which paludrine was subsequently evolved.

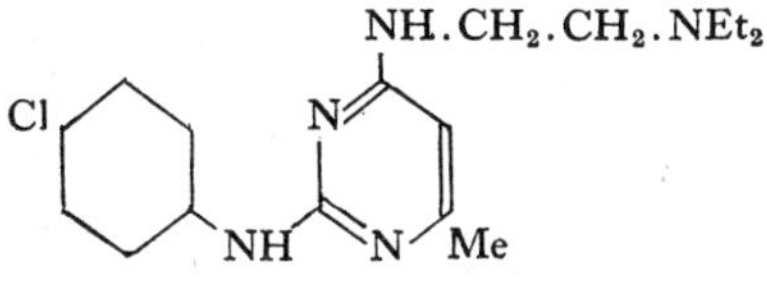

(I) (2666)

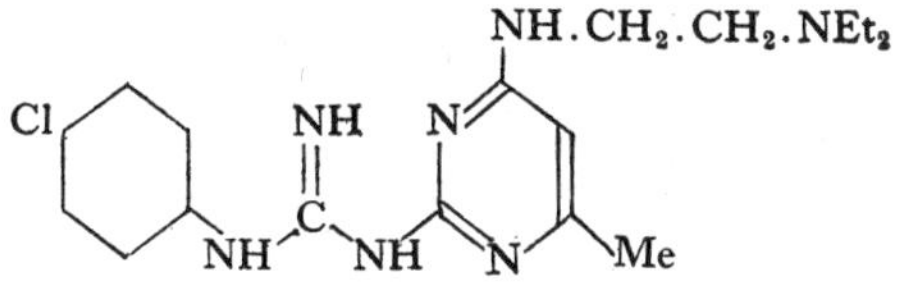

(II) (3349)

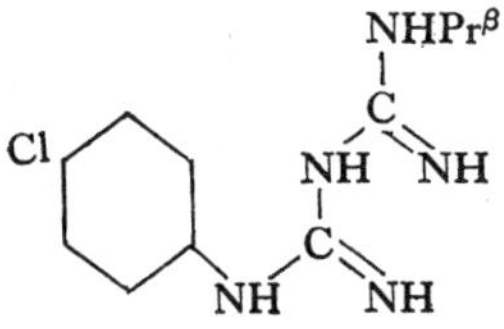

(III*a*) ('Paludrine')

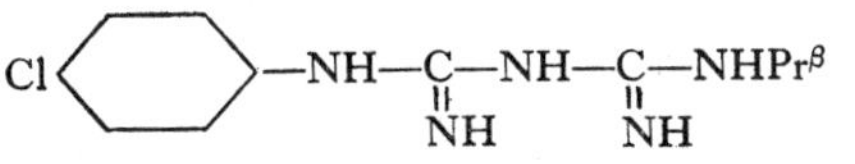

(III*b*) ('Paludrine'—conventional formulation)

Appreciation of the chemical interrelationship which exists between these three substances is best assisted by a summarized recapitulation of the chemical line of thought which led to their successive synthesis.

The chemical basis for the evaluation of the anilinopyrimidine (I) has been reported fully elsewhere (Curd & Rose, 1946). Briefly, it was the first member to show activity in experimental malaria (Curd, Davey & Rose, 1945*a*) of a small series of simple derivatives of pyrimidine in which were incorporated some of the salient features of the then known antimalarial drugs, in particular, mepacrine (IV).

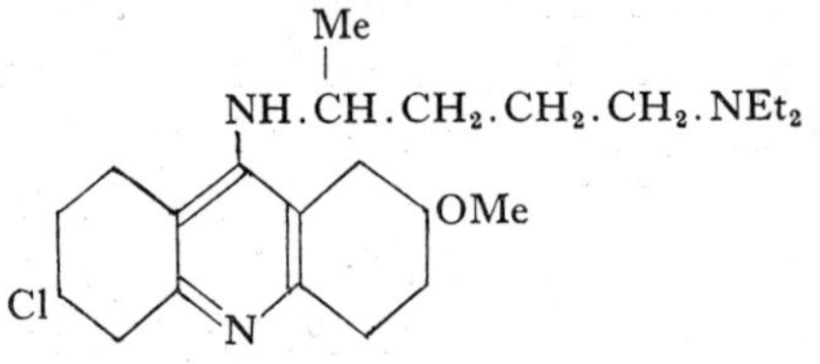

Me
NH.CH.CH_2.CH_2.CH_2.NEt_2
OMe
Cl
N

(IV) (Mepacrine)

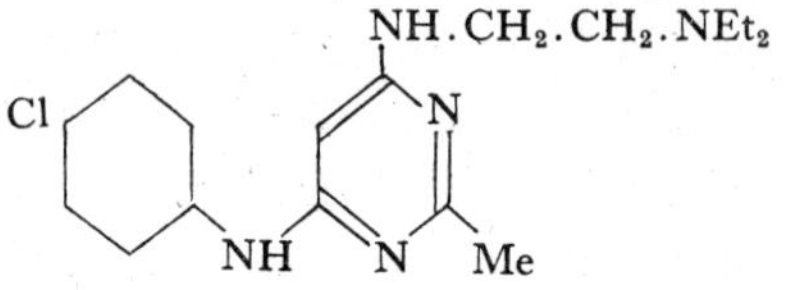

NH.CH_2.CH_2.NEt_2
Cl
N
NH N Me

(V)

These features included the halogen-substituted benzene ring, a 'basic side-chain' of the dialkylaminoalkylamino class, and the existence of tautomeric potentialities between a ring-nitrogen atom and side-chain amino group. The lack of antimalarial activity in the isomeric substance (V) and the enhancement of activity observed in (II) through replacement of the *p*-chloroanilino group of (I) by *p*-chlorophenylguanidino, stressed the importance of the tautomeric possibilities inherent in the active structures, and in particular of those tautomers found only in the active substances in which it was possible to trace conjugation through alternate carbon and nitrogen atoms from the anilino group to the basic side-chain as in (I*a*) and (II*a*), corresponding to (I) and (II) respectively.

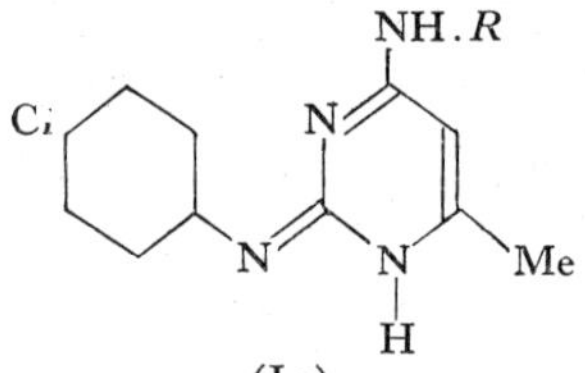

NH.*R*
Cl
N
N N Me
H

(I*a*)

NH.*R*
Cl
NH_2
C
N
N N N Me
H

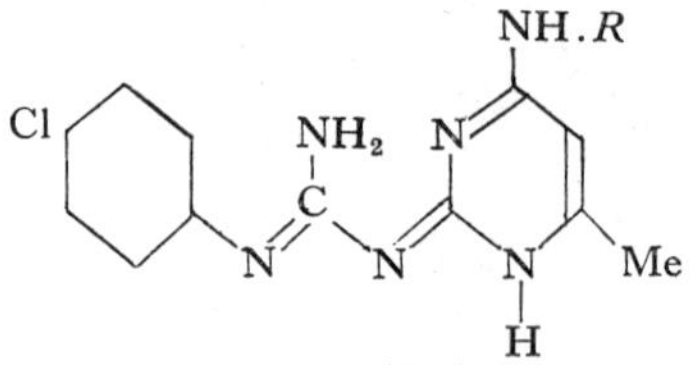

(II*a*)

Further considerations of these implications led to the abandonment of the pyrimidine ring and the substitution for it of the closely related acyclic diguanide system, and although the compounds strictly analogous to (I*a*), in which *R* was dialkylaminoalkyl, were devoid of antimalarial action, the corresponding unsubstituted alkyl derivatives typified by the *iso*propyl homologue (III*a*) (paludrine) were even more active than their progenitors (I) and (II).

II. ABSORPTION AND BLOOD CONCENTRATIONS IN EXPERIMENTAL ANIMALS AND MAN

(a) *In experimental animals*

These factors are directly concerned with the therapeutic effects of the drugs, since they control the relative availability of the latter to the parasite in the infected host, and must therefore be taken into account in any discussion on comparative activity *in vivo*. This aspect is of particular importance in a disease such as malaria, in which the assessment of intrinsic antiprotozoal activity by direct means is a matter of some difficulty.

Absorption data, following administration of the drugs by mouth, are available for a number of different subjects, but in these laboratories the most exhaustive investigations have been concerned with the rat, and Table 2 gives the blood and plasma concentrations for (II), (III), (IV) and quinine at intervals after oral administration of nearly similar doses of these substances as soluble salts, to this animal.

Data are also available relating to the concentrations attained by these drugs in a wide variety of solid tissues. These are not recorded here in detail, but Table 2 carries a footnote giving the approximate range of these concentrations as a multiple of the corresponding plasma levels.

In addition, since actual measurement of antimalarial activity is made in the chick, the concentrations given by (III) in the blood of this species following oral dosing is included in the table.

(b) *In man*

Blood and plasma concentrations expressed as mg./l. are given in Table 1 for the drugs (III) (Maegraith, Tottey, Adams, Andrews & King, 1946) and (IV) (Army Malaria Research Unit, Oxford, 1946).

Table 1. *Blood and plasma concentrations in humans*

Drug ... Dose to adult human ...	(III) 100 mg.		(IV) 100 mg.	
Conc. in mg./l. as read from experimental curves in ...	Blood	Plasma	Blood	Plasma
Time after dosing (hr.)				
1	2–3 × plasma conc.	—	0·026	0·002
2	(vide Maegraith	0·017	0·032	0·006
8	*et al.* 1946)	0·034	0·017	0·006
24		0·002	0·006	Tr.

Table 2. *Blood and plasma concentrations in experimental species*

Drug ... Dose ...	(II) 100 mg./kg. (rat)		(III) 80 mg./kg. (rat)		(IV) 80 mg./kg. (rat)		Quinine 80 mg./kg. (rat)		(III) 40 mg./kg. (chick)	
Conc. (mg./l.) in ...	Blood	Plasma	Blood	Plasma	Blood	Plasma	Blood	Plasma	Blood	Plasma
Time after dosing										
20 min.	0·278	0·0725	—	—	—	—	—	—	—	—
30 min.	—	—	1·2	0·29	—	—	4·5	1·93	0·43	0·25
50 min.	0·303	0·143	—	—	—	—	—	—	—	—
1·5 hr.	0·346	0·15	1·58	0·73	0·29	0·165	3·25	1·61	1·83	0·88
1·75 hr.	—	—	1·98	0·76	—	—	—	—	—	—
2·5 hr.	0·83	0·23	1·26	0·59	0·515	0·232	—	—	2·20	0·51
3·5 hr.	0·61	0·13	—	0·47	0·7	0·38	3·1	1·42	2·02	0·95
5 hr.	0·56	0·04	—	—	0·635	0·286	—	—	2·07	0·86
6 hr.	—	—	—	—	—	—	1·95	0·8	—	—
7 hr.	0·45	0·16	0·7	0·24	0·485	0·217	—	—	1·76	0·74
18 hr.	—	—	—	—	—	—	0·07	Tr.	—	—
24 hr.	0·354	Tr.	0·58	0·12	0·298	0·139	—	—	0·47	0·24
Approx. range of tissue conc. as multiple of plasma conc.	100–300		10–100		200–2000		50–300		10–100	

III. TOXICITY IN EXPERIMENTAL ANIMALS AND IN MAN

(a) *In experimental animals*

The results of experiments on the acute and chronic toxicity of (I), (II), (III) and (IV), following the administration to white mice (approximately 20 g. weight) of aqueous solutions of neutral water-soluble salts, are given in Table 3.

The dosage figures are adjusted to correspond to the amount of base used in each case.

Table 3. *Toxicity in the mouse*

Drug	L.D. 50 in the mouse, as mg./kg.			
	Oral (acute)	Oral (chronic; b.i.d. × 5)	Intraperitoneal (acute)	Intravenous (acute)
(I)	300–350	150	65	45
(II)	750–1000	250–450	—	—
(III)	60–80	25	20–30	20–30
(IV)	750–1000	375	—	40

Comparative acute toxicity figures for (III) are available for the chick, mouse, rat and rabbit (Butler, Davey & Spinks, 1947) and are recorded in Table 4.

Table 4. *Toxicity in several species*

Experimental animal	L.D. 50 of (III) as mg./kg.		
	Oral	Intraperitoneal	Intravenous
Chick (50 g.)	400–600	—	60–8
Mouse (18–22 g.)	60–80	20–30 (delayed deaths)	20–30 (delayed deaths)
Rat (100 g.)	100–150	40 (delayed deaths)	40 (delayed deaths)
Rabbit (1·5 kg.)	*c.* 150	—	*c.* 50

(b) *In man*

L.D. 50 figures are obviously not available, but it is possible in each instance to quote approximate threshold dosages which when exceeded in adult humans produce untoward symptoms attributable to the drug. These symptoms differ so widely in character and severity from drug to drug that they clearly cannot be related in any way to the lethal properties of the several compounds, and any comparison with observations made in laboratory animals would be unwise. Nevertheless, for the sake of completeness, the order of magnitude of tolerated doses and the nature of toxic effects produced when these are exceeded are given in Table 5.

Table 5. *Tolerated doses in adult humans*

Drug	Approximate dose for adult humans at which toxic symptoms appear	Nature of toxic effects
(I)	40 mg. t.i.d.	Headache, marked mental depression
(II)	100–200 mg. t.i.d.	Colic, diarrhoea, frontal headache
(III)	1–1·5 g. daily	Nausea, vomiting
(IV)	200 mg. t.i.d.	Gastro-intestinal discomfort, mental disturbances, skin-staining

IV. ANTIMALARIAL ACTIVITY IN EXPERIMENTAL ANIMALS AND MAN

(*a*) Plasmodium gallinaceum *in chicks*

The experimental infections used in these laboratories have been described in detail elsewhere (Davey, 1946; Curd *et al.* 1945*a*, *b*). For the screening of drugs for potential use either in the clinical treatment and clinical prophylaxis of malaria, or for causal prophylaxis and radical cure, the standard infection was that produced by *Plasmodium gallinaceum* in chicks. For the former purpose, 6-day-old chicks, weighing 45–55 g., were injected intravenously with approximately 50 million parasitized cells drawn from heavily infected chicks. The drug was then given orally, either in solution or as a suspension, twice daily for the next 3 days. Blood smears were taken on the fifth day, and the density of the residual infection expressed as the number of parasitized corpuscles in a random sample of 500. The minimum effective dose (M.E.D.) was that at which the drug exerts what is materially its maximum effect, or stated in another manner, the figure at which the density of infection-dose curve becomes parallel to the base-line.

Table 6 shows the comparative action of the three selected drugs, together with controls based on mepacrine and quinine in the therapeutic test. Data relating to a homologue (VI) corresponding to (I) in which the diethylaminoethyl side-chain of the latter is replaced by the homologous diethylaminopropyl grouping is included since this substance was at a later date examined for certain purposes more extensively than its prototype. The 'fractions' quoted indicate the number of parasitized cells found per 500 in the drug-treated chicks compared with the infected but untreated controls of the same experiment.

Table 7 shows corresponding results obtained by administration of the drugs as single intravenous doses on each of the first 3 days following infection.

The causal prophylaxis test aims at detecting drugs which prevent parasites, presented to the host as sporozoites obtained from the salivary glands of the insect vector *Aëdes aegypti*, from developing as far as the

Table 6. *Antimalarial activity (oral) against* Plasmodium gallinaceum *in the chick*

	Activity expressed as no. of parasitized cells per 500 in treated/control birds						Minimum effective dose (mg./kg.)
Dose (mg./kg.) ...	80	40	20	10	5	2	
Drug							
(I)	17/331	237/306	—	—	—	—	>80
(VI) (homologue of (I))	1/373	9/441	44/441	—	—	—	40
(II)	3/265	15/302	273/302	—	—	—	60
(III)	—	1/332	2/332	2/400	1/398	250/398	5
(IV)	—	2/302	120/302	—	—	—	40
Quinine	—	2/422	69/422	—	—	—	40

Table 7. *Antimalarial activity (intravenous) against* Plasmodium gallinaceum *in the chick*

	Activity expressed as no. of parasitized cells per 500 in treated/control birds						Minimum effective dose (mg./kg.)
Dose (mg./kg.) ...	28	24	20	16	12	8	
Drug							
(VI) (homologue of (I))	31/276	112/276	178/385	—	—	—	>28
(II)	—	—	6/354	31/376	127/376	—	16–20
(III)	—	2/401	2/401	—	11/340	—	12
(IV)	—	—	—	1·3/382	18/385	94/410	12
Quinine	6/293	17/293	43/293	116/383	202/381	—	24

Table 8. *Antimalarial activity in sporozoite-induced infections of* Plasmodium gallinaceum *in chicks*

Drug	Dose (mg./kg.)	Result
(I) (II)	Up to maximum tolerated during life of host	No apparent effect on mortality rate
(III)	60 b.i.d. × 5	Cure
	40 b.i.d. × 5	Some cures
	20 b.i.d. × 5	Marked delay in course of infection
	5 b.i.d. × 6	Death delayed some days
	2 b.i.d. × 6	Death delayed some days in some chicks
(IV)	160 b.i.d. as long as possible	No apparent effect on mortality rate
Pamaquin	40 b.i.d. × 6	Death delayed a few days: no cures
	20 b.i.d. × 6	No apparent effect on mortality rate

erythrocytic stages, by an action against either the sporozoites themselves or, since separate experiments have shown the resistance *in vitro* of these to many drugs, almost certainly against the first generation of exoerythrocytic forms which follow. Davey (1946), working in particular with drugs of the diguanide class, has drawn attention to the susceptibility of the latter, and to the inhibition of reproduction of the later (or secondary) exoerythrocytic forms, which, contemporaneous with the parasites in the red cells, normally increase in number as the infection develops and through blockage of the brain capillaries result in the ultimate death of the chick.

Table 8 indicates the degree of action of the several speculative and standard drugs against a sporozoite-induced infection of *Plasmodium gallinaceum* in chicks, the drugs being administered on the day of infection.

(*b*) *Antimalarial activity against other experimental infections*

A similar range of compounds has been examined by Davey (1946) against experimental Protozoa other than *P. gallinaceum*, for example, against *P. lophurae* in chicks and *P. cathemerium* in canaries. *P. lophurae* has been widely employed elsewhere in the duck as host, but in this instance the chick was used to facilitate comparison with the results obtained in infections due to *P. gallinaceum*. The chicks were similarly selected, but since *P. lophurae* is less virulent to this host than the latter species, the infecting inoculum of parasitized red cells was higher, and to accommodate the longer period of time needed to reach the peak value of the parasitaemia, the chicks were dosed on the day of inoculation and on each of the four succeeding days.

P. cathemerium was introduced intravenously into canaries (18–22 g.) using 8–12 million parasitized red cells drawn from a canary at the height of its infection. The drugs were given orally, first, 4 hr. after inoculation, and then twice daily for 3 or 4 days. The density of the residual infection was estimated by a count of the number of parasites per microscope field, working under clearly defined conditions.

The results obtained are indicated in abbreviated form in Table 9 in which the M.E.D. against *P. gallinaceum* in the chick is included to assist comparison.

Table 9. *Antimalarial activity against several experimental infections*

Drug	Minimum effective dose (mg./kg.) against		
	P. gallinaceum in chicks	*P. lophurae* in chicks	*P. cathemerium* in canaries
(VI) (homologue of (I))	40	—	*c.* 50
(II)	60	Inactive	*c.* 50
(III)	5	*c.* 5	25
(IV)	40	40	10
Quinine	40	*c.* 40	—

Limited experimentation has been made to determine the activity of the drugs against the exo-erythrocytic forms of other species of parasite. Thus *P. cathemerium* injected into canaries in the sporozoite phase rapidly disappear from the blood but induce a parasitaemia in the course of some days from which the birds subsequently die. Of the several drugs examined, only (III) and pamaquin have prevented this appearance of the parasites in the blood stream, but the effect is only temporary and parasites reappear a few days after dosing has ceased. *P. relictum* in canaries behaves similarly. The course of a sporozoite-induced infection of *P. lophurae* in canaries, chicks or ducklings does not permit a clear assessment of drug action, but of the experiments conducted by Davey, the more important indicate the inability of (IV) at 160 mg./kg., b.i.d. × 4, to prevent the appearance of the parasites of this species in the blood of chicks, a similar failure with (III) at 40 mg./kg., b.i.d. × 4, but complete protection with the latter at 60 mg./kg., b.i.d. × 4.

(*c*) *Antimalarial activity in man*

No single dose, or even dose range, can adequately define the clinical response consequent upon the administration of an antimalarial drug. Variability may exist, not only in the virulence of the many species of the causal parasite, but in the reaction of the host to the disease, and again the same clinical end-effect may result from the interference of different drugs at different points in the life cycles of the Protozoa.

For the purpose of this communication, it is proposed to effect a rough comparison between the several drugs in their action on the clinical phases of the disease due to *P. falciparum* (malignant tertian) and *P. vivax* (benign tertian). In other words, the drugs are compared on the basis of:

(1) Their action as causal prophylactics (action on pre-erythrocytic phase).

(2) Their value as therapeutic agents (schizonticidal action).

(3) Their action on later forms of the parasite (e.g. gametocytes and/or later hypothetical exoerythrocytic forms).

This information (from a number of sources), summarized in Table 10, is given in terms of dosage for adult humans, and represents no more than an attempt, bearing in mind the many variable factors, to extract an approximate mean value from the data available.

V. ANTIBACTERIAL ACTIVITY

Dr A. R. Martin, in these laboratories, has examined the several drugs *in vitro*, and to a limited extent *in vivo*, against a range of bacteria. Limiting concentrations which just inhibit the growth of the organisms *in vitro* are given in Table 11. It is not proposed to describe the experimental conditions in detail, since these were standard for any given test organism, enabling a fair comparison to be made between the drugs in each case. The results obtained using the well-known antibacterial substance acriflavine are included in the table.

Table 10. *Antimalarial activity in humans*

Drug	Clinical effects					
	Action on pre-erythrocytic forms		Action on erythrocytic forms. Approx. lowest daily dose fully effective in controlling symptoms		Action on later forms of parasite	
	P. vivax	*P. falciparum*	*P. vivax*	*P. falciparum*	*P. vivax*	*P. falciparum*
(I)	(Not examined)	(Not examined)	Inactive at highest permissible dose	Inactive at highest permissible dose	(Not examined)	(Not examined)
(II)	(Not examined)	(Not examined)	200 mg. t.i.d.	200 mg. t.i.d.	No clear action on relapse rate. No action on gametocytes	No clear action on relapse rate. No action on gametocytes
(III)	At 100 mg. every 3–4 days no parasites appear in blood, but eradication not complete	Complete causal prophylactic at 100 mg. every 3–4 days. 50–100 mg. up to 5 days from infection affords complete protection. 10–25 mg. effective in some cases	25 mg. b.i.d.	25 mg. b.i.d.	Therapeutic doses give proportion of radical cures. No action on gametocytes. Sterilizes gut infection in mosquito	Therapeutic dose gives radical cure. No action on gametocytes. Sterilizes gut infection in mosquito
(IV)	No action	No action	100 mg. t.i.d.	100 mg. t.i.d.	Some influence on relapse rate. Slight gametocidal action	Therapeutic dose gives radical cure
Quinine	No action	No action	300–600 mg. t.i.d.	600 mg. t.i.d.	No effect on relapse rate	Not completely effective in giving radical cures

So far as examination *in vivo* is concerned, drugs (I), (II), (III) and (IV) have all been tested against streptococcal and tubercular infections in mice. In only one instance, that of (I) against the latter infection, was any activity observed, and that but slight (Hoggarth & Martin, 1948).

Table 11. *Antibacterial activity* in vitro

Drug	Lowest concentrations which just inhibit growth *in vitro* of			
	Strep. pyogenes	*Strep. agalactiae*	*Staph. aureus*	*B. coli*
(I)	1/3000–1/9000	1/9000	1/3000	1/1000
(II)	1/3000–1/9000	1/27,000	1/3000	1/000–1/3000
(III)	1/9000–1/27,000	—	1/9000	1/3000
(IV)	1/80,000	—	1/3000–1/9000	1/1000
Acriflavine	1/240,000	1/240,000	1/80,000	1/27,000

Drug	Lowest concentrations which just inhibit growth *in vitro* of			
	Cl. welchii	*Ps. pyocyanea*	*C. pyogenes*	*M. tuberculosis*
(I)	1/1000–1/9000	—	1/9000	1/3000
(II)	1/3000–1/9000	> 1/1000	1/27,000	1/1000
(III)	1/3000	1/3000	—	1/3000
(IV)	1/80,000	—	—	1/1000
Acriflavine	1/80,000–1/240,000	1/3000–1/9000	1/240,000	—

VI. ACTIVITY AGAINST COCCIDIOSIS AMOEBIASIS AND TRYPANOSOMIASIS

Compounds (II), (III) and (IV) have been examined in these laboratories by Mr J. S. Steward for their activity against an infection due to *Eimeria acervulina* in chickens. In no case was any effect observed.

A technique for the production of an infection due to *Entamoeba histolytica* in young rats has been described by Jones (1946) who also developed a means of assessing the influence of drugs on the course of the infection. Compounds (II), (III) and (IV) have been examined in this way, but only the last named showed any appreciable activity, and that at the high single oral dose of 1 g./kg.

Numerous compounds of the same chemical types as (I), (II) and (III) have been examined by Dr D. G. Davey in these laboratories against experimental trypanosome infections, but none has shown positive activity.

VII. DISCUSSION

The data given for the selected drugs in the preceding sections can be classified on the following broad basis:

(1) Absorption properties in experimental animals and man.

(2) General toxic effects in these species.

(3) Antimalarial activity against a range of plasmodia.

(4) Activity *in vitro* and *in vivo* against a number of other micro-organisms.

(*a*) *Toxicity*

A complete discussion in a comparative manner of effects observed upon administration of a series of drugs to the living animal is a difficult if not impossible task. This is particularly so when considering toxic action, unless precise information is available relating to the influence of the drugs on a great many biological and biochemical processes. While the selected drugs have in some instances been subjected to fairly detailed pharmacological examination, the data still remain insufficient for useful comparison to be made, and such discussion as is possible will therefore be confined to overall toxic effects as defined by median lethal dose. Clearly (1) (above) is of very considerable importance in governing toxicity. The capacity of a drug to undergo absorption, for example, from the gut, and to pass into and remain in the tissues is a complex function of, among many other things, physico-chemical properties, degradation, excretion, combination with plasma proteins, specific affinities for specialized tissues, etc. Despite the chemical similarity underlying the structures of drugs (II), (III) and (IV), Table 1 shows that paludrine (III) is distinct in its capacity to concentrate in the blood stream of the rat following oral administration. Not only is the maximum reached in a shorter period of time, but the absolute value attained is nearly three times that of mepacrine (IV). A similar relationship is found in man (Table 2).

Examination of the toxicity data for rats (Table 4) and man (Table 5) in the light of the concentrations of the drugs achieved at toxic doses reveals, however, a considerable difference between the test species. Thus, a dose of 50 mg./kg. of (III) daily in rats produces a peak of about 1·4 mg./l. (Butler *et al.* 1947). This dose is toxic and does, indeed, lead to a proportion of deaths. Maegraith *et al.* (1946) have recorded *plasma* concentrations of about 0·5 mg./l. of (III) 12 hr. after doses of 500 mg. b.i.d. to man, which must correspond to blood concentrations between 1 and 1·5 mg./l. All available evidence indicates that these values, and even higher, are readily tolerated by the subjects.

The positions of (II) and (IV) are less favourable. Thus it is generally accepted that mepacrine (IV) at 200 mg. b.i.d. in man produces toxic effects in a high percentage of subjects. Workers of the Army Malaria Research Unit, Oxford (1946), have shown that the whole-blood concentration of (IV) in a group of healthy volunteers taking a therapeutic regime of 800 mg. daily for 2 days followed by 300 mg. for 10 days, ranged between 0·300 and 0·400 mg./l. These figures are about half of the values obtained in the rat following the administration of approximately one-tenth of the acute oral dose.

Unfortunately, blood concentrations for (I) are not available, but from consideration of the several tables, and assuming absorption characteristics similar to those of the closely related (II), then it would follow that toxic effects in man were discernible with this drug at even lower relative blood concentrations than was the case with (IV).

The concentrations attained in the chick by (III) (Table 1) require comment. Single doses of 80 mg./kg. in the rat and 40 mg./kg. in the chick achieve similar blood and plasma concentrations. Repeated daily dosing in the rat does not cause any significant 'build-up', but in the chick, on the fifth day after giving 60 mg./kg. b.i.d., a peak blood level of 12 mg./l. has been reported (Spinks, 1948).

As Butler *et al.* (1947) have pointed out, a fundamental difference clearly exists between the metabolism of paludrine (III) in rats (and mice), and its metabolism in chicks and man. A similar, and perhaps related, contrast is discernible when the toxicities of the drugs (I), (II) and (IV), as observed in rats and man, are compared with (III). Since, as is discussed below, these distinctions also occur in the therapeutic and enzyme-inhibitory behaviour of these drugs, it is conceivable that despite their common chemical ancestry, their ultimate action on living tissue, whether of animal or micro-organismal origin, is fundamentally different.

(*b*) *Antimalarial activity*

When a therapeutic effect is obtained by the direct action of a drug upon the causal micro-organism, it is frequently possible to analyse the course of the response in an infected host on the basis of (*a*) intrinsic antimicro-organismal action, and (*b*) accessibility of the drug to the micro-organism. Consideration of antimalarial action in a quantitative manner along these lines is a matter of extreme difficulty. Not only is the precise determination of intrinsic antiplasmodial effect almost impossible at the present time, but the complex and as yet incompletely understood history of the parasite in the animal host, with the probability of differential action at the several phases, preclude more than the roughest guess as to the influence of the pharmacological properties of the drug on the course of the infection. Because of these difficulties, the discussion under this heading must be qualitative rather than quantitative. Even so, some amazing and perplexing differences stand out in considering the action of the selected drugs both in experimental and human malarias.

Thus, reference to Table 9, shows that whereas (II) and (VI) (=homologue of II) are about equi-effective against *Plasmodium gallinaceum* and

P. cathemerium, yet the former organism is some five times as susceptible as the latter to (III). With mepacrine (IV), on the other hand, the susceptibilities are completely reversed in approximately the same ratio. These results were obtained with the parasite infecting different hosts, the chick and canary, respectively, and it might be argued, particularly since *P. lophurae*, also in the chick, resembles *P. gallinaceum*, that the different reaction of these host species to the two drugs, for example, by way of absorption, distribution, metabolism, etc., could provide the underlying cause. The experimental evidence available is as yet insufficient to test this possible explanation, but since *P. lophurae* proved quite unaffected by (II) it must be concluded that a marked intrinsic differential sensitivity amongst the several species of plasmodia is very probable.

For reasons that have been explained in detail by Davey (1946), the main experimental infection on which most work in these laboratories has been based was that due to *P. gallinaceum* in chicks. Not least among the reasons was the abundance of exoerythrocytic forms of the parasite in this host and the opportunity which the occurrence of these forms provided for the study of drugs in the causal prophylaxis and radical cure of malaria. In this way some striking differences can be demonstrated between the selected drugs (I)–(IV). Of these, only (III) can be said to have any action on the exoerythrocytic (e.e.) forms of *P. gallinaceum*, and here it is necessary to draw some distinction between activity against the primary e.e. forms, that is, those which must immediately follow the sporozoite phase, and the secondary e.e. forms, which appear simultaneously with the erythrocytic phase, and which become progressively more difficult to influence as the disease is allowed to run its course. Thus, it has been shown by Davey (1946) that (III), in common with all other drugs, is without action on the sporozoites of *P. gallinaceum*, yet as indicated in Table 8, this substance in doses of 3 mg./50 g. chick b.i.d. for 5 days can be used to cure all chicks initially infected with sporozoites, provided that treatment does not commence later than about 80 hr. afterwards. When treatment has been delayed for 4 days no cures can be achieved. The infection can be held in check by continued medication, but withdrawal of the drug leads to the ultimate death of the host. It is of interest to note that certain sulphanilamide drugs are capable of acting in the same way, but in much higher doses. Experiments with e.e. forms of other species of experimentally used parasites do not lend themselves to such precise analysis as *P. gallinaceum*, but the results indicate that again the effects obtained with (III) are unique.

In man, the distinctive antimalarial activity of (III) is also apparent. The toxicity of (I) was such that the dose which it was anticipated should be therapeutically effective could not be given, but there is no reason to believe, even if such a dose had been permissible, that the results obtained would differ qualitatively from those produced with (II), (IV) or quinine. That is, in both *P. vivax* and *P. falciparum* infections, it would have had no marked action other than that of a schizonticide. (III), on the other hand, exhibits not only a powerful effect of this kind, but exerts a definite lethal action on the pre-erythrocytic forms of *P. falciparum* (almost certainly against the tissue forms and not the sporozoites), and with a suitable dose regime affords complete protection. In *vivax* malaria, (III) acts only as a partial causal prophylactic, and asexual parasites fail to gain access to the blood so long as the drug is being given, suggesting an inhibitory action on the e.e. forms.

Comparison of the influence of the drugs on the sexual forms (gametocytes) of *P. vivax* and *P. falciparum* and the later development of these in the insect vector provides further contrasts. Mepacrine (IV) and quinine are attributed with weak gametocidal activity. The limited clinical trials with (II) (Adams & Sanderson, 1945) provided no evidence of any such activity on the part of this drug, nor does (III) exert any obvious primary effect on either the number or microscopical appearance of gametocytes in the blood of a carrier (Fairley, 1946). Yet so long as the drug is being administered, that is, so long as more than a certain limiting concentration is present in the blood, sterilization of the infection occurs in the gut of a mosquito vector fed on a *falciparum* or *vivax* gametocyte carrier. This result provides additional support for the contention that the basis of the biochemical interference of (III) with the life processes of the malaria parasite is fundamentally different from that of its progenitors.

(c) *Other activities of the drugs*

For the purposes of comparative discussion we are concerned with the action of the drug series on micro-organisms in general, not so much from the viewpoint of the practical value of the results obtained, but rather in so far as they throw light on the mechanism by means of which the drugs produce their several effects. The first conclusion to be drawn from the experimental findings recorded above is that the therapeutic activity of the substances under consideration is directed almost specifically against the malaria parasite. Either, then, the latter is intrinsically more susceptible to the drugs, or alternatively, or additionally, the distribution of the drugs in the tissues of the host is such as to provide a high local concentration in the vicinity of the parasite, e.g. within the red blood cells where the erythrocytic phase is concerned. Certainly, the distribution factor can be held to account for the inability of the compounds to deal with a septicaemia of bacterial origin. For each of the organisms listed in Table 11, the concentrations necessary to inhibit growth are higher than those which could be attained by the drugs in the plasma of the test animals when administered at doses approaching the toxic limit. On the other hand, the characteristic partition coefficients exhibited by these compounds in favour of the cellular components of the blood might well provide concentrations within these of a similar order to those which would inhibit the growth of bacteria when present in the culture media. These arguments cannot, however, be pressed far, since the ultimate lethal action of a drug on an intracellular parasite might require consideration as a surface rather than a bulk phenomenon, that is, adsorption on to some common enzyme system might be the determinant factor, as distinct from concentration in the cytoplasm.

The significance of the distribution characteristics of the drugs in leading specifically to antimalarial activity receives support from consideration of the overall antibacterial activities of (I), (II) and (IV). The general picture presented in Table 11 is that the intrinsic antimicro-organismal action of these substances increases in the order in which they are enumerated, and this is also the order of ascending antimalarial activity. The position of (III) is again anomalous.

Little is to be gained from a study of the remaining therapeutic effects shown by the test drugs. The chemotherapy of amoebiasis and coccidiosis will again require in each case specific pharmacological features to be present in useful drugs, features about which little or nothing is known at the present time. It is proposed to do no more with regard to these two disease conditions than to point out that the latter is caused by a protozoan parasite having a complex life cycle in which a sporozoite phase, an asexual multiplication within the cells (epithelial) of the host, and differentiation into male and female elements (gametocytes) all occur, closely reminiscent of some aspects of the life cycle of the malaria parasite. Yet from the present results, and also from the wider experience gained in these laboratories, it is clear that the two species have little in common in regard to their sensitivity to chemotherapeutic agents.

(*d*) *Other experimental results*

From the preceding discussion it will have become apparent that despite their points of chemical resemblance, the three drugs (I), (II) and (IV) differ as a group from the diguanide (III) in the reaction to them of many forms of organized life, from that of the animal host to that of a unicellular parasite. Further understanding of these effects, and differences in effect, requires some knowledge of the mode in which these substances interfere with vital processes. At any early stage in the chemical development of these compounds (Curd & Rose, 1946) it was appreciated that the molecules of (I), (IV), and to some extent (II), when formulated as planar structures as above, showed some structural resemblance to the important component of many respiratory enzymes, riboflavin (VII):

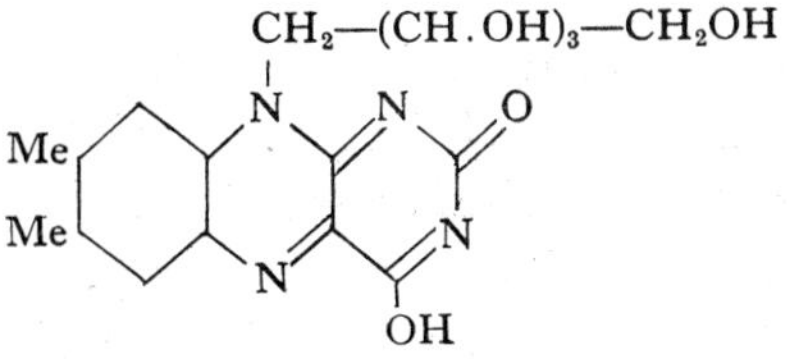

CH_2—$(CH.OH)_3$—CH_2OH

Me Me N N O N N OH

(VII)

ribose N N N N NH_2

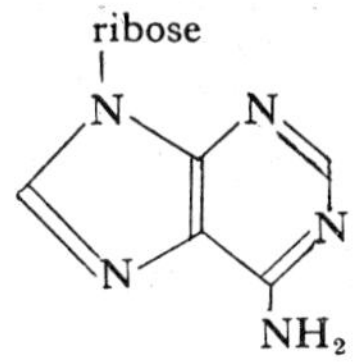

(X)

The points of resemblance and difference between riboflavin and the several drugs are perhaps not patently obvious, and it may be helpful to indicate what are regarded as the significant features. (IV), like riboflavin, possesses a fused three-ring skeleton, the left-hand ring in each case being benzene. The centre nuclei in both molecules are heterocyclic, being pyrazine and pyridine, respectively. The pyrimidine ring of riboflavin corresponds with the right-hand benzene ring of the drug, and this may be of some special significance, since it is known that a similar substitution in the vitamin aneurin results in an antagonistic substance. Finally, it has been pointed out by Madinaveitia (private communication) that the ribityl group of the growth factor, which in the mononucleotide will be phosphorylated, i.e. will carry a terminal *acidic* grouping, is in the drug simulated by a carbon chain carrying a terminal *basic* group.

(I) does not contain a trinucleate system, but exhibits benzene and pyrimidine rings joined by a single nitrogen, in contrast to the double linkage through nitrogen of the same ring systems seen in riboflavin. The basic side-chain is attached to the molecule at the nearest point equivalent to the missing *meso* position.

(II) is probably the drug least obviously related to riboflavin, and little more can be said for it than that it is closely derived from (I). With the aid of models it can be shown that the benzene ring of (II) corresponds to the *ortho*-methyl groups of riboflavin. The biological equivalence of such structural features is well known in the case of the polycyclic carcinogens.

It has been suggested (Curd *et al.* 1945*c*) that the antimicro-organismal (and perhaps also the toxic) effects of these three substances is due to their interference with the functioning (or even the synthesis) of one or more flavin enzyme systems. Direct experimental evidence in the case of malaria is not available, but Madinaveitia (1946*a*), using *Lactobacillus casei* as test organism, has shown that the inhibition of growth produced by the addition of (I), (II) or (IV) to cultures of this organism is reversed by the addition of riboflavin. The same worker has shown (private communication) that a strain of *B. coli* made resistant to (IV) is also resistant to (I) and (II).

(III) was derived from (I) by a chain of chemical reasoning which indicated the retention of a similar formal structure in both cases (compare (III*a*) and (I)). The most notable difference from a purely chemical view-point is the replacement of the dialkylaminoalkyl side-chain of the latter by the unsubstituted alkyl side-chain of the former. However, the loss of basicity due to the removal of the more complex side-chain is almost exactly balanced by the strongly basic character of the diguanide system *per se* as compared with the corresponding diaminopyrimidine system found in (I) (see below). It might confidently be expected, therefore, that the capacity of (I) to function as a riboflavin antagonist would transfer to the structurally and (physico-) chemically related (III). In point of fact, Dr J. Madinaveitia, again using the *Lactobacillus casei* technique, has shown in these laboratories that (III) exhibits no such property. Further, he has shown that *B. coli* resistant to (I), (II) and (IV) is not resistant to (III).

Even making the reasonable assumption that the antimalarial and anti-bacterial activities of (I), (II) and (IV) are, indeed, linked with riboflavin, it does not follow conclusively from the evidence just adduced that the biological properties of (III) are due to some entirely different connexion. It is still possible that some metabolite of III is the ultimate effective agent (see later), and that this metabolite, if it could be isolated, would exhibit riboflavin antagonism.

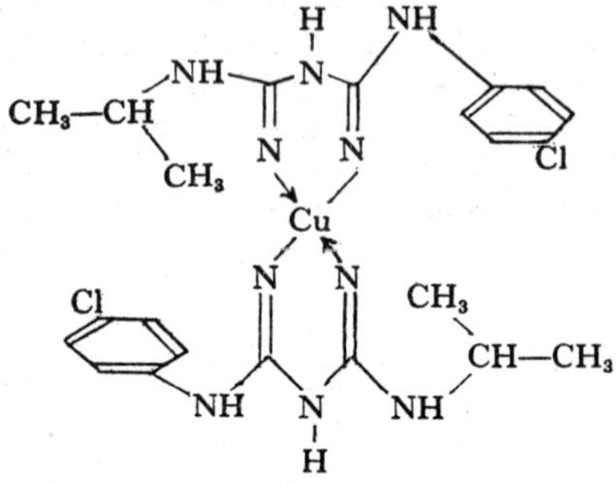

(VIII) 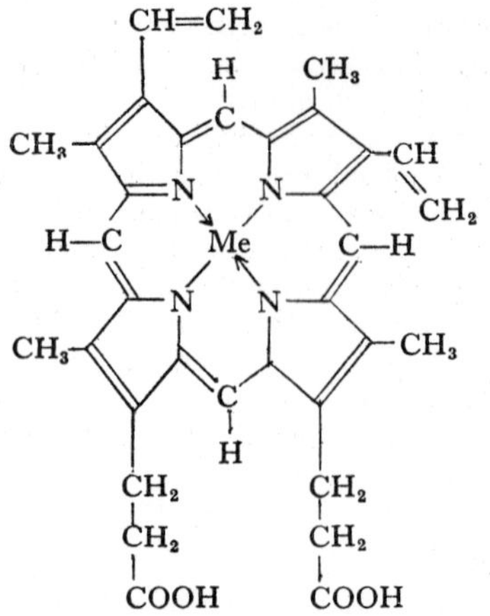 (IX)

An alternative explanation has been put forward by Curd & Rose (1947) based solely on chemical reasoning. This drew attention to the formal resemblance between an assumed structure for the characteristic metal complexes of (III), e.g. the copper complex (VIII), and a typical metal porphyrin (IX), and suggested that the diguanide drug exerted its biological effects through interference, not with a riboflavin enzyme system, but with some hypothetical porphyrin-containing respiratory system. Attempts to establish the validity of this view by researches involving the use of typical porphyrin enzyme systems, and of the bacterium *H. influenzae* which requires a haematin source for its culture, have been made here by

Dr J. Madinaveitia and Mr J. Francis, respectively, but have produced only negative results.

Reverting to the matter of the base strength of the drugs referred to above, Mr J. C. Gage in these laboratories has undertaken a thorough investigation of these properties in relation to antimalarial effect. These results are now in course of publication and detailed discussion elsewhere, but it is of interest to note the actual values for the several drugs with which we are now concerned. As already indicated, these substances are very strong bases. This would seem to be the first requirement for antimalarial activity, and is probably closely connected with the access of the drugs to the malaria parasite. In all cases, except (III), the most strongly basic group in the molecule is that of the trialkylamino residue of the basic side-chain, the corresponding pK_b values being 10·0, 10·2, and 10·1, for (I), (II) and (IV), respectively. Further analysis reveals a second base dissociation constant associated with the hetero portions of the molecules for the same three compounds, respectively, 6·5, 7·5 and 7·5. It will be noted that these values are in the order of increasing antimalarial (and antibacterial) activity, and are such that at physiological pH the relevant portions of the molecules will exist largely in the cationic form. (III) is unique in that the special characteristics of the diguanide system provide a base strength (pK 10·4) comparable with that of the trialkylamino groups of the other drugs as part of the intrinsic physical make-up of the hetero system of the molecule, and it is attractive to consider that these fortuitous circumstances are of special significance. However, the importance of these factors must not be over-emphasized, since it is possible to quote many instances in which molecules, while differing in other structural features, have yet the seemingly critical base-strength properties, but are devoid of antimalarial activity.

Finally, two items remain to be mentioned. The first relates to the development in chicks by Williamson & Lourie (1947) of a strain of *Plasmodium gallinaceum* resistant to (III). This strain was resistant to another diguanide derivative closely related to (III), but not to (I), (II) or (IV), again indicative of an entirely different mode of action. The second item concerns an observation made by Raventos (1947) working on the pharmacological action of adenosine (*X*) (which is, of course, both chemically and biochemically related to riboflavin), who found that the typical auriculo-ventricular heart block produced by intravenous administration of this substance to anaesthetized guinea-pigs is of shorter duration in animals receiving (I), (II) or (IV) whether orally or intravenously. On the other hand, the effect was only produced by (III) when given orally, a further indication of the prior formation of a metabolite. Further work is in hand aimed at the correlation of these effects and will be discussed elsewhere, but in this connexion, an observation recorded in Table 7 is suggestive. This shows that (III) is less effective in experimental malaria when given intravenously than by the oral route. More recently, Hawking (1947) had adduced evidence which indicates that whereas this drug is without action on the e.e. forms of *P. gallinaceum* grown in tissue culture, inhibition occurred when the serum of an uninfected monkey receiving (III) (intramuscularly) was added to the preparation.

VIII. CONCLUSION

The experimental data available are clearly insufficient to do more than point the direction in which the search for the explanation of the biological effects produced by the several compounds might be pursued. To the chemist, the alteration in chemical structure in passing from the drugs (I) and (II) to (III) would be considered gradual, and he would accept the acyclic diguanide grouping of (III) as fundamentally equivalent to the pyrimidine nucleus of the former compounds. Yet in almost every respect, the change in the biological behaviour between these two types is abrupt. Either, then, the continuance of antimalarial activity is purely fortuitous, or else the susceptible processes within the malaria parasite (and other micro-organisms on which these drugs act), if different in each case, might still be related in a structural sense, or additionally, be concomitant in some biochemical reaction series.

The difficulties that would be associated with the direct approach to these problems are obvious, and it is important therefore that full consideration be given to circumstantial evidence such as that displayed in this contribution.

REFERENCES

ADAMS, A. R. D. & SANDERSON, G. (1945). *Ann. Trop. Med. Parasit.* **39**, 173.
ARMY MALARIA RESEARCH UNIT, OXFORD (1946). *Ann. Trop. Med. Parasit.* **40**, 472.
BUTLER, R., DAVEY, D. G. & SPINKS, A. (1947). *Brit. J. Pharmacol.* **2**, 181.
CURD, F. H. S., DAVEY, D. G. & ROSE, F. L. (1945*a*). *Ann. Trop. Med. Parasit.* **39**, 139.
CURD, F. H. S., DAVEY, D. G. & ROSE, F. L. (1945*b*). *Ann. Trop. Med. Parasit.* **40**, 208.
CURD, F. H. S., DAVEY, D. G. & ROSE, F. L. (1945*c*). *Ann. Trop. Med. Parasit.* **39**, 157.
CURD, F. H. S. & ROSE, F. L. (1946). *J. Chem. Soc.* p. 343.
CURD, F. H. S. & ROSE, F. L. (1947). *Nature, Lond.*, **158**, 707.
DAVEY, D. G. (1946). *Ann. Trop. Med. Parasit.* **40**, 52, 453.
FAIRLEY, N. H. (1946). *Trans. R. Soc. Trop. Med. Hyg.* **40**, 105.
HAWKING, F. (1947). *Nature, Lond.*, **159**, 409.
HOGGARTH, E. & MARTIN, A. R. (1948). *Brit. J. Pharmacol.* (in the Press).
JONES, W. R. (1946). *Ann. Trop. Med. Parasit.* **40**, 130.

Madinaveitia, J. (1946*a*). *Biochem. J.* **40**, 373.
Madinaveitia, J. (1946*b*). *Biochem. J.* **40**, i.
Maegraith, B. G., Tottey, M. M., Adams, A. R. D., Andrews, W. H. H. & King, J. D. (1946). *Ann. Trop. Med. Parasit.* **40**, 493.
Raventos, J. (1947). Communication to the Pharmacological Society, Manchester (1947). To be published.
Spinks, A. (1948). *Brit. J. Pharmacol.* (in the Press).
Williamson, J. & Lourie, E. M. (1947). *Ann. Trop. Med. Parasit.* **41**, 278.

THE STUDY OF ENZYMES IN RELATION TO SELECTIVE TOXICITY IN ANIMAL TISSUES

By R. A. PETERS

Department of Biochemistry, University of Oxford

I. INTRODUCTION

From time to time it is useful to try to see in perspective the lines of thought which have led to the present position in science. The problem of selective toxicity is so fundamental that it has excited attention for at least half a century; Ehrlich's idea of chemoreceptors in protoplasm was a bold and early conception. In this article, the development of the idea of selective toxicity to enzymes will be examined in relation to experimental evidence; no attempt will be made to give an inclusive review, but rather to quote work which seems to the writer to be significant in the whole development. It is interesting to note that up to about 1940, studies of selective toxicity *in vivo* were based mainly upon the events in vitamin deficiency and upon agents combining with active groups in the enzyme; studies of selective toxicity based upon the idea of substrate competition waited for the crystallization of thought by the studies of Fildes (1940) and Woods (1940). This work upon bacteria will be reviewed by D. D. Woods.

There is apt to be confusion at the outset if a clear distinction is not drawn between the problem of permeating a cell, and that of more specific toxic action. Though a truism, even now, there is sometimes slight confusion upon this point. The writer remembers being confronted with this in World War I in connexion with the action of the war gases. Phosgene and chloropicrin, for instance, both give rise to a fatal lung oedema, due to the escape of fluid into the lung caused by alteration of capillary permeability. At first it was thought that this was due to a physico-chemical change, perhaps an alteration of colloids, and that it might be reversed by the giving of calcium salts. This proved to be untrue. Bacq & Goffart (1941) have found that the lung changes occur equally well if the chloropicrin is injected into the circulation, so that it is not even necessary that the agent should enter the lung tissue through the air passage. A prominent theory then advanced was simple; the changes were due to intracellular liberation of acid. But some chemical substances used in gas warfare show prominent selective toxicities even when acid liberation is out of the question. For instance the chlorarsines hydrolyse, in an aqueous medium, to toxic oxides;

it was found that the oxide $(C_6H_5)_2AsOH$, formed from diphenylchlorarsine, is 10 times more toxic to Protozoa than that formed from phenyldichlorarsine, C_6H_5AsO (McCleland & Peters, 1919), and diphenylchlorarsine is equally more irritant to mucous membranes. It seems clear that these substances penetrate as the chlor compounds and hydrolyse to toxic products within the cells. It is almost a commonplace to point out that as we do not yet believe in action at a distance, one condition of toxicity must be that a substance can reach a site where it can exert activity. Such a view is embodied in the Overton-Meyer theory of narcotic action. After this first step other factors become involved which depend more upon structural chemistry.

Purely physico-chemical explanations of these activities became less unsatisfactory intellectually when the conceptions of surface structure in monomolecular layers were given to us by W. B. Hardy, Langmuir, Harkins (1912–19) which have been developed further by Adam (1938). The theory that surface structures can have elaborate molecular and atomic patterns at once gave some chemical basis for a belief in the possibility that relatively small amounts of a toxic substance could interfere with cell activity, provided that changes in a small part of the surface structure can influence a whole living cell. In this way a better explanation could be advanced for such phenomena as the oligodynamic action of traces of copper, which were such a puzzle to the older physiologists; we also now know that algae will concentrate the copper (Eichholtz, 1934; Clark, Straub, Peters, Quastel, Ing, Gaddum, Yorke, Danielli, 1936). The conception is now reasonable that interference with the surface structure of a cell is one way in which a small amount of a chemical substance can exert toxicity (Clark, 1937).

In the last years, experimental evidence has been growing for the different idea that damage to a cell can be caused by the selective toxicity of a poison upon some enzyme concerned with intermediary metabolism in the tissue cell. This view may be held to have developed out of studies upon the action of inhibitors on enzymes, which have been in progress for at least 50 years. They will generally be found to be reviewed in books on enzymes (Euler, 1920; Haldane, 1930) under the heading of enzyme poisons. To this category may belong fluoride and cyanide. As an instance of this, enzymes have been inactivated quantitatively with heavy metals and reactivated by removal of the metal with H_2S; an early example was that of Euler & Svanberg (1920) using saccharase and mercury salts. Among such studies there were also many devoted to the specific effect of inhibitors. It is hard to quote an example without being invidious, but one may mention the studies of Quastel (1931, 1932) on urease and fumarase, of Leloir & Dixon (1937) on pyrophosphate, and the fundamental work of the Stedmans (1931) on eserine and acetylcholine. Most of these studies were made with the object of throwing light upon the enzyme action itself.

Some recognized the need for connecting such enzyme investigations more definitely with the biological object of investigating the problems of toxicity in living tissue. An example of this arose in chemical warfare. Flury (1921) and Rona and György (1920) studied the effects of arsenicals on some enzymes upon the general idea that substances active in such small concentrations must be poisoning enzymes; for instance, 0·01 mM. (10^{-5}) methylarsenoxide produced about 50% inhibition of urease. The relation of the enzymes even to arsenicals proved complex, and, though interesting as studies in specificity, no intracellular enzyme was implicated at this stage. Somewhat later Peters & Walker (1923–5) could obtain no support for the enzyme theory using chemical warfare agents. Equally it may be said that, though Onaka (1911) in Warburg's laboratory linked the poisonous action of arsenic with tissue respiration in nucleated red blood cells, and though Dresel (1926) showed that the rates of respiration of several tissues were powerfully inhibited by arsenite, up to 1929 no satisfactory evidence was produced to connect cell toxicity with cell enzyme inhibition. An exception was the proof by Keilin (1925) that cyanide inhibited cytochrome oxidase activity, and the association of this with Warburg's work on the 'respiratory enzyme'; but as Dixon (1929) pointed out this classical work on oxidation mechanisms was not then linked with that on intermediary metabolism. In a more specialized physiological category, but of great importance as connecting *in vivo* with *in vitro* events, was the correlation of acetylcholine and the action of eserine with an esterase in heart extract by Loewi & Navratil (1926; see also Dale, 1934).

Meanwhile, the discovery and isolation of glutathione by Hopkins (1921) gave an impetus to studies of sulphydryl groups. Early in this work, in addition to the soluble glutathione, Hopkins & Dixon (1922) had shown the presence of a so-called 'fixed' SH group in proteins; this gave a nitroprusside reaction upon proteins such as boiled muscle tissue and reacted with the soluble SH group in glutathione after oxidation. Though it was many years before a better understanding of the SH group in proteins began to appear, some studies made in 1923–5 were significant. Two sets of workers independently found a connexion between SH groups and arsenicals. Peters & Walker (1924–5) found that several vesicants, including diphenylchlorarsine, abolished the fixed SH reaction in muscle and skin slices; in a series of papers, acting upon the theory of Voegtlin & Smith (1920) that the active principle of arsphenamine was arsenoxide formed by oxidation, Voegtlin, Dyer & Leonard (1923; Voegtlin, 1925) connected the toxic action of therapeutic arsenicals with SH-containing substances in the tissues. The latter workers found a limited power of protection of rats against the toxic effect of arsenoxide by previous injection of thioglycollate ($HS.CH_2.COOH$), and also protection of trypanosomes *in vivo* by sulphydryl compounds. This action was confirmed for Protozoa by Walker (1928) using diphenylchlorarsine, and some temporary reversal was found with monothioethyleneglycol, $HS.CH_2CH_2.OH$. Though suggestive, these studies of SH groups produced no compelling evidence for enzyme theories; they might have been explained equally well by the idea that some more non-specific attack upon an SH group in cell proteins was involved.

Arsphenamine

As ═ As

H_2N NH_2

OH OH

↓

Arsenoxide

As=O

NH_2

OH

Steps towards more definite evidence of the connexion between enzymes and pathological processes were taken about 1929 in researches where enzymes were studied *in vitro* in parallel with *in vivo* studies. Kinnersley & Peters (1929) initiated a series of researches upon the intermediary metabolism in the brain of the vitamin B_1-deficient pigeon in opisthotonus, which were continued by Gavrilescu & Peters (1931) and others (see Peters, 1936*a*); these latter found a specific action of vitamin B_1 *in vitro* upon the brain tissue from the thiamin-deficient animal. Increases in oxygen uptake due to addition of vitamin *in vitro*, with lactate as substrate, went parallel with the recovery of the animal *in vivo* after dosing with the vitamin. About this time also, Lundsgaard (1930) discovered the fact that muscle poisoned with iodoacetate contracted without production of lactic acid; such muscles develop a peculiar contracture after stimulation, the appearance of which can be delayed by a lactate medium (Mawson, 1932). Another *in vivo*, *in vitro* study of interest was made by Quastel & Wheatley (1932*a*), who connected a reversible inhibition by narcotics of brain-tissue respiration with the intermediary metabolism of glucose (Quastel, 1939).

These studies were all useful in their connexion of abnormal effects in tissues with enzymes; in some ways the most definite case is still that of vitamin B_1 deficiency (aneurin, thiamin). Continued study over 10 years, mainly in the Oxford Biochemistry Department, has given the final proof not only that the particular enzyme system concerned was the pyruvate oxidase system (Banga, Ochoa & Peters, 1939), but also that the deficient entity was the thiamin pyrophosphate, cocarboxylase, which was isolated from yeast in 1937 by Lohmann & Schuster. This coenzyme is also responsible for the essential step in alcoholic fermentation in which pyruvic acid is converted to acetaldehyde and carbon dioxide. In the animal there is not a simple decarboxylation, but an oxidative decarboxylation, and in deficiency of this vitamin pyruvate accumulated in the blood. The pyruvate oxidase system was shown in 1939 to be composed of a battery of enzymes, and needs not only the pyruvate dehydrogenase component, but also cocarboxylase, phosphate, magnesium, adenylic acid, and an acid of the 4-carbon di-COOH series, e.g. fumarate. In more modern terms this suggests catalysis by the Krebs tricarboxylic cycle. Though there is still much to be learnt, this phenomenon demonstrates clearly a close connexion between the reversible and pathological abnormality of a tissue and lack of a coenzyme necessary for pyruvate oxidation in an enzyme system. It should be emphasized that the abnormality in the cells leading to the convulsive opisthotonus condition precedes changes which are visible with the microscope. In order to crystallize such thoughts, the term 'biochemical lesion' was proposed for this and still appears useful.

Connected with these studies of vitamin B_1 it should be noted that the isolation of vitamin C, and the identification of riboflavin in 1932–3 (for literature see György, 1942) as part of an enzyme system, and later of nicotinic acid (Elvehjem, Maddon, Storey & Woolley, 1937) as the antipellagra factor have been important in the development of thought. But it must also be recognized as a stimulus for further work that even now it cannot be said definitely which enzyme is inhibited before the induction of the known pathological lesions in ariboflavinosis and pellagra. Pyridoxin is in a somewhat different position, because, apart from the work upon bacteria (Gale, 1945), recently Blaschko, Carter, O'Brien & Sloane-Stanley (1947) in Oxford have found that pyridoxin deficiency decreases dopa and cysteic acid oxidation in liver, and that in some cases reactivation is induced by addition of pyridoxal + ATP. A whole new chapter connected with thiamin and peripheral neurophysiology has been recently developed by von Muralt and colleagues (for review see von Muralt, 1948), and is still to be finally appraised.

Biochemical lesions should be distinguished from the more permanent biochemical changes involving genes; they can be classified as shown in Table 1.

Table 1

Biochemical lesions due to
- (*a*) Deficiencies (nutritional)
- (*b*) Toxic agents:
    - Metals (As, Hg, etc.)
    - Vesicants and war gases
    - Temperature changes
    - Narcosis
    - Metabolic antagonists
    - Gases, such as oxygen, carbon monoxide

II. SH GROUPS AND THEIR SIGNIFICANCE

During part of this work upon the biochemical lesions in thiamin deficiency, a deeper knowledge of the role of sulphydryl groups in biochemistry began to appear. A connexion was made between the action of SH groups and iodoacetate by the finding that this would combine with a compound like glutathione (Dickens, 1933; Quastel & Wheatley, 1932*b*; Rapkine, 1933). In 1935,* the writer made an observation which to him reopened the whole question of vesication as due to an initial attack upon an enzyme (Peters, 1936*b*). It had been found earlier that iodoacetate had a selective action upon the pyruvate component in the lactate oxidase system in brain (Peters, Rydin & Thompson, 1935); now it was observed that mere traces of the vesicant (0·00009 M) mustard gas sulphone, had a similar effect. It was also found that arsenite, which Krebs (1933) had found to stop oxidation of α-keto acids in high concentration, was selectively toxic at the same stage in minute concentrations (5 μg./3 c.c.). Since it was known that iodoacetic acid was a mild vesicant, the connexion seemed possible and was then advanced; further, the reactions of arsenicals and of iodoacetate with —SH led to the idea that an SH component of the pyruvate system was involved.

In connexion with the relation between SH groups and enzymes, in 1933, Hellerman, Perkins & Clark proved that the enzyme urease was influenced by oxidation and reduction, and that when poisoned by metals it could be reactivated by —SH. From 1936 onwards, evidence accumulated which could be interpreted upon the idea that some enzymes contained an essential SH group. Particularly important was the work by Hopkins and his colleagues on succinodehydrogenase (e.g. Hopkins & Morgan, 1938). As early as 1937, Dixon proved clearly that there was a high selectivity in action of iodoacetic acid on enzymes. The alcohol dehydrogenase of yeast and the triosephosphate dehydrogenase of yeast were highly sensitive to iodoacetate; the succinic dehydrogenase, on the other hand, is comparatively insensitive to iodoacetate though susceptible to —SS— reagents and to the action of maleate. Such discrepancies very properly made critical observers

* This work was in progress at the same time as work by Berenblum *et al.* (see below).

(Dixon, 1939) reluctant to accept *in toto* the view that these so-called SH enzymes necessarily contained an essential SH group, especially because the evidence was indirect, and because some (Michaelis & Schubert, 1934) claimed that iodoacetate could also combine with NH groups though less easily. After all the further work done, a modern biochemical view would be that the discrepancies are due to the existence in proteins of SH groups in more than one state of activation; this indicates that as the next approximation we shall have to classify such enzymes according to their sensitivity to various SH reagents.

III. BIOCHEMICAL ATTACK UPON CHEMICAL WARFARE SUBSTANCES

An attack on the problem of the biochemical lesions produced by the poison gases during World War II in the hope of finding antidotes was intensively pursued in this country and later in the U.S.A. Independently, a somewhat similar study was carried out by Bacq and his colleagues in Belgium (see 1946, 1947), the results of which were not known until the end of the war. The basic idea of the researches in Oxford and Cambridge was that the poisons attack intracellular enzymes, and that this initial attack starts the train of pathological disturbances leading, for instance, to vesication. It involves also the notion that the attack is such that the cells cannot readily turn to an alternative metabolic path for their energy. This war work has been recently reviewed (Peters, Stocken & Thompson, 1945; Waters & Stock, 1945; Dixon & Needham, 1946; Gilman & Phillips, 1946; Peters, 1947); it raises so many points vital to our subject that some reference to the more important features is necessary. The question may be asked: How far does this war work advance the conception that a biochemical lesion due to a poison may be induced by damage to some enzyme?

IV. TRIVALENT ARSENICALS AND SOME OTHER METALLIC SUBSTANCES

It is convenient to start with the trivalent arsenical, lewisite

$$(CH.Cl{:}CH.AsCl_2).$$

The attack in the Oxford Biochemistry Department by the team under the writer's direction is reviewed in the article by Peters *et al.* (1945); it started in October 1939 with two basic assumptions: (1) that the arsenical combined with some SH group, and (2) that this SH grouping formed part of the pyruvate oxidase system. It was soon found that the pyruvate dehydrogenase system was sensitive to two tests used for identifying SH enzymes; it was easily inhibited by maleate, and it was reversibly oxidized and reduced by SS and SH groups respectively. At the same time the earlier evidence for selective action of arsenite upon the pyruvate component of the lactate oxidation system was confirmed and extended; and it was made clear that the pyruvate oxidase system in brain was much more sensitive to arsenicals than several other enzymes; the sensitive component was not the coenzyme. Table 2 shows some of these results. One interesting point is that the total

Table 2. *Comparison of the effects of trivalent arsenicals upon the pyruvate and succinate oxidase systems in brain (pigeon)*

Preparation	Arsenical	Conc. (mM.)	% inhibition	
Brei	Arsenite	0·034	75	13
	Lewisite	0·017	89	5
Dispersion (homogenate)	Arsenite	0·04	55	5

The pyruvate oxidase system is also more easily poisoned than the α-glycerophosphate and lactate dehydrogenase (Peters, Sinclair & Thompson, 1940, 1946).

pyruvate oxidase system is more sensitive to arsenic than the pyruvate dehydrogenase component. Though not sufficiently extensive to prove a specificity, the data gave very strong ground for belief in a selective action. This was confirmed by showing that the injection of arsenicals into animals gave an increase of pyruvic acid in the blood rather like that in thiamin deficiency, suggesting again that the pyruvate oxidizing system is the most readily attacked. In view of the earlier work on arsenic and SH groups it might be thought that this problem could have been readily solved by the use of a monothiol as antidote. However, though large excess of monothiols might reverse the effects of arsenoxide (Eagle, 1939), no monothiol could even protect either *in vitro* or *in vivo* from the toxic effects of lewisite (Sinclair, 1940). This is now known to be due to the biochemical existence of the reversible reaction of monothiol substances discovered by Cohen, King & Strangeways (1931) and embodied in equation (1):

$$R.\mathrm{AsO}+2R'\mathrm{SH}\rightleftharpoons R\mathrm{As}(\mathrm{S}R')_2+\mathrm{H_2O}. \qquad (1)$$

At alkaline pH the thioarsenite dissociates and will give the nitroprusside reaction for SH compounds. Even at neutral pH there is a strong tendency to dissociate. From this it was clear that there was some grouping in the enzyme protein concerned, making a more stable compound with the arsenical than that made with a monothiol. After some 13 months the problem was finally resolved by the analysis by our two colleagues, Stocken & Thompson (1940, 1946), of lewisite and arsenite compounds of kerateine (Goddard & Michaelis, 1934), the reduced product of hair keratin. When treated even to excess with these arsenicals, at pH 7·0, the kerateine takes up a definite but limited amount of arsenic; the average content of arsenic for lewisite-kerateine was 0·52%, and for arsenite-kerateine 0·41%.

Lewisite-kerateine is rather stable upon reprecipitation and dialysis. The SS form, metakeratin, on the other hand, takes up negligible amounts, so that the arsenical is combining with SH groups. Most important of all, the amount of arsenic taken up by the kerateine is less than the number of SH groups disappearing. This led them to the realization that the more stable form of arsenic in the enzyme must be a ring form in which two sulphur atoms are combined with one arsenic, as follows:

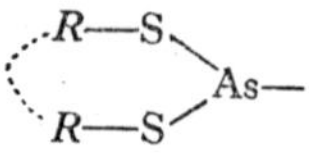

$$\begin{matrix} \cdot\cdot R\text{—S} \\ \vdots \qquad\qquad \text{As—} \\ \cdot\cdot R\text{—S} \end{matrix}$$

Porphyrindin titration of the —SH disappearing, as compared with the arsenic taken up, indicated that about 75% of arsenic was combined in the ring form. This would explain why combination with the component of the enzyme was more stable than that with a monothiol:

Formula of BAL

$$\begin{array}{c} H_2C.SH \\ | \\ HC.SH \\ | \\ H_2C.OH \end{array}$$

The new compound 2:3-dimercaptopropanol was synthesized by Stocken; it was the third compound to be tried and proved to have the desired action; it was called British anti-lewisite (BAL) in the U.S.A. BAL, and also some other dithiols tried, had the protective effect desired against lewisite for the test enzyme system, the pyruvate oxidase of brain; some monothiols were entirely ineffective (Table 3). It is still more important that the compound reverses a poisoning of the enzyme after it has been thoroughly established. This effect is shown in Table 4. This result strongly supports the idea that the antidote acts by taking advantage of a slightly greater dissociation of the protein-arsenic ring. The order of stability is: most stable lewisite —BAL, next lewisite-pyruvate enzyme, and then with a different order of stability lewisite-monothiol. This was confirmed in experiments using porphyrindin as an indicator.

The proof of the pudding is in the eating! These facts of reversal obtained *in vitro* were confirmed *in vivo* in a striking way. Both in Oxford and in Porton, it was found that the pathological damage visible in a developing lewisite blister could be reversed even after 2 hr. by application of BAL to the skin, and the compound would save an eye poisoned with a trivalent arsenical; no compound other than a dithiol is known which will do this. Equally the systemic effects produced by the presence of arsenicals in the whole animal rather than in the skin also responded to the antidote.

Table 3. *Protective action of dithiols for the brain pyruvate oxidase system (brei from pigeon) poisoned by lewisite oxide* (0·016 mM.)

(Data of Stocken & Thompson, 1941.)

	mM.	Inhibition of O_2 uptake Respiration periods			
		Lewisite oxide		Lewisite oxide + thiol	
		0–30 min.	30–90 min.	0–30 min.	30–90 min.
Dithiol compounds:					
Toluene-3:4-dithiol	(0·21)	67	64	0	0
Ethane-1:2-dithiol	(0·36)	56	51	0	0
2:3-Dimercaptopropanol (BAL)	(0·27)	72	46	0	0
1:3-Dimercaptopropanol	(0·27)	45	44	0	2
1:3-Dimercaptopropane	(0·31)	49	51	0	0
2:3-Dimercaptopropylamine	(0·27)	45	48	0	13
Keratеine 0·3%		65	53	10	12
Monothiol compounds:					
2-Mercaptoethanol	(0·43)	57	55*	54	55*
Cysteine (HCl)	(0·27)	48	47	51	50

* Period 30–60 min.

Table 4. *Ability of 2:3-dimercaptopropanol (BAL) to reverse the toxic effect of lewisite oxide on the brain pyruvate oxidase system (brei)*

Conc. BAL (mM.)	Average of exp. nos.	Poison antidote	Inhibition of O_2 uptake			
			15–45 min.	45–75 min.	75–105 min.	105–135 min.
0·27	2	Lewisite alone	47	52	44	—
		Lewisite + BAL added after 15 min.	↓ 7	2	0	—
0·14	3	Lewisite alone	56	51	54	56
		Lewisite + BAL added after 45 min.	56	↓40	23	16
0·27	1	Lewisite alone	56	57	53	52
		Lewisite + BAL added after 48 min.	49	↓48	26	27

↓ represents the time at which the BAL was added.

Injection of lewisite and some other trivalent arsenicals into animals such as rats rapidly produces acute diarrhoea and death. If BAL is injected sufficiently soon, these effects disappear. This forms the basis of the extension of the work to the clinical use of BAL in the accidental arsenic poisoning which may occasionally follow therapy with arsphenamines, due presumably to the arsenoxide formed, and which has given a remarkable success (*M.R.C. Report*, 1947).

Even though this dithiol is less toxic than any 'skin-penetrant' dithiol of similar chemical nature and has been clinically effective, it is still too toxic to use in very large doses. This is desirable for treatment, and the difficulty has been met for the treatment of systemic arsenical poisoning on the experimental level by the synthesis of a glucoside known as BAL-intrav, which was designed on theoretical grounds (Danielli, Danielli, Mitchell, Owen & Shaw, 1946). BAL-intrav is so water-soluble that it remains in the blood stream and is therefore much less toxic. It is hoped that it will prove successful in the clinic.

BAL is also most useful in mercury poisoning, as has been shown especially in human cases by Longcope & Luetscher (1946) and by Stocken (1947) in rats, and Gilman, Allen, Phillips & St John (1946) in dogs. In the case of mercury, the theoretical basis is not so certain as it is for the arsenicals. It is known that the poisoning of the pyruvate oxidase system by mercurials may be reversed by small amounts of BAL (Stocken, Thompson & Whittaker, 1947); but differing from the trivalent arsenic, the monothiol glutathione has a similar action in much larger amounts. In addition, it is also not yet clear which is the enzyme involved in the poisoning. As a matter of fact it has long been known that sulphides would reverse the action of mercuric salts; Rapkine (1931) showed that the division of sea-urchin eggs was inhibited by traces of mercuric salts and that this could be reversed by glutathione; and much earlier Chick (1908) found that the disinfectant power of mercuric salts could be reversed by ammonium sulphide.

These practical effects of 2:3-dimercaptopropanol surely remove possible doubts as to the correctness of the underlying theory. The idea that some component of the pyruvate oxidase system has two thiol groups so arranged in space that they can form a ring with trivalent arsenic can be known as Stocken & Thompson's dithiol theory; the stability resembles that of an 8-membered thioarsenite ring (Whittaker, 1947). Proof of the dithiol nature of any enzyme component in this sense will lie in the failure to reactivate with a monothiol in arsenical poisoning. Dixon and his colleagues (Dixon & Needham, 1946) have found that the poisoning of pyrophosphatase and of succinodehydrogenase by lewisite can only be reversed with a dithiol, whereas hexokinase, which we shall see to be important in another connexion, after inhibition by lewisite responds to a monothiol such as cysteine; choline esterase also responds to a monothiol (Thompson, 1948). These facts make it clear that the enzyme poisoned in skin is a 'dithiol' enzyme like the pyruvate oxidase; it cannot be hexokinase alone which is affected in lewisite poisoning. A large number of enzymes were examined by Barron & Singer (1945), both as regards their response to a reactivating thiol and their behaviour to various arsenical compounds. Many of these could be reactivated by monothiols. It should be mentioned that trivalent arsenical compounds vary somewhat in their activity. For instance, mapharside is less toxic at a given concentration than lewisite and forms a slightly more dissociable compound with BAL. The high toxicity of secondary aromatic arsenicals *in vivo* has not yet been explained, and there are grounds for thinking that monothiols may reverse their action (Peters & Stocken, unpublished experiment). A theory of the influence of the substituent groups in arsenicals upon the dissociation of thioarsenites formed with monothiols is badly needed.

This work on lewisite has been discussed at some length, because it demonstrates in a clear way the view that vesication may be due to an initial attack upon the pyruvate oxidizing system. It is surely important for theory that an attack upon this system can lead to reversible pathological effects, both when made negatively by removal of its coenzyme in a dietary deficiency or positively by the introduction of a toxic agent.

In summary, it does not seem too much to claim that these arsenical researches have established that one mode of attack upon tissue cells can be a selective effect upon the SH groups of an intracellular enzyme. In this sense it forms a direct proof of Ehrlich's theory of chemoreceptors, and it is interesting that Ehrlich himself (1909) suggested somewhat prophetically as possibilities that OH or SH groups might be arsenic receptors. Armed with this evidence, it is now possible to examine more intelligently whether the action of other substances fits into such a simple scheme. Before doing this, the susceptibility of the pyruvate system, and some points about vesication, may be briefly considered.

V. PYRUVATE OXIDASE SYSTEM

It has been mentioned previously that this is a co-ordinated complex of enzymes; as such, it might be expected to be sensitive to toxic agencies. The system does not stand freezing without loss of most of the activity; in the pigeon the brei preparation, in which the nuclei are largely intact, is relatively stable *in vitro* (see Peters, 1940); but after fracture of the nuclei by grinding to form a homogenate (dispersion), the system loses activity in 30 min. at 38° C. An active dispersion of rat brain can never be prepared. This relation to fracture of the nucleus is suggestive. The pyruvate oxidase has been considered to be implicated in oxygen poisoning (Dickens, 1946; Mann & Quastel, 1946) and in narcosis (Quastel, 1939); the system seems therefore to be most important from the pharmacological point of view. It is therefore relevant that the lethal doses for chemical warfare substances found by Fell & Allsopp (1939, 1946) upon tissue cultures correspond rather closely with the amounts needed to poison this same enzyme system in brain brei experiments; they are for lewisite oxide 11–18μM., and for the mustard gas sulphone 75 μM. If it is not the pyruvate oxidase system which is attacked, it must be one which is practically as sensitive.

Note added Oct. 1948. We have recently concluded (Coxon, Liébecq & Peters, 1948) that the pyruvate oxidase system in brain includes the Krebs tricarboxylic acid cycle when fumarate is added. Since this is well known to be a complex of enzymes, the susceptibility of the system to various toxic agents is well explained.

Quite recently, Massart, Peeters & Vanhoucki (1947) have studied the influence of acridines upon the respiration of yeast; they found a progressive diminution of this respiration which was reversed by ATP, or adenylic, or nucleic acid. They think that the ribonucleoproteins form adsorption complexes with acridines. At the writer's suggestion early in the war, Manifold (1941) investigated the effect of some acridines upon the pyruvate oxidase system in brain (human, pigeon, rabbit) and found toxic effects with low concentrations, 1/4000 and less. There were interesting differences, e.g. the 2:7 diamino compound was less toxic than acriflavine itself. The correlation of toxicity to this enzyme system with nucleic acid, and with the instability caused by fracture of the nucleus is certainly most suggestive (compare also McIlwain, 1941; Dickens, 1936).

VI. VESICATION

A discussion of the complex primary and secondary phenomena of vesication is outside the scope of this article. Arsenic attacks SH groups in an enzyme and this appears to be the initial step. From this it might be thought that vesication is always induced by an attack upon an —SH enzyme; in fact this is the assumption upon which the war work on mustard gas started in Oxford which was not justified in this case by the researches. But there is no evidence yet that the vesication so rapidly caused by heat is initiated by damage to an enzyme (Peters, 1945), so that it is wrong to think that enzyme damage must necessarily be a first step. However, it may often be so; Danielli (1941) found that a concentration of cyanide (M/1000) which only inhibits respiration produced no vesication in the perfused frog, whereas larger amounts which would also inhibit glycolysis (M/100), would produce vesicles, suggesting some damage to the enzymes in the carbohydrate series. The liberation of a skin proteinase has been considered by Beloff & Peters (1945) to be a necessary stage in the separation of the skin layer.

VII. SH ENZYMES AND LACHRYMATORS

Lachrymators are somewhat allied to vesicants, because in rather high concentration the 'tear gases' will also produce vesicles. During the war, in Cambridge, Dixon's team studied the action of lachrymators and in

Belgium independently of the British work, Z. Bacq and his colleagues made an intensive study of the effects of these on SH proteins, from which Bacq (1946) has developed the pharmacological conception of 'substances thioloprivés'. From this angle, several apparently unrelated substances which have the same pharmacological effect, can be brought into a unitary scheme. This generalization is undoubtedly useful as a first approximation; but closer inspection shows that the facts are less simple. For instance, the pyruvate oxidase system in brain which is so sensitive to arsenicals and to the mustard gas sulphone is only of intermediate sensitivity to iodoacetic acid at pH 7·0–7·3, and is not nearly as easily poisoned by mustard gas. The varied views which can still be taken are shown from the fact that Dixon regards lachrymators as the test agents *par excellence* for SH enzymes, whereas Barron considers that the supreme agents are arsenicals. Bacq thinks that both vesicants and lachrymators owe their power to their capacity for altering SH groups whether it be by oxidation or by combination; whereas, in an American statement (Gilman & Phillips, 1946), based largely upon experiments upon the sulphur and nitrogen mustard gases, there was not much stress laid upon enzymes or SH groups.

Considering the facts, it is now generally agreed both by ourselves, by Dixon and colleagues and by Bacq that the war gas substances including allyl*iso*thiocyanate, etc. mostly interfere with SH groups on proteins. The 'fixed' SH groups in skin and muscle tissue are destroyed; lachrymators irreversibly inhibited all enzymes which are believed to require an SH group for their activity (Mackworth, 1941, 1948); the reaction is as follows, and can be followed manometrically:

$$\underset{\text{Chloracetophenone}}{C_6H_5CO.CH_2Cl} + HS.R = C_6H_5CO.CH_2S.R + HCl.$$

The molar concentration required to inhibit was of the same order as that needed to poison enzymes. Owing to this action and to the presence of essential SH groups in the glycolytic system, e.g. the triosephosphate dehydrogenase, the lachrymators poison powerfully both cell respiration and glycolysis. The workers in Belgium have added many points of interest. Chloropicrin attacks both the free and masked SH groups in native albumin without changing the molecular weight (Desreux, Fredericq & Fischer, 1946). The speed of action under standard conditions is greater for some compounds like chloropicrin than for mustard gas sulphone. Furthermore, practically all these compounds give the 'Lundsgaard' (iodoacetate) effect with frog's muscle. All this work gives strong support to the theory connecting enzyme inhibition with an action upon the cell. For the case of the lachrymators, Dixon has pointed out that his theory of reaction with

SH groups will explain the lachrymatory activity not only of substances containing a positive halogen, but also of those containing a —C=C— link such as acrolein. It was further shown that the site of action was the corneal nerve endings; we do not yet know whether the sensory stimulation in the eye is an action upon an enzyme or upon some less specialized surface protein there.

VIII. MUSTARD GAS

The remarks here will be mainly confined to 'sulphur' mustard gas $S(CH_2CH_2Cl)_2$. This chemical compound is so wayward that it has defied the efforts of biochemists of two continents to find an antidote during the last war. It forms an exception to the general rule that SH groups are attacked preferentially and does not fit the enzyme ideas without reservations. It is now generally agreed by workers that the main sting of mustard gas is not due to the liberation of acid; this was shown clearly by Peters & Walker in 1923. It is also generally agreed that the compound does not combine preferentially with SH groups in proteins, though it combines well with a soluble SH compound such as glutathione or cysteine and will stop the action of glyoxalase in this way. The effects of mustard gas upon living organisms are so diverse that it is necessary to limit discussion rather strictly to those aspects concerned with our main argument, its relation to enzymes. Quite early in the war, members of the Oxford team worked out the kinetics of mustard gas action (Ogston, Holiday, Philpot & Stocken, 1940, 1948). It is an alkylating reagent capable of combining with a variety of substances even at room temperature. Put very briefly, the rate of formation of any compound of mustard gas, as for instance with cysteine, depends in the first place upon the rate of conversion to an activated form, believed now to be the cyclic 'onium' ion

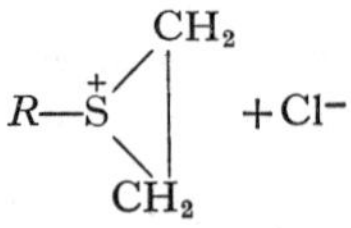

$$R\text{—}\overset{+}{S}\begin{matrix} \diagup CH_2 \\ | \\ \diagdown CH_2 \end{matrix} \quad + Cl^-$$

and then upon the competition between water molecules and cysteine for the activated mustard gas. In a mixture the proportion in which a particular reagent reacts depends on its concentration and characteristic velocity content and on those of all the other reagents (Table 5). Among a series of SH compounds prepared, all were not equally able to compete for the activated mustard gas. As an example, ethanedithiophosphonate competes 100 times more actively than cysteine; the SH present is differently activated. The development of the competition theory in Oxford (by Ogston *et al.* 1940, 1948), at Porton (by Ogston & Powell, 1944) and in the U.S.A. (by Doering & Linstead, 1942) showed that mustard gas also combined with groups such as —COOH (especially poly-carboxylic acids) and $—NH_2$. In addition, hydrolysed mustard gas solutions may contain sulphonium salts. Some of these facts are fundamental to any understanding of the way in which mustard gas attacks biochemical substances and enzymes. When proteins or protein products are treated with mustard gas, the predominant reaction is not combination with —SH (Banks, Boursnell, Francis, Hopwood & Wormall, 1946; Peters & Wakelin, 1941, 1947*b*; Northrop, 1942; Bergman, Fruton *et al.* 1942; see also Moore, Stein & Fruton, 1946). Though some mustard gas combines with the SH in kerateine, unlike the arsenicals much combines with other groups.

Table 5. *Some reactions of mustard gas forming basis of competition theory*

SN_1 activation	$R.Cl \rightarrow R.Cl^*$
Hydrolysis	$R.Cl^* + H_2O \rightarrow ROH + HCl$
Inhibition by Cl^-	$R.Cl^* + Cl^- \rightarrow R.Cl + Cl^-$
Substitution where B is a base	$R.Cl^* + B \rightarrow R.B + Cl^-$
Substitution where A and B are conjugate acid and base	$R.Cl^* + A \rightarrow RB + HCl$

Data of Ogston, Holiday, Philpot & Stocken (1940, 1948): cf. also Doering & Linstead (by communication, 1942).

In regard to enzymes, in 1936 Berenblum, Kendall & Orr reported that mustard gas was more toxic to glycolysis than to respiration in a brain brei; this was suggestive though not related to individual enzymes. Some observations, which have been made in the Oxford laboratory from 1939 onwards upon the pyruvate oxidase system are collected in Table 6. They show that mustard gas is toxic, but far less so than the arsenicals or

Table 6. *Concentrations of various substances which inhibit the uptake of oxygen (pyruvate oxidase system) by a pigeon brain brei under standard conditions* (pH 7·3 *etc.*) (mM., 10^{-6} M)

		mM.
Arsenicals	Lewisite	16
	Arsenite	34
Non-arsenical	Iodoacetamide	60
	Divinylsulphone	67
	Mustard gas sulphone	110
	Allyl*iso*thiocyanate	242
	2:2′:2″-Trichlortriethylamine	750
	Mustard gas	1000

Data of Peters and Wakelin (1947*a*, 1940)

divinylsulphone. The only enzyme proteins known to be affected by mustard gas are hexokinase, the pyruvate oxidase system, some proteinases, creatine and pyruvate phosphokinase (enzymes concerned in phosphate transfer), and inorganic pyrophosphatase. Choline oxidase, acetylcholine esterase from brain (Gilman & Phillips, 1946) and pyruvate oxidase (Peters, Thompson & Wakelin, 1942*b*) were inactivated by mustard gas nitrogen. The action of mustard gas on the pyruvate oxidase system is also much potentiated by the presence of diethyldithiocarbamate (Peters & Wakelin, 1947*a*). No cell compound is yet known which could exert such an effect but it is evidently conceivable. Dixon & Needham have advanced the theory, that vesication is due to inhibition of hexokinase; the first stage in the intermediary metabolism of glucose would therefore be cut out. Although this theory should not be pressed too far (compare p. 46), nevertheless they have produced an impressive correlation between vesication and inhibition of hexokinase under their conditions in the case of some fifty substances tried. The teams of Dixon and of Wormall have investigated, with radioactively marked mustard gas, the quantitative aspects of combination of mustard gas with proteins. In this way (Bailey & Webb, 1944, 1948, with Boursnell, Francis & Wormall) it has been found that 6–7 molecules of H were combined with a molecule of hexokinase when this was completely inactivated. Calculation shows that this means a very specific attack upon specialized groups, less than 1 H group in 100 amino acid residues. Since the amounts of H which combined with the protein ovalbumin were very similar, it must be concluded that the H specially attacks the active groups in hexokinase. Some is certainly combined with —COOH groups in proteins.

In spite of these facts there is still not complete conviction that the biochemical lesion in vesication by mustard gas is due to an initial attack upon hexokinase. One of the main difficulties is that whereas some of the actions of mustard gas are immediate, as for instance the liberation of leucotaxin, the poisoning of hexokinase in skin has a delay period of one to two hours. This has been met by Dixon by the hypothesis that the mustard gas only attacks the hexokinase after combination with some more reactive protein has ceased.

It should be emphasized that these observations are concerned with the first action of relatively small amounts of mustard gas; it follows from the competition theory that with larger amounts there will be a gradual reaction with a wider circle of groups in increasing numbers of biochemical substances. Hence the biochemical lesion in mustard gas poisoning must be graded and complex, depending upon the amount and conditions; the phenomenon resembles in this respect that in thermal burns. The most significant happening will be that taking place initially with the smaller concentrations. It is therefore relevant to point out that concentrations of 'the mustards' below those which affect either respiration or glycolysis have been found to produce fundamental changes in mitotic activity (see Gilman & Phillips, 1946). Hence it is important that mustard gas was shown early in the war by Berenblum & Schoental (1940) to combine readily with nucleoprotein. This was confirmed and extended by Young & Campbell (1947) and by Elmore, Gulland, Jordan & Taylor (1948). The Canadian workers found that under their conditions twice as much mustard gas combined with thymus nucleoprotein as with globin, 10 times as much with yeast nucleoprotein, and 26 times as much with guanine. Undoubtedly the nuclear effects of both S and N mustard are most important as shown by their cytotoxic action and influence upon genetics. From this it would seem that further study of the action of mustard gas upon nucleoproteins and upon enzymes concerned with these will be fruitful.

The writer felt that the delay in poisoning of hexokinase in skin was an insuperable difficulty and has fallen back to the idea that the first change produced by the mustard gas is irreversible damage of some component in the surface of the cell, leading as a first step to an alteration of permeability. There is direct evidence for this in thermal burns because skin proteinase leaves the skin. With mustard gas the evidence was not so clear; but there was experimental support for the view that the action of mustard gas on skin led to a disappearance of proteinase which was greater than could be accounted for by any direct attack of the mustard gas upon the enzyme itself (Beloff, Peters & Wakelin, 1945). Since Brachet (1942) has shown that ribosenucleic acid is present in the cytoplasm, this might of course be an attack upon such nucleoproteins in the cell surface. Whether the first attack be upon the permeability or not, it must be still regarded as controversial (and therefore not yet proven) that mustard gas makes its initial attack upon an enzyme. In this sense it is in a different category from the substances previously discussed.

IX. SOME OTHER SUBSTANCES

Before concluding this account, brief reference can be made to some other cases where selective toxicity is being studied. It should be stressed that all enzyme inhibitors *in vitro* are not necessarily active *in vivo*; the problem of permeability and accessibility to the enzyme arises, as well as the possibility of diversion from the active centre by biochemical conjugation. The writer encountered one clear instance of this in the war work; the compound of mustard gas and diethyldithiocarbamate (Peters & Wakelin, 1947*a*) was toxic to the pyruvate oxidase system *in vitro* and quite non-toxic *in vivo*. The opposite effect is found with pentavalent arsenicals, which are toxic *in vivo* because of conversion to the trivalent form.

Fluorine compounds. The two classes of compounds which have been receiving most attention are the fluoroacetates and the fluorophosphonates.

Fluoroacetates. Of this series, McCombie & Saunders (1946*b*) considered that $CH_2F.CO$ was the potent group; the action of Na fluoroacetate ($CH_2F.COONa$) may be discussed, first prepared by Swarts (1896) and now used as a rat poison. Even as little as 50 μg./kg. when given to dogs leads to death with convulsions in about two hours (Chenoweth & Gilman, 1946); until the terminal stages, the animals appear to be remarkably well. The question is still open as to whether this is due to an action *in vivo* upon some enzyme. Most enzymes are relatively unaffected by fluoroacetate *in vitro* (Webb, 1943); it was found in Oxford (Peters & Wakelin, 1943, unpublished) that the pyruvate oxidase system in brain was insensitive to large concentrations (0·02 M) of fluoroacetate; in view of the number of enzymes involved in pyruvate oxidation, this latter fact considerably narrows the field. It seems to be inconsistent with the suggestion made (Phillips, 1946) that 'a possible pathway in the oxidative disposal of pyruvate by mammalian tissues may involve a direct oxidation to acetate and CO_2 followed in turn by oxidation of the formed acetate'. Bartlett & Barron (1947) consider that fluoroacetate is concerned with this oxidation of acetate for which they believe it to be a competitive inhibitor. This evidence would be more convincing if the doses *in vitro* were not so much greater than those needed *in vivo*. Further, though fluoroacetate upsets the functioning of brain tissue, there is little evidence that acetate is a normal metabolite in brain.

Note added October 1948. Since writing the above, Liébecq & Peters (1948) have concluded from experiments with guinea-pig kidney tissue that fluoroacetate in addition to any competitive effects with acetate interferes directly in the reactions of the tricarboxylic acid cycle, so that citrate accumulates. This effect can be induced with comparatively small concentration of the poison, much nearer to those causing effects *in vivo*.

Fluorophosphonates. These compounds were intensively studied in Cambridge and elsewhere during the war and the facts about them have been well reviewed (McCombie & Saunders, 1946*a*; Adrian, Feldberg & Kilby, 1946; Mackworth & Webb, 1948; Bodansky & Mazur, 1946). The standard substance is d*iso*ipropylfluorophosphonate (DFP). Unlike eserine, the fluorophosphonates poison choline esterase irreversibly in very low concentration, are powerful constrictors of the pupil and are very toxic. The brain choline esterase is less susceptible to DFP than the less specific serum choline esterase, 10^{-7} M as against 10^{-9} M (Mendel & Hawkins, 1947). Recently, Adams & Thompson (1948) have found that nitrogen mustard gas (DDM) has a relatively more powerful action on the brain choline esterase. Nevertheless, DFP will reduce the amount of choline esterase present in the brain, and animals die when it is reduced to 5% of the normal value. Though DFP has action on other enzymes, much higher concentrations are required, and the case seems a strong one for specificity. The views of the importance of choline esterase in nerve conductivity receive support from these facts, but Nachmansohn's hypothesis of the relation of acetylcholine to the activity of the nervous system is still somewhat controversial.

Non-fluorine compounds. Another most interesting substance and illustrative of a class which there is no time to discuss here is another rat poison, alpha naphthyl thiourea (ANTU) (Richter, 1945; Dubois, Holm & Doyle, 1946). In fact our colleagues in the U.S.A. under the leadership of Dr Winternitz & Dr Rhoads are developing with much enterprise the practical applications of this biochemical field. To our American colleagues, we owe the trials of nitrogen mustards in Hodgkin's disease and of di*iso*propylfluorophosphonate in myasthenia gravis. Apart from the classical example of acetylcholine and eserine, which was mentioned earlier, there are also many important researches of much pharmacological interest, which do not yet connect so definitely with the problem under discussion; there may be mentioned the competitive inhibition by ephedrine of the oxidation of adrenaline by amine oxidase (Blaschko, Richter & Schlossman, 1937), and the antihistamine substances. The antimalarial substance atabrine inhibits cytochrome reductase (Wright & Sabine, 1944; Haas, 1944).

X. FINAL REMARKS

A review of the field as a whole has shown that, based upon vitamin deficiency studies, convincing cases now exist in which the action of poisons can be directly related to selective effects upon enzymes; the arsenicals and lachrymators fit into this category, and also fluorophosphonates; mustard gas still remains somewhat of an enigma. This study of selective toxicity *in vivo* is still relatively little advanced, but enough has been done to form a firmer ground for future studies, and there is no doubt that the future will show big advances. It will not be out of sight that this type of interference with cell metabolism should do much to clarify the bearing of '*in vitro*' studies of enzymes upon the normal functioning of the tissues. It is not entirely fanciful to apply Sherringtonian principles of integrative action to the chemical components which make up the co-ordinative biochemistry of the organism, especially if the writer's belief in a cytoskeleton is shared. The views given in this account have separated rather sharply toxicity to enzymes from toxicity to a surface. At the same time, it must be pointed out that this may be merely a stage in intellectual evolution. The most damaging effect of all would be that upon an enzyme oriented in the surface of a cell or nucleus. There is evidence that some enzymes live at the surface, so that ultimately we may look for some higher unity.

Thanks are due to Miss J. Allen for Secretarial help.

REFERENCES

ADAM, N. K. (1938). *Physics and Chemistry of Surfaces*. Oxford Univ. Press.
ADAMS, D. H. & THOMPSON, R. H. S. (1948). *Biochem. J.* **42**, 170.
ADRIAN, E. D., FELDBERG, W. & KILBY, B. A. (1946). *Nature, Lond.*, **158**, 625.
BACQ, Z. M. (1946). *Summaries Experientia*, **2**.
BACQ, Z. M. (1947). *Actualités biochimiques. Travaux recents sur les toxiques de guerre*. Liege: Editions Désœur.
BACQ, Z. M. & GOFFART, M. G. (1941). *Acta Belg.* **1**, 33.
BAILEY, K. & WEBB, E. C. (1944). Report to Ministry of Supply by Dixon, no. 30.
BAILEY, K. & WEBB, E. C. (1948). *Biochem. J.* **42**, 60.
BANGA, I., OCHOA, S. & PETERS, R. A. (1939). *Biochem. J.* **33**, 1109.
BANKS, T. E., BOURSNELL, J. C., FRANCIS, G. E., HOPWOOD, F. L. & WORMALL, A. (1946). *Biochem. J.* **40**, 745.
BARRON, E. S. G. & SINGER, T. P. (1945). *J. Biol. Chem.* **157**, 221.
BARTLETT, G. R. & BARRON, E. S. G. (1947). *J. Biol. Chem.* **170**, 67.
BELOFF, A. & PETERS, R. A. (1945). *J. Physiol.* **103**, 461.
BELOFF, A., PETERS, R. A. & WAKELIN, R. W. (1945). Report to Ministry of Supply and M.R.C. by Peters, no. 89.
BERENBLUM, I., KENDALL, L. P. & ORR, J. W. (1936). *Biochem. J.* **30**, 709.
BERENBLUM, I. & SCHOENTAL, R. (1940). Report to Ministry of Supply.
BERENBLUM, I. & SCHOENTAL, R. (1947). *Nature, Lond.*
BERGMANN, M. & FRUTON, J. S. *et al.* (1942). By communication.
BLASCHKO, H., CARTER, C. W., O'BRIEN, J. R. P. & SLOANE-STANLEY, G. H. (1947). *J. Physiol.* **107**, 18 P.
BLASCHKO, H., RICHTER, D. & SCHLOSSMAN, H. (1937). *Biochem. J.* **31**, 2187.
BODANSKY, O. & MAZUR, A. (1946). *Fed. Proc.* **5**, 123.
BRACHET, J. (1942). *Arch. Biol., Paris*, **52**, 307.
CHENOWETH, M. B. & GILMAN, A. (1946). *J. Pharmacol.* **87**, 90.
CHICK, H. (1908). *J. Hyg., Camb.*, **8**, 92.
CLARK, A. J., STRAUB, F. B., PETERS, R. A., QUASTEL, J. H., ING, H. R., GADDUM, J. H., YORKE, & DANIELLI, J. F. (1936). *Proc. Roy. Soc.* B, **121**, 580.
CLARK, A. J. (1937). Heffter, *Handb. Exp. Pharmakol.* **4**, 1. Springer.
COHEN, A., KING, H. & STRANGEWAYS, W. I. (1931). *J. Chem. Soc.* p. 3043.
COXON, R. V., LIÉBECQ, C. & PETERS, R. A. (1948). (Unpublished)
DALE, H. H. (1934). *Proc. Roy. Soc. Med.* **28**, 319.
DANIELLI, J. F. (1941). Quoted by Dixon & Needham (1946).
DANIELLI, J. F., DANIELLI, M., MITCHELL, P. D., OWEN, L. N. & SHAW, G. (1946). *Nature, Lond.*, **157**, 217.
DESREUX, V., FREDERICQ, E. & FISCHER, P. (1946). *Bull. Soc. Chim. Biol., Paris*, **28**, 493.
DICKENS, F. (1933). *Biochem. J.* **27**, 1141.
DICKENS, F. (1936). *Biochem. J.* **30**, 1233.
DICKENS, F. (1946). *Biochem. J.* **40**, 145.
DIXON, M. (1929). *Biol. Rev.* **4**, 352.
DIXON, M. (1937). *Nature, Lond.*, **140**, 806.
DIXON, M. (1939). *Amer. Rev. Biochem.* p. 1.

DIXON, M. & NEEDHAM, D. M. (1946). *Nature, Lond.*, **158**, 432.
DOERING, W. E. & LINSTEAD, R. P. (1942). By communication.
DRESEL, K. (1926). *Biochem. Z.* **178**, 70.
DUBOIS, K. P., HOLM, L. W. & DOYLE, W. L. (1946). *J. Pharmacol.* **87**, 53.
EAGLE, H. (1939). *J. Pharmacol.* **66**, 423, 436.
EHRLICH, P. (1909). *Ber. dtsch. chem. Ges.* **42**, 17.
EICHHOLTZ, R. (1934). Heffter, *Handb. Exp. Pharmakol.* **3**, 1942.
ELMORE, D. T., GULLAND, J. M., JORDAN, D. O. & TAYLOR, H. F. W. (1948). *Biochem. J.* **42**, 308.
ELVEHJEM, C. A., MADDON, R. J., STOREY, F. M. & WOOLLEY, D. W. (1937). *J. Amer. Chem. Soc.* **59**, 1767.
EULER, H. VON (1920). *Chem. der Enzyme*, **1**. München u. Wiesbaden: J. Bergman.
EULER, H. VON & SVANBERG, O. (1920). *Fermentforschung*, **3**, 330.
FELL, H. B. & ALLSOPP, C. B. (1939). Report to Ministry of Supply (December).
FELL, H. B. & ALLSOPP, C. B. (1946). *Brit. J. Exp. Path.* **27**, 305.
FILDES, P. (1940). *Lancet*, **1**, 955.
FLURY, F. (1921). *Z. ges. exp. Med.* **13**, 523.
GALE, E. F. (1945). *Biochem. J.* **39**, 46.
GAVRILESCU, N. & PETERS, R. A. (1931). *Biochem. J.* **25**, 2150.
GILMAN, A., ALLEN, R. P., PHILLIPS, F. S. & ST JOHN, E. (1946). *J. Clin. Invest.* **25**, 549.
GILMAN, A. & PHILLIPS, F. S. (1946). *Science*, **103**, 409.
GODDARD, D. R. & MICHAELIS, L. (1934). *J. Biol. Chem.* **106**, 605.
GYORGI, P. (1942). *Biological Actions of the Vitamins*, p. 54. Chicago Press, U.S.A.
HAAS, E. (1944). *J. Biol. Chem.* **155**, 321.
HALDANE, J. B. S. (1930). *Enzymes*. London: Longmans Green and Co.
HELLERMAN, L., PERKINS, M. E. & CLARK, W. M. (1933). *Proc. Nat. Acad. Sci., Wash.*, **19**, 855.
HOPKINS, F. G. (1921). *Biochem. J.* **15**, 286.
HOPKINS, F. G. & DIXON, M. (1922). *J. Biol. Chem.* **54**, 527.
HOPKINS, F. G. & MORGAN, E. J. (1938). *Biochem. J.* **32**, 611.
KEILIN, D. K. (1925). *Proc. Roy. Soc.* B, **98**, 312.
KINNERSLEY, H. W. & PETERS, R. A. (1929). *Biochem. J.* **23**, 1126.
KREBS, H. A. (1933). *Hoppe-Seyl. Z.* **217**, 191.
LELOIR, L. E. & DIXON, M. (1937). *Enzymologia*, **2**, 81.
LIÉBECQ, C. & PETERS, R. A. (1948). *Biochim. et Biophys. Acta.* (in Press).
LOEWI, O. & NAVRATIL, E. (1926). *Pflüg. Arch. ges. Physiol.* **214**, 678.
LOHMANN, K. & SCHUSTER, P. (1937). *Biochem. Z.* **294**, 188.
LONGCOPE, W. T. & LUETSCHER, J. A. (1946). *J. Clin. Invest.* **25**, 557.
LUNDSGAARD, E. (1930). *Biochem. Z.* **217**, 162.
MCCLELAND, N. & PETERS, R. A. (1919). *J. Physiol.* **53**, xxiv.
MCCOMBIE, H. & SAUNDERS, B. C. (1945*a*). *Nature, Lond.*, **157**, 287, 776.
MCCOMBIE, H. & SAUNDERS, B. C. (1946*b*). *Nature, Lond.*, **158**, 382.
MCILWAIN, H. (1941). *Biochem. J.* **35**, 1311.
MACKWORTH, J. F. (1941). Report to Ministry of Supply by Dixon, nos. 4 and 13.
MACKWORTH, J. F. (1948). *Biochem. J.* **42**, 82.
MACKWORTH, J. F. & WEBB, E. C. (1948). *Biochem. J.* **42**, 91.
MANIFOLD, M. (1941). *Brit. J. Exp. Path.* **22**, 111.
MANN, P. J. G. & QUASTEL, J. H. (1946). *Biochem. J.* **40**, 139.
MASSART, L., PEETERS, G. & VANHOUCKI, A. (1947). *Arch. int. Pharmacodyn.* **75**, 1210.
MAWSON, C. A. (1932). *J. Physiol.* **75**, 201.
MEDICAL RESEARCH COUNCIL REPORT (1947). *Lancet*, **2**, 497.

Mendel, B. & Hawkins, R. D. (1947). *Biochem. J.* **41**, xxii.
Michaelis, L. & Schubert, M. P. (1934). *J. Biol. Chem.* **106**, 331.
Moore, S., Stein, W. H. & Fruton, J. S. (1946). *J. Organ. Chem.* **11**, 675.
Muralt, A. von (1948). *Vitamins and Hormones*, **5**, 93. Acad. Press. Inc.
Northrop, J. H. (1942). By communication.
Ogston, A. G., Holiday, E. R., Philpot, J. St L. & Stocken, L. A. (1948). *Trans. Faraday Soc.* p. 511 (in the Press), and (1940) Report to Ministry of Supply by Peters, no. 7.
Ogston, A. G. & Powell, E. O. (1944). Report to Ministry of Supply by Peters, no. 88.
Onaka, M. (1911). *Hoppe-Seyl. Z.* **70**, 433.
Peters, R. A. (1936*a*). *Lancet*, **1**, 1161.
Peters, R. A. (1936*b*). *Nature, Lond.*, **138**, 327.
Peters, R. A. (1940). *Chem. & Ind.* **59**, 373.
Peters, R. A. (1945). *Brit. Med. Bull.* **3**, 81.
Peters, R. A. (1947). *Nature, Lond.*, **159**, 149.
Peters, R. A., Rydin, H. & Thompson, R. H. S. (1935). *Biochem. J.* **29**, 63.
Peters, R. A., Sinclair, H. M. & Thompson, R. H. S. (1946). *Biochem. J.* **40**, 516, and 1940.
Peters, R. A., Stocken, L. A. & Thompson, R. H. S. (1945). *Nature, Lond.*, **156**, 616.
Peters, R. A., Thompson, R. H. S. & Wakelin, R. W. (1942). Report to Ministry of Supply by Peters, no. 60.
Peters, R. A. & Wakelin, R. W. (1941). Report to Ministry of Supply by Peters, no. 39.
Peters, R. A. & Wakelin, R. W. (1947*a*). *Biochem. J.* **41**, 545, and (1940) Report to Ministry of Supply by Peters, nos. 2 and 8.
Peters, R. A. & Wakelin, R. W. (1947*b*). *Biochem. J.* **41**, 550.
Peters, R. A. & Walker, E. (1923). *Biochem. J.* **17**, 260.
Peters, R. A. & Walker, E. (1924–5). Government Reports.
Phillips, F. S. (1946). *Fed. Proc.* **5**, 297.
Quastel, J. H. (1931). *Biochem. J.* **25**, 898.
Quastel, J. H. (1932). *Biochem. J.* **26**, 1685.
Quastel, J. H. (1939). *Physiol. Rev.* **19**, 135.
Quastel, J. H. & Wheatley, H. M. (1932*a*). *Proc. Roy. Soc.* B, **112**, 60.
Quastel, J. H. & Wheatley, H. M. (1932*b*). *Biochem. J.* **26**, 2169.
Rapkine, L. (1931). *Ann. Physiol. Physicochim. biol.*, **7**, 382.
Rapkine, L. (1933). *C.R. Soc. Biol., Paris*, **112**, 790.
Richter, C. P. (1945). *J. Amer. Med. Ass.* **129**, 927.
Rona, P. & György, P. (1920). *Biochem. Z.* **111**, 115.
Sinclair, H. M. (1940). Quoted by Peters, Progress Report to Ministry of Supply.
Stedman, E. & Stedman, E. (1931). *Biochem. J.* **25**, 1147.
Stocken, L. A. (1947). *Biochem. J.* **41**, 358.
Stocken, L. A. & Thompson, R. H. S. (1940). Report to Ministry of Supply by Peters, no. 20.
Stocken, L. A. & Thompson, R. H. S. (1941). Report to Ministry of Supply by Peters, no. 33.
Stocken, L. A. & Thompson, R. H. S. (1946). *Biochem. J.* **40**, 529.
Stocken, L. A., Thompson, R. H. S. & Whittaker, V. P. (1947). *Biochem. J.* **41**, 47.
Swarts, F. (1896). *Bull. Acad. roy. Belg.*, III[e] S., **31**, 675.
Thompson, R. H. S. (1948). *Biochem. J.* **42**, xxvi.
Voegtlin, C. (1925). *Physiol. Rev.* **5**, 63.

VOEGTLIN, C., DYER, H. A. & LEONARD, C. S. (1923). *U.S. Publ. Hlth Rep.* **38**, 1911.
VOEGTLIN, C. & SMITH, H. W. (1920). *J. Pharmacol.* **15**, 475.
WALKER, E. (1928). *Biochem. J.* **22**, 292.
WATERS, L. L. & STOCK, C. (1945). *Science*, **102**, 601.
WEBB, E. C. (1943) See DIXON & NEEDHAM (1946).
WHITTAKER, V. P. (1947). *Biochem. J.* **41**, 56.
WOODS, D. D. (1940). *Brit. J. Exp. Path.* **21**, 74.
WRIGHT, C. I. & SABINE, J. C. (1944). *J. Biol. Chem.* **155**, 315.
YOUNG, E. G. & CAMPBELL, R. B. (1947). *Canad. J. Res.* **25**, 37.

THE DESIGN OF BACTERIAL INHIBITORS MODELLED ON ESSENTIAL METABOLITES

By H. N. RYDON

Department of Chemistry, Birkbeck College, London

I. INTRODUCTION

In 1940 Fildes (1940*a*) put forward a plea for 'A Rational Approach to Chemotherapy', in which he advocated research directed towards the preparation of modified essential metabolites which should be sufficiently closely related to the essential metabolites on which they were based as to fit the same enzyme, but sufficiently different as to be themselves devoid of essential metabolic activity. This idea was taken up widely, and fruitfully, but often without due regard to the factors making for true similarity between metabolite and inhibitor; the present paper is an appeal for such a truly rational approach to chemotherapy to be based on a proper appreciation of the structural factors on which such similarity must be founded.

Ehrlich was the first to attempt to put chemotherapy on a rational basis (cf. Ehrlich, 1909). Modelling his ideas on Witt's (1876) theory of dyestuffs, he postulated the necessity for the presence of two specific groupings in an effective chemotherapeutic agent, viz. a 'haptophore' group, through which the agent becomes attached to the parasite, and a 'toxophore' group by means of which it exerts its toxic action. Ehrlich's view of the matter was definitely chemical, and he pictured the haptophores as chemical groupings capable of reacting with other suitable chemical groupings (receptors) in the parasite. He was, however, unable to determine the precise chemical nature of the haptophores and receptors; this is not surprising since, in Ehrlich's day, little attention was paid to the secondary, non-covalent, binding forces which, we are coming to believe, play a predominant part in such interactions. Partly because of this insufficiency of the theoretical chemical background and partly because of their indefiniteness, Ehrlich's views did little to remove research on chemotherapy from empiricism. Starting either with some fairly promising agent discovered by chance, or with some substance obtained by introducing a toxophore group into a dyestuff known to stain the parasite under attack, chemical modifications were made, more or less at random, until a satisfactory agent was found; hundreds of substances might need to be made and tested before anything satisfactory was obtained, e.g. Ehrlich and Bertheim's '606' or Salvarsan (1912) and Bayer '205' or Fourneau '309' (Fourneau, Tréfouel &

Vallée, 1924). Such work was slow, inefficient and costly, and some new guiding principle was clearly needed.

This new principle was provided by Fildes's paper (1940*a*). Directly based on Woods's (1940) discovery of the antagonism of *p*-aminobenzoic acid and sulphanilamide and his own work (Fildes, 1940*b*) on the relationship between mercurial inhibitors and —SH compounds, Fildes's idea also had roots in the earlier work of Voegtlin (1925) on arsenicals and the purely biochemical work on the competitive inhibition of enzymes, such as succinoxidase (Quastel & Wooldridge, 1928; Potter & Elvehjem, 1937; Hopkins, Morgan & Lutwak-Mann, 1938). In enunciating Fildes's principle I cannot do better than paraphrase his own summary:

'Antibacterial substances function by interfering with an essential metabolite (i.e. a substance which takes an essential part in a chain of reactions necessary for bacterial growth). The interference may be (1) by oxidizing a substance which requires to be reduced, (2) by molecular combination forming an inactive product or (3) by competition for an enzyme associated with that essential metabolite. It is claimed that sulphanilamide acts as in (3), the essential metabolite being *p*-aminobenzoic acid. Class (3) inhibitions require an inhibitor so closely related in structure to the essential metabolite that it can fit the same enzyme and is sufficiently unrelated to be devoid of essential metabolic activity. It is suggested that research in chemotherapy might reasonably be directed to making such modifications of known essential metabolites as will have these characteristics.'

This was not only the required new principle but it was also put forward at the right moment in the development of chemistry, when the structures of the vitamins were falling rapidly to the attack of the organic chemists, while the biochemists were fast establishing the details of their functions in living processes. Accordingly, Fildes's idea was seized upon by many workers and a whole host of potential inhibitors, modelled on essential metabolites, were prepared and tested. Roblin (1946) has reviewed the subject very fully and I need quote only a few typical examples (pp. 62–4).

It must be admitted that most of these analogues have not proved of practical value as chemotherapeutic agents. An exception to this statement is to be found in the antimalarial drugs; here the analogy is, to say the least of it, rather remote and was clearly based on the observed interactions; however, the idea that such drugs acted in virtue of their relationship to riboflavin played some part in the development of the very successful paludrine (Curd, Davey & Rose, 1945).

A major criticism of the work which has been based so far on Fildes's principle is that the analogies between metabolite and inhibitor have been

Table 1

Metabolite	Inhibitor	References
$R\text{—}CH(NH_2)CO_2H$ α-Amino-carboxylic acids	$R\text{—}CH(NH_2)SO_3H$ α-Amino-sulphonic acids	McIlwain (1941)
$CH_3.S.CH_2.CH_2.CH(NH_2)CO_2H$ Methionine	$CH_3.CH_2.S.CH_2.CH_2.CH(NH_2)CO_2H$ Ethionine	Dyer (1938); Harris & Kohn (1941)
	$CH_3.O.CH_2.CH_2.CH(NH_2)CO_2H$ Methoxinine	Roblin *et al.* (1945)
Thiamine (Aneurin)	Pyrithiamine	Tracy & Elderfield (1941); Robbins (1941); Woolley & White (1943*a*, *b*); Woolley (1944*a*)
Nicotinic acid	Pyridine-3-sulphonic acid	McIlwain (1940)

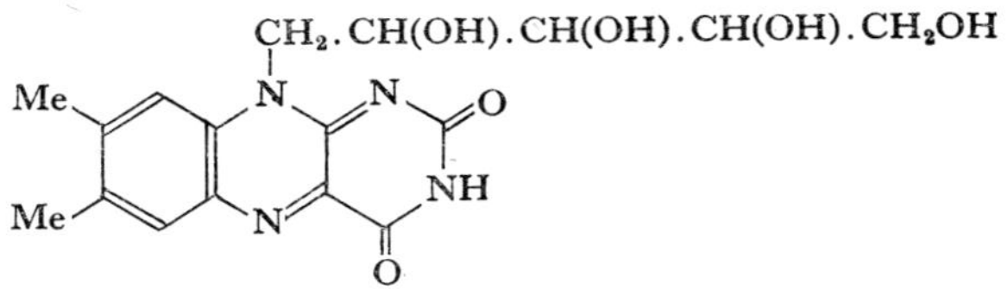

Compound	Formula labels	Reference
Riboflavin	$CH_2.CH(OH).CH(OH).CH(OH).CH_2OH$; Me, Me; N, N, N; O, O; NH	
Dichloroflavin	$CH_2.CH(OH).CH(OH).CH(OH).CH_2OH$; Cl, Cl; N, N, N; O, O; NH	Kuhn *et al.* (1943)
Methylene blue	Cl^-; S^+; Me_2N; NMe_2; N	Oesterlin (1936); Madinaveitia (1946)
Mepacrine	$NH.CH(CH_3).CH_2.CH_2.CH_2.NEt_2$; MeO; Cl; N	Oesterlin (1936); Madinaveitia (1946)
Quinine	CH(OH); MeO; N; CH_2, CH_2; $CH{=}CH_2$	Oesterlin (1936); Madinaveitia (1946)
Paludrine	Cl; NH; C=NH; NH; $CHMe_2$	Curd *et al.* (1945)

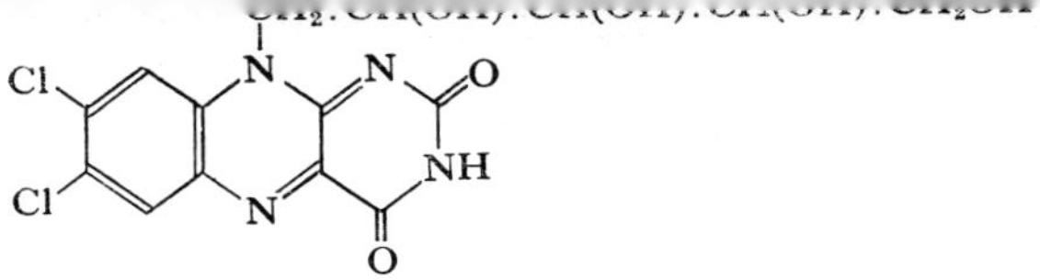

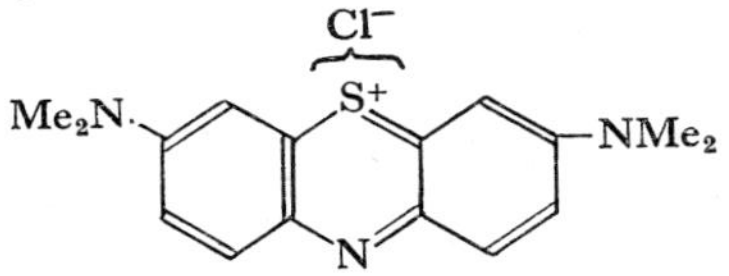

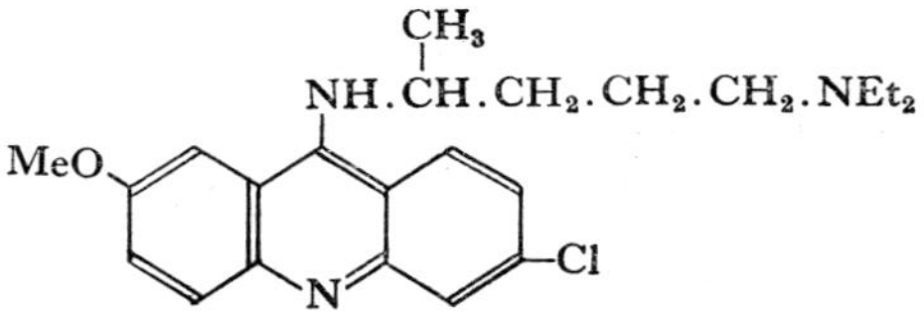

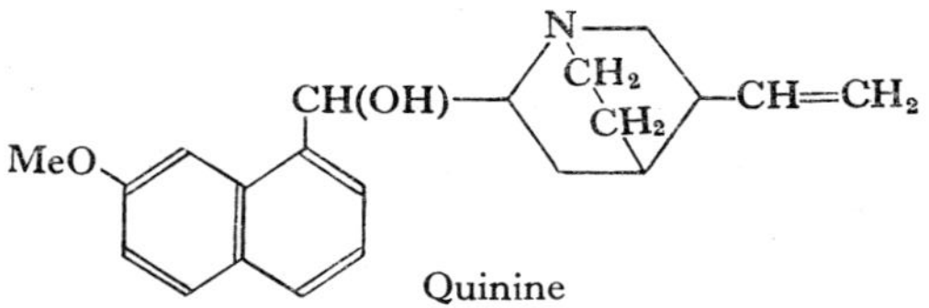

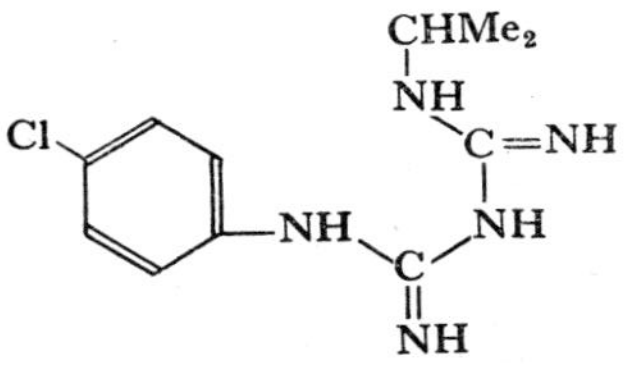

Table 1 (*cont.*)

Metabolite	Inhibitor	References
$HOCH_2.CMe_2.CH(OH).CO.NH.CH_2.CH_2.CO_2H$ Pantothenic acid	$HOCH_2.CMe_2.CH(OH).CO.NH.CH_2.CH_2.SO_3H$ Pantoyltaurine	Snell (1941 *a*, *b*); McIlwain (1942)
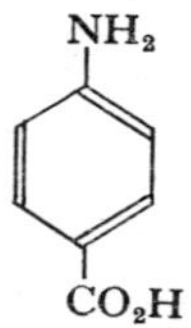 NH_2, CO_2H *p*-Aminobenzoic acid	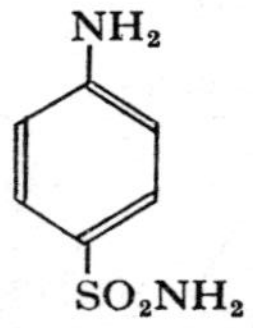 NH_2, SO_2NH_2 Sulphanilamide	Woods (1940)
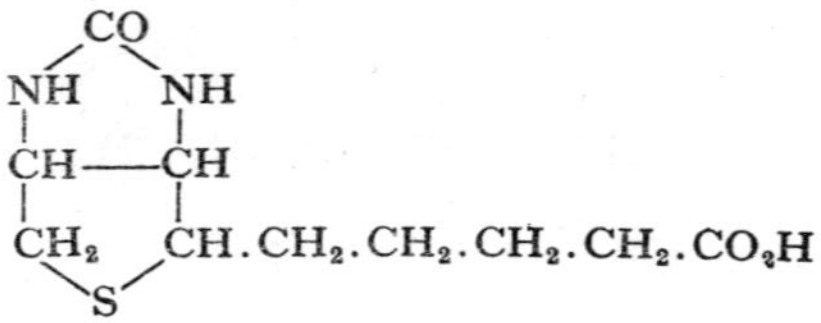	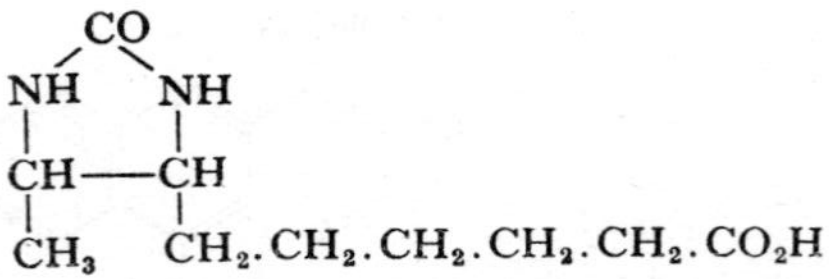 CO, NH, NH, CH—CH, CH_3, $CH_2.CH_2.CH_2.CH_2.CH_2.CO_2H$ Desthiobiotin	Dittmer *et al.* (1944); Lilly *et al.* (1944); Rubin *et al.* (1945)
CO, NH, NH, CH—CH, CH_2, $CH.CH_2.CH_2.CH_2.CH_2.CO_2H$, S Biotin	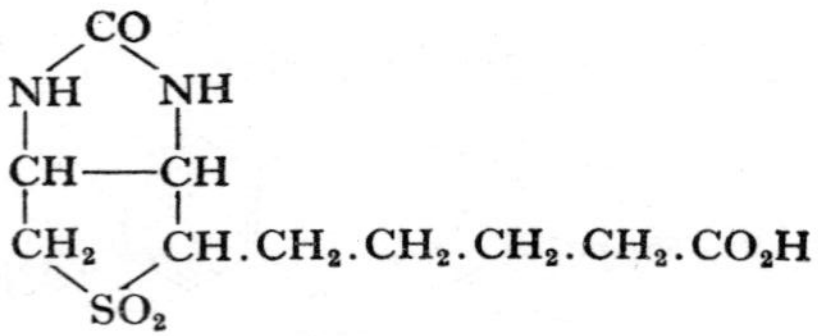 CO, NH, NH, CH—CH, CH_2, $CH.CH_2.CH_2.CH_2.CH_2.CO_2H$, SO_2 Biotin sulphone	Dittmer *et al.* (1944)
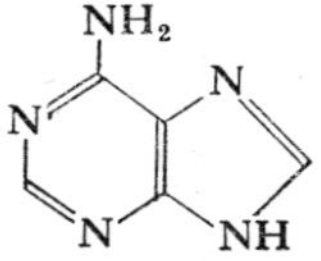 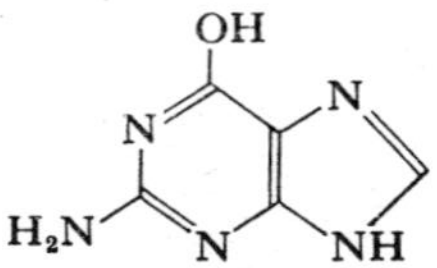 NH_2, N, N, N, NH Adenine OH, N, N, N, NH, H_2N Guanine	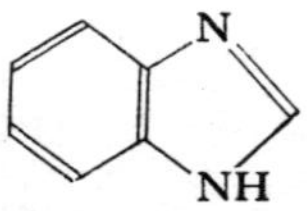 N, NH Benzimidazole	Woolley (1944 *b*)

'pictorial' in nature, the inhibitor having simply been made to 'look like' the metabolite. A notable exception, to which we shall return shortly, is the dichloroflavin of Kuhn, Weygand & Möller (1943). In my opinion Fildes's principle is unlikely to be very fruitful if we continue to confine ourselves to such naïve pictorial analogues, which are both not enough and too much. Pictorial analogy is not enough because an analogue may look like a metabolite and yet lack the chemical groupings necessary for combination with the appropriate enzyme; it leads us to overemphasize the geometry of the molecule and underestimate the importance of the right sort of combining centres. Pictorial analogy is too much, since there may be a great deal in the structure of the metabolite which is not directly concerned in its attachment to the enzyme, which is our only concern in designing an inhibitor; it leads us to overload our inhibitors with possibly unnecessary complications.

It is essential, then, that we should attempt to design inhibitors on a more rational basis. In order to do so we must know two things: the structure of the metabolite we are imitating, and the precise manner in which this metabolite is attached to the enzyme to which it is related as substrate, coenzyme or product.

The first of these requirements should give us little trouble. The work of the last two decades has resulted in such a wealth of knowledge of the structure of vitamins and other essential metabolites that there can be no lack of starting points for imitations. However, if we are interested in a particular disease, it may still be necessary to begin by studying the mode of action of known effective agents in order to determine with what essential metabolite they interfere; such knowledge will always be of the greatest value in designing improved agents. Moreover, success or failure in the chemotherapy of a given infection may depend on a wise choice of the essential metabolite to be attacked.

Regarding the second point, however, we have little, if any, information, and it is with this aspect of the matter that the present paper is mainly concerned. We may begin by considering the possible modes of linkage of essential metabolites to enzymes; attention is drawn to the reviews of Pauling (1945) and Bateman (1945). Following Michaelis & Menten (1913), enzyme-substrate combination is regarded as a readily reversible reaction; this means that the forces holding enzyme and substrate together must be such that the union can be easily broken. Although this does not altogether exclude a covalent linkage, cases where this is sufficiently easily reversed are likely to be rare; possibilities are the reversible oxidative union of thiol groups:

$$\text{—SH} + \text{HS—} \rightleftharpoons \text{—S—S—} + 2\text{H},$$

and sulphonium salt formation at methionine side-chains:

$$-CH_2-X + Me-S-CH_2-\ldots \rightleftharpoons -CH_2-\overset{\displaystyle Me}{\overset{|}{\underset{\underset{\displaystyle X^-}{\smile}}{\overset{+}{S}}}}-CH_2-\ldots$$

Many more possibilities arise when consideration is given to what may be called 'secondary valencies', which range from the strong electrostatic interaction between poles (ionic centres) to the weak interaction between transitory dipoles known as van der Waals forces. It is noteworthy that such secondary forces are of great importance in determining the arrangement of molecules in crystals and the analogy between enzyme-substrate combination and crystallization is striking and likely to be fruitful. As Adam (1941) has pointed out, a parallel can be drawn between the attachment of the substrate to specific groupings in an enzyme and the preferential adsorption of dyestuffs on particular faces of certain crystals. Pauling (1947) has drawn a similar analogy in the case of antigen-antibody combination, while Lettré (1937, 1943) regarded both antigen-antibody and enzyme-substrate combination as analogous to the formation of partial racemates (mixed crystals) by such pairs as *d*-tartaric and *l*-malic acids. Kuhn (1942) has also emphasized the importance of isomorphism of inhibitor and metabolite, showing that the *N*-acetyl derivatives of the methyl esters of sulphanilic and *p*-aminobenzoic acids form mixed crystals, whereas those of *o*- and *p*-aminobenzoic acids do not; the deliberate preparation by Kuhn *et al.* (1943), of dichloroflavin as an inhibitor modelled on riboflavin, on the basis of the known isomorphism of, and mixed crystal formation by, methyl- and chloro-derivatives in the benzene series, is an important instance of the successful application of this analogy. For the present purpose, we may classify these secondary forces into three groups, viz. (*a*) interaction of poles and dipoles, (*b*) hydrogen bonding and (*c*) van der Waals forces, which we may now discuss briefly.

The interaction of poles and dipoles is electrostatic in character. The strongest attractive forces will be between pairs of oppositely charged poles, e.g. enzyme-$\overset{+}{N}H_3$ and $\overset{-}{O_2}C$-substrate; suitable ionic centres in the enzyme may be afforded by the amino-groups of lysine, arginine and histidine residues, the carboxyl groups of aspartic and glutamic acid residues, the phenolic hydroxyl groups of tyrosine residues and, possibly, the thiol groups of cysteine residues. It is to be noted that the extent to which these groupings will assume the necessary ionized forms will depend very markedly on pH, and there seems little doubt that the influence of pH on enzyme-substrate combination is to be ascribed, in part, to such effects. A weaker electrostatic attraction is set up between permanent dipoles, e.g.

$$\overset{\longleftarrow\!\!\!+}{\text{N—H}} \text{ (with CO) and } \overset{\longleftarrow\!\!\!+}{\text{O=C}} \text{ (with NH)} \qquad \text{enzyme-}\overset{\longleftarrow\!\!\!+}{\text{O—H}} \text{ and } \overset{\longleftarrow\!\!\!+}{\text{Cl—CH}_2}\text{-substrate}$$

In the enzyme the —CO— and —NH— groups of the peptide backbone and all the side-chains, except those derived from glycine, alanine, valine, leucine, isoleucine and phenylalanine, offer opportunities for this type of interaction. Electrostatic interaction can also occur between a pole and a permanent dipole, e.g.

$$\text{Enzyme-}\overset{+}{\text{NH}_3}\ \ \overset{\longleftarrow\!\!\!+}{\text{Cl—CH}_2}\text{-substrate}; \quad \text{Substrate-}\overset{-}{\text{CO}_2}\ \ \overset{+\!\!\longrightarrow}{\text{H—N}}\text{=Enzyme},$$

and between poles (or dipoles) and dipoles induced by them in neighbouring molecules. These interactions fall off increasingly rapidly with distance as we pass from pole-pole interactions, through pole-dipole and dipole-dipole interactions, to those involving induced dipoles.

In the hydrogen bond a single hydrogen atom serves to unite two electronegative atoms, of which one carries a hydrogen atom and the other a pair of unshared electrons:

$$X\text{—H}: Y, \quad \text{e.g.} \quad R\text{—C}\begin{matrix}\text{=O- -H—O}\\ \text{—O—H- -O=}\end{matrix}\text{C—}R$$

At one time (cf. Sidgwick, 1937) hydrogen bonding was ascribed entirely to quantum-mechanical resonance between structures such as

$$R\text{—C}\begin{matrix}\text{=O H—O}\\ \text{—O—H O=}\end{matrix}\text{C—}R \quad \text{and} \quad R\text{—C}\begin{matrix}\text{—O—H O=}\\ \text{=O H—O—}\end{matrix}\text{C—}R$$

However, it is not possible to differentiate strictly between hydrogen bonding of this type and electrostatic attraction between poles, e.g.

$$\overset{+}{\text{N}}\text{H}_2\text{—H- -O=}\overset{\bar{\text{O}}}{\text{C}}\text{—} \quad \text{and} \quad \overset{+}{\text{N}}\text{H}_2\text{—H- -}\bar{\text{O}}\text{—}\overset{\text{O}}{\overset{\|}{\text{C}}}\text{—} \quad \text{and} \quad \overset{+}{\text{N}}\text{H}_2\text{—H}\ \ \bar{\text{O}}\text{—}\overset{\text{O}}{\overset{\|}{\text{C}}}\text{—}$$

or dipoles, e.g.

$$R\text{—C}\begin{matrix}\text{=O- -H—O}\\ \text{—O—H- -O=}\end{matrix}\text{C—}R \quad \text{and} \quad R\text{—C}\begin{matrix}\text{=O}\ \ \overset{+\!\!\longrightarrow}{\text{H—O}}\\ \text{—}\underset{\longleftarrow\!\!+}{\text{O—H}}\ \ \text{O=}\end{matrix}\text{C—}R$$

and at the present time (e.g. Davies, 1946) there is a tendency to regard the electrostatic aspect as of greater importance than the resonance aspect. It seems probable that there is in fact a continuous gradation between extreme cases in which the bonding is purely electrostatic and others in which it is due solely to resonance. It is of interest to quote the opinions of two authorities on the biochemical importance of the hydrogen bond; Pauling (1940) regards it as more significant 'than any other single structural feature', whereas Jordan-Lloyd (1940) thought it 'no more than an attractive possibility'; current general opinion seems to incline towards Pauling's estimate. Groups suitable for hydrogen bonding abound in proteins; as hydrogen donors we have the —NH— of the backbone, —OH of serine, threonine, tyrosine and hydroxy-proline side-chains, $—CO_2H$ of aspartic and glutamic acid side-chains; —SH of cysteine, $—NH_2$ of arginine and lysine, =NH of arginine, proline, hydroxy-proline, histidine and tryptophan; as electron donors we have all these and, in addition, the backbone —CO—, the —SMe of methionine and the =N— of histidine.

Attraction by what are known as dispersion or van der Waals forces (London, 1930) results in two molecules (more strictly their van der Waals 'envelopes') being brought into contact. The attraction falls off with the seventh power of the distance between the atoms concerned and is thus only of comparatively short range; London (1942) has, however, suggested that long-range van der Waals forces, falling off with the third power of the distance, may operate in macromolecules, but only when these contain conjugated double bond systems. The magnitude of van der Waals forces also depends on the atomic weight of the atoms involved and may, for our purposes, be neglected for light atoms such as hydrogen. In general, van der Waals attraction will only be reasonably strong if several atoms in each of the molecules concerned can be brought into apposition. This type of interaction demands, then, that the interacting molecules have geometrically complementary structures so that they can be fitted together in such a way that several atoms in each can come into contact; we are strongly reminded of Fischer's 'lock and key' simile for enzyme action. On general grounds it would seem that flatness in the molecules involved might give the best chance for van der Waals attraction, and the importance of the flatness of aromatic systems from this point of view has been emphasized by recent workers (e.g. Albert, Rubbo, Goldacre, Davey & Stone, 1945). However, enzyme-substrate combination by van der Waals forces with such flat substrates also requires a suitable flat area on the enzyme molecule, and it is worth while looking into the implications of this aspect of the matter. The uniform L-configuration of the amino-acid residues in the normal proteins implies that in any reasonable structure (e.g. Astbury, 1941) the peptide backbone will be masked by side-chains protruding alternately above and below it. In order to get the necessary flatness we must either have an area in which there is a group of side-chains of equal length or an area in which the absence of side-chains bares the peptide backbone. In the former case the relatively great separation of the side-chains would seem to imply that only a small proportion of the atoms in a flat substrate or inhibitor could come into contact with atoms in the enzyme side-chains, and we could only expect appreciable van der Waals attraction with relatively large molecules; the steroids and carcinogenic hydrocarbons may possibly fall into this category. In the latter case we can bare the backbone by having an area composed exclusively of glycine residues; in this event the comparatively close packing of the C, O and N atoms would permit van der Waals interaction with almost all of the non-hydrogen atoms of a flat substrate or inhibitor, and it seems to the author that van der Waals interaction of enzymes with small flat molecules, such as the acridine derivatives of Albert *et al.* (1945), is most likely to involve such side-chain-free glycine areas. It may be remarked that complete complementariness of side-chains may well occur between pairs of macro-molecules (e.g. antigen and antibody; protein and nucleic acid) and lead to very considerable van der Waals interaction without any prerequisite of flatness. To sum up we may say that van der Waals attraction demands a degree of specialization in the molecules concerned which seems unlikely to be common; probably, then, the majority of enzyme-substrate or enzyme-inhibitor interactions are to be ascribed to the less exacting mechanisms (polar and dipolar association, hydrogen bonding) discussed earlier, although it is as well to point out that van der Waals attraction may well reinforce and be itself directed by these other types of attraction (cf. Pauling, 1945).

II. THE DESIGN OF INHIBITORS

Having outlined the theoretical background we may return to the task of designing an effective bacterial inhibitor, modelled on some essential metabolite; this involves, as a first step, determining as precisely as possible how the metabolite is attached to the bacterial enzyme to which it is related. It is suggested that this may be done by introducing substituents of known electropolar properties into the metabolite molecule and studying the inhibitory (or metabolic) activities of the resulting compounds. Differences in the inhibitory powers of the members of such a series of anti-metabolites may be due to differences in their affinities for the enzyme concerned or to differences in their powers of penetration into the bacterial cell or to a combination of both factors. Ideally, it would be best to work with isolated enzymes, thus avoiding any complications due to permeability differences; isolated bacterial enzymes are rare, however, and we must usually do the best we can with living bacteria. To avoid gross differences in permeability the inhibitors studied should be as similar as possible and comparisons should probably be restricted to isomerides. The suggested procedure, then, is to prepare one or more complete series of isomeric anti-metabolites and determine their activities as bacterial inhibitors; differences in their inhibitory powers may resonably be ascribed to different affinities for the bacterial enzyme and, by correlating the results with the known polar and steric properties of the substituents, it may be possible to deduce, in some detail, the nature of the combination of the metabolite and its related bacterial enzyme. If we succeed in this task, it should not be too difficult to devise other modifications of the metabolite structure leading to substances having maximal affinity for the enzyme; in this way conscious design might supplant the empiricism which still governs the preparation of metabolite analogues, and we could claim to have a really rational approach to chemotherapy.

I propose to illustrate this modified method of approach by three examples, viz. substituted indoles, substituted tryptophans and substituted *p*-aminobenzoic acids.

A study of all the isomeric methyl-indoles as inhibitors of *Salmonella typhi* (Fildes & Rydon, 1947) has shown three of them, viz. the 1-, 2- and 3-methyl compounds, to be devoid of inhibitory activity; the other four inhibit certain strains of the organism by interfering with the conversion of indole into tryptophan, the order of decreasing inhibitory activity being 4-Me > 6-Me > 7-Me > 5-Me. It is noteworthy that the non-inhibitory methyl-indoles comprise all of those in which the methyl group is attached to the pyrrole ring; this suggests that the combination of indole with the enzyme closely involves this ring, which contains both the —NH— group, the most reactive grouping in the indole molecule, and the carbon atom, C_3, which is concerned in the conversion of indole into tryptophan; it is not surprising that substitution in this ring causes, perhaps for stereochemical reasons, a complete failure of the modified metabolite to combine with the enzyme. On the other hand, the inhibitory activity of the compounds containing methyl groups attached to the benzene ring indicates that substitution here does not prevent combination of the substituted indole with the enzyme, although it does profoundly influence reactivity towards the enzyme. The alternating effect observed in the inhibitory activities at once suggests that the underlying mechanism is polar in nature.

Now, differences in the reactivity, including that towards enzymes, of substituted indoles must depend on differences in the contributions of the two extreme forms, (A) and (B), of the indole nucleus to the resonance hybrid:

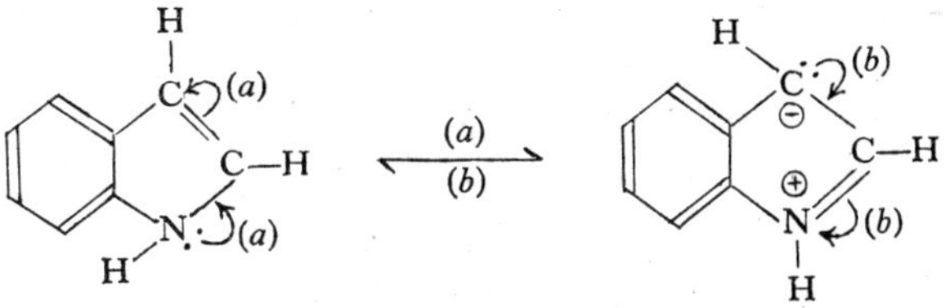

(A) $\underset{(b)}{\overset{(a)}{\rightleftarrows}}$ (B)

Substitution tending to induce electron shifts favouring process (*a*) and opposing (*b*) will result in an increased contribution of (B) in the substituted indole, as compared with indole itself, while substitution favouring (*b*) will conversely result in an increased contribution from (A).

The state of affairs in the methyl-indoles is as follows (Me→ is used to include all forms of electron release from the methyl group):

	Effect on process (*a*)	Effect on process (*b*)	Effect on contributions of (A) and (B) to hybrid
4-Me	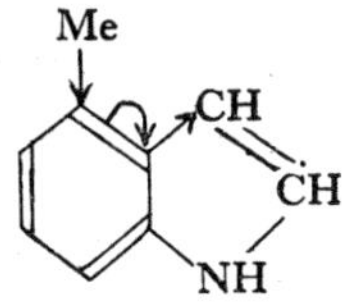 Opposed	Favoured	More (A)
5-Me	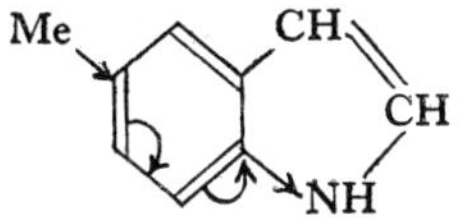 Favoured	Opposed	More (B)
6-Me	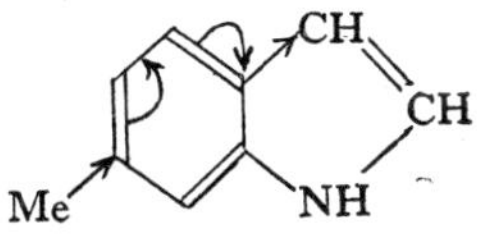 Opposed	Favoured	More (A)
7-Me	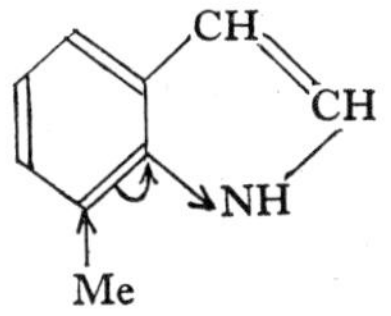 Favoured	Opposed	More (B)

Comparing these conclusions with the experimental findings, it is seen that the two most inhibitory compounds (4-Me and 6-Me) are those in which (A) preponderates. The results thus indicate that the nature of the combination of indole and the methyl-indoles with the enzyme must be such that it is favoured by structure (A) rather than structure (B).

Of the possible ways in which indole might be attached to an enzyme, hydrogen bonding (or the dipole interaction into which it may merge) at the —NH— group seems the most likely; electrostatic attraction of poles is only possible for structure (B), while both (A) and (B) are equally flat and so equally suitable for van der Waals interaction with an enzyme molecule. However, there is a very real difference between (A) and (B) in their ability to form hydrogen bonds. For, whereas (B) can form only one such bond, (A) can enter simultaneously into two:

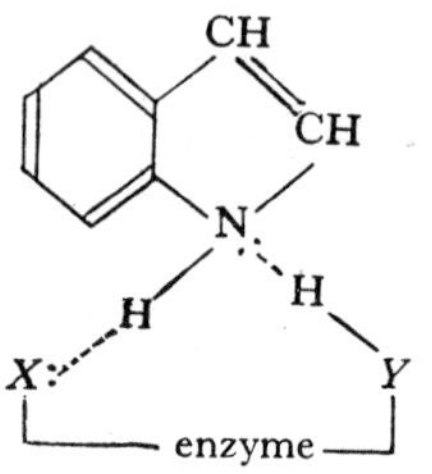

CH CH N H H X Y enzyme

(A)

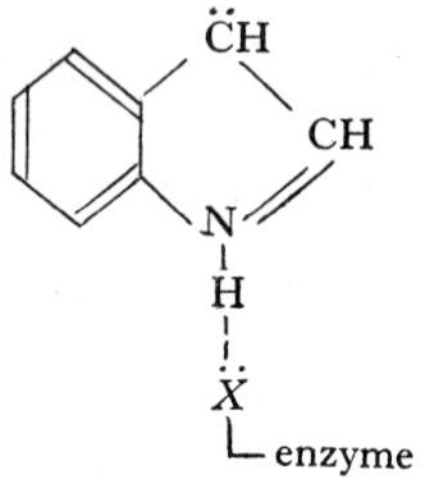

CH CH N H X enzyme

(B)

The results suggest, therefore, that the combination of indole with the enzyme is of this second type, involving attachment by two hydrogen bonds to two side-chains of the enzyme molecule; such double attachment, besides being stronger than attachment by a single hydrogen bond, would result in a rigid anchoring of the indole molecule into position for further reaction. Assuming a tetrahedral angle between the hydrogen bonds and a hydrogen bond length of 2·9 A. (reasonable for either N—H—N or N—H—O), we arrive at the following picture:

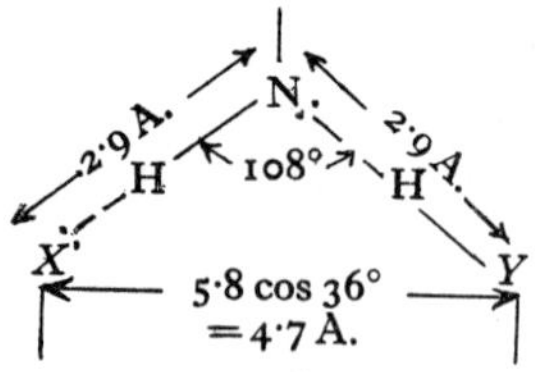

N 2·9 A. 2·9 A. H 108° H X Y 5·8 cos 36° = 4·7 A.

leading to a distance of 4·7 A. between the two enzyme side-chains with which the indole combines. Considering the uncertainties in the calculation, this distance is remarkably close to the 5·1 A. spacing postulated in Astbury & Bell's (1941) folded structure for α-keratin and α-myosin, which also affords a reasonable model for the globular proteins. It may be, then, that the indole molecule fits into a fold of such a structure by double hydrogen bonding to suitable atoms in two of the side-chains. Scale drawings (Fig. 1) of such an arrangement show that the carbon atom, C_3, involved in the formation of tryptophan is thus brought into reasonable reacting distance of the corresponding pair of side-chains in the corresponding fold of the adjacent peptide chain. The drawings also suggest how methylation on N_1, C_2 or C_3 may cause steric inhibition of the fitting of indole into the enzyme. Such conclusions are, of course, highly speculative, but it is encouraging to find that they are by no means unreasonable

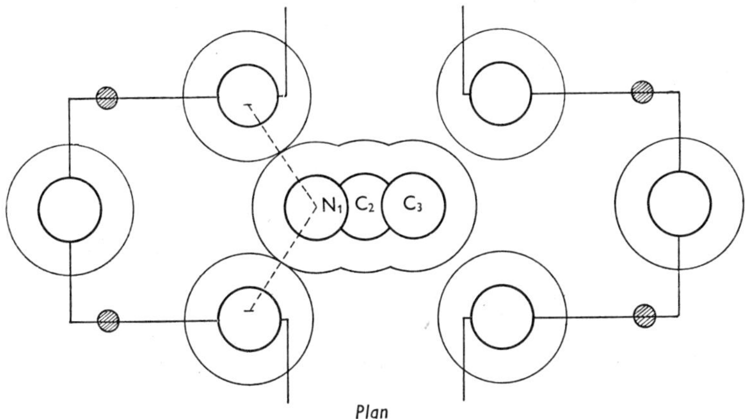

Plan

(The broken lines from N_1 are the projections of 2·9 A. hydrogen bonds. The small shaded circles indicate the positions of side-chains projecting below the peptide backbone, which is shown diagrammatically by the unbroken straight lines. Only N_1, C_2 and C_3 of the indole molecule are shown.)

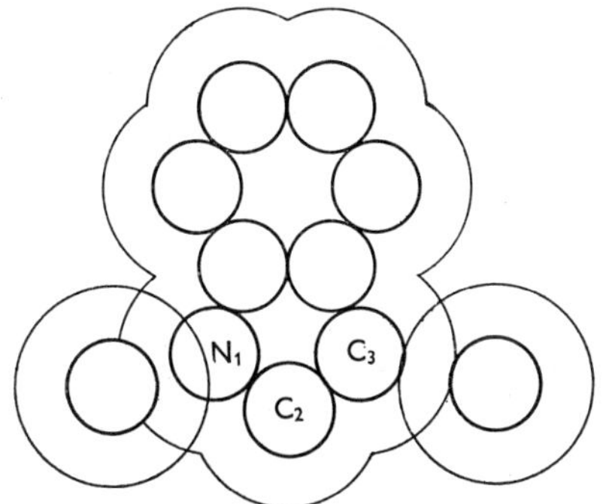

Elevation

(Covalent radii, thick circles, have been taken as 0·7 A. and van der Waals radii, thin circles, as 1·5 A. Hydrogen atoms have been omitted.)

Fig. 1. Indole and an α-folded protein.

as a first attempt to picture the details of the combination between indole and its related bacterial enzyme.

Further work on substituted indoles as inhibitors of *Salmonella typhi* is being carried out by the author and Mr C. A. Long. In order to test the above hypothesis, substituents of different electropolar character (e.g.

halogens) are being introduced in place of methyl on C_4, C_5, C_6 and C_7, while more powerfully electron-releasing groups are being introduced into the 4-position in the hope of obtaining still more inhibitory compounds.

Fildes & Rydon (1947) also studied the five nuclear-C-methylated tryptophans as inhibitors of *S. typhi*; these substances act by preventing the utilization of tryptophan, the order of decreasing potency being 4-Me > 5-Me > 6-Me > 7-Me with the 2-Me compound inactive. This order is quite different from that observed with the methyl-indoles, and it thus seems that tryptophan does not combine with its enzyme by double hydrogen bonding on the —NH— group of the indole nucleus; presumably the $—CH(NH_2)CO_2H$ grouping of the side-chain is concerned, but studies with other types of substituent are required before any details of the combination can be suggested and before any conscious planning of inhibitors more potent than 4-methyl-tryptophan can be undertaken; such work is also being carried out in collaboration with Mr C. A. Long.

As a last illustration we may discuss the substituted *p*-amino-benzoic acids, whose inhibitory action on various bacterial species has been studied by several groups of workers (Wyss, Rubin & Strandskov, 1943; Johnson, Green & Pauli, 1944; Martin & Rose, 1945). Excluding doubtful cases, the following series of compounds, arranged in order of decreasing inhibitory activity, emerge from their results:

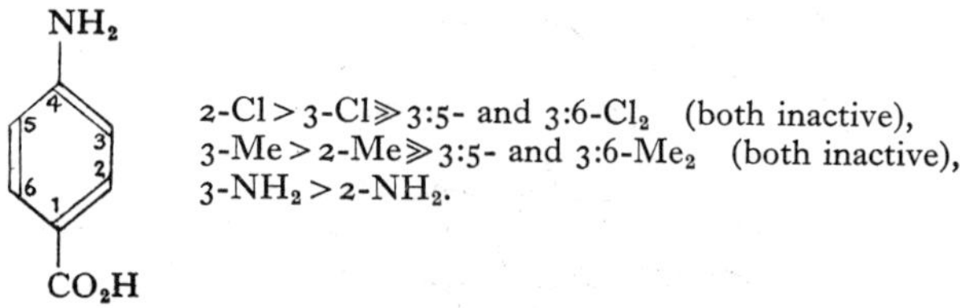

It is seen at once that polar effects are probably operative here, too, since the two substituents with electron releasing effects (methyl and amino) show the same order of inhibitory activities, which is reversed when the substituent is changed to one (chloro) with an electron-attracting effect.

Both the amino and carboxyl groups in *p*-aminobenzoic acid can enter into hydrogen bonds and both can do so in two ways, contributing either an electron pair (mechanism A) or a hydrogen atom (mechanism B) to the bond (shown on p. 75). Of these possibilities, mechanism A will be far the most probable for the amino and mechanism B for the carboxyl group, since it is these two which merge into polar interactions. Mechanism A will be facilitated by a drift of electrons in the molecule towards the reacting group and mechanism B by an electron drift in the opposite direction.

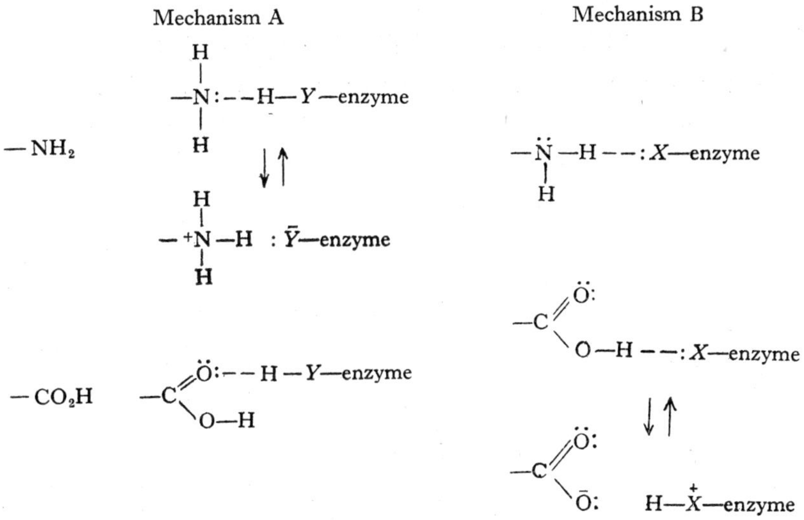

	Mechanism A	Mechanism B
$-NH_2$	$-\overset{H}{\underset{H}{N}}:--H-Y-$enzyme $\rightleftharpoons$ $-\overset{H}{\underset{H}{\overset{+}{N}}}-H\ \ :\bar{Y}-$enzyme	$-\underset{H}{\ddot{N}}-H--:X-$enzyme
$-CO_2H$	$-C(=\ddot{O}:--H-Y-$enzyme$)-O-H$	$-C(=\ddot{O}:)-O-H--:X-$enzyme $\rightleftharpoons$ $-C(=\ddot{O}:)-\bar{\ddot{O}}:\ \ \ H-\overset{+}{X}-$enzyme

The effect of substitution on hydrogen bonding by these two mechanisms depends on the electropolar nature of the substituent as follows:

Type of substituent	Position	Electron drift	Effect on hydrogen bonding by NH_2: Mech. A	NH_2: Mech. B	CO_2H: Mech. A	CO_2H: Mech. B
$+I$	2–	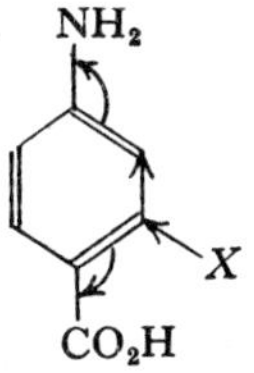 NH_2 / X / CO_2H	Weakly facilitated	Weakly opposed	Facilitated	Opposed
	3–	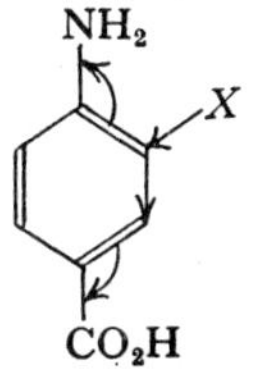 NH_2 / X / CO_2H	Facilitated	Opposed	Weakly facilitated	Weakly opposed
$+T$	2–	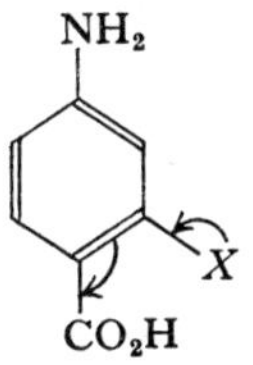 NH_2 / X / CO_2H	Little effect		Facilitated	Opposed
	3–	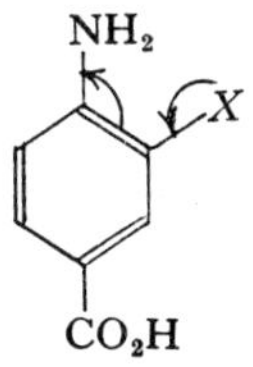 NH_2 / X / CO_2H	Facilitated	Opposed	Little effect	

(The effects of $-I$ and $-T$ substituents will be the reverse of these.)

We may now draw up a table showing the anticipated effect of various types of substituent on the relative capacity of the two isomeric monosubstituted *p*-aminobenzoic acids for hydrogen bonding:

Substituent	Position most favourable for hydrogen bonding on				Observed most inhibitory isomer
	NH_2 by		CO_2H by		
	Mech. A	Mech. B	Mech. A	Mech. B	
$+I$	3	2	2	3	—
$-I$	2	3	3	2	—
$+T$	3	2	2	3	—
$-T$	2	3	3	2	—
Me ($+I$; $+T$)	3	2	2	3	3
NH_2 ($+T$)	3	2	2	3	3
Cl ($-I$)	2	3	3	2	2
Cl ($+T$)	3	2	2	3	

There is a clear correlation between bacterial inhibitory power and facilitation of hydrogen bonding by the most probable mechanisms, viz. A for the amino and B for the carboxyl group; it appears that, as in aromatic substitution, the $-I$ effect of chlorine overpowers its $+T$ effect. We are thus led to the conclusion that *p*-aminobenzoic acid combines with the bacterial enzyme of which it is the substrate in one or both of these two ways.

From a consideration of these monosubstitution products alone it is not possible to conclude whether both mechanisms are involved, or only one, but the operation of both is suggested by the results for the disubstituted compounds. The inactivity of the 3:5-dimethyl- and -dichloro-compounds, contrasted with the activity of the 3-monosubstitution products, suggests that steric hindrance is involved, as in the chemical reactions of many similar compounds (for references see Branch & Calvin, 1941). The similar inactivity of the 3:6 compounds suggests that combination takes place at both ends of the *p*-aminobenzoic acid molecule, the configuration of which is such that, in these substitution products, rotation into position for minimally hindered reaction at one end automatically leads to maximum steric hindrance at the other:

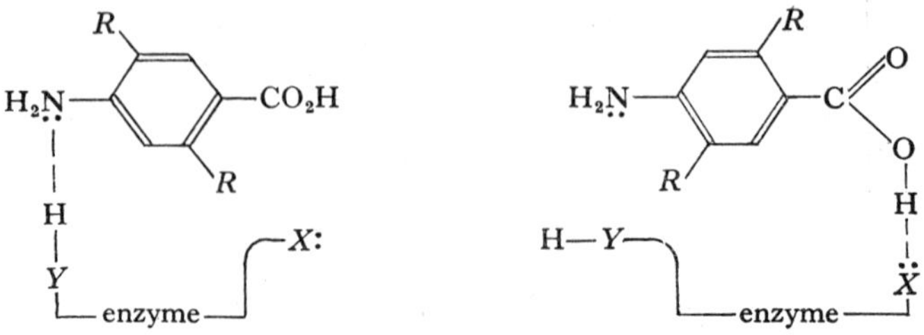

We are thus led to conclude that the attachment of *p*-aminobenzoic acid to its bacterial enzyme involves hydrogen bonding at both the amino and the carboxyl group, thus:

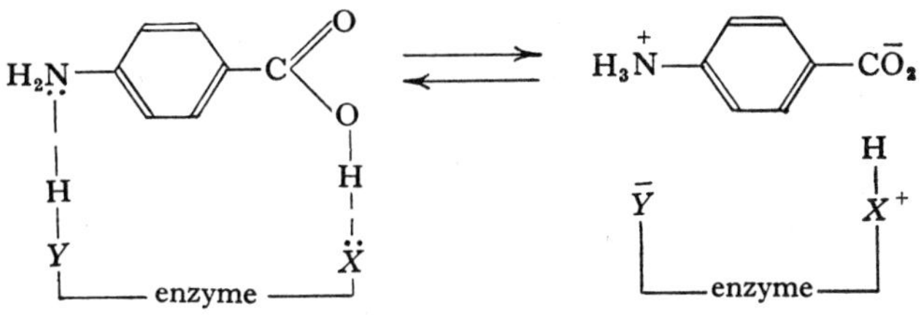

$H_2\ddot{N}$ — C — O — O — H — Y — H — $\ddot{X}$ — enzyme ⇌ $H_3\overset{+}{N}$ — $C\bar{O}_2$ — H — $\bar{Y}$ — X^+ — enzyme

The existence of free rotation in the two single bonds of the carboxyl prevents us from assigning any definite value to the distance between the two enzyme side-chains concerned; appropriate rotation around the single bonds would allow fitting to any suitable pair of side-chains between 5 and 10 A. apart.

III. DISCUSSION

Apart from the possible utilitarian value of studies such as those we have just been considering they are also of value in helping to elucidate the nature of enzyme-substrate combination in general. Similar studies using isolated enzymes are less likely to be upset by permeability considerations and should provide a background for the more practical task of improving chemotherapeutic agents; there is no doubt that the two lines of investigation are complementary and mutually helpful and should proceed side by side. Similar work on antigen-antibody interactions seems less likely to be immediately relevant; here no chemical reaction succeeds combination of the two entities, and hence the approach distance may be greater and the forces involved rather different in type.

It may be worth while drawing attention to the finding that our reasoning, from both the methyl-indoles and the substituted *p*-aminobenzoic acids, has led us to postulate attachment of the substrate to two side-chains of the enzyme. It seems not improbable that this double attachment may prove to be general; it would certainly lead to greater specificity (cf. Rydon, 1948). If, as suggested for indole, the double attachment occurs across a fold of the peptide chain, this circumstance may help to explain several of the phenomena of enzyme chemistry, e.g. the protection of enzymes by their substrates which may keep them folded in opposition to the unfolding which constitutes denaturation; the function of bivalent metals, such as magnesium and zinc, as coenzymes which may assist folding into the right pattern; the production of adaptive enzymes in the presence of the substrate, which may impress the right type of folding in the polypeptide chain from the beginning of its formation.

So far we have considered only competition for an enzyme between a normal substrate and an inhibitor modelled on it and capable of combining with the same groupings in the enzyme. There seems to be no valid reason why there should not equally well be competition for a metabolite between an enzyme and an inhibitor modelled on the enzyme and capable of combining with the metabolite in the same way as the enzyme. The polypeptide nature of such antibiotics as the gramicidins, aerosporin and, in a modified way, of penicillin may be significant in this connexion; perhaps these substances are enzyme analogues which compete with enzymes for some metabolite just as the sulphonamides and the like are metabolite analogues which compete with metabolites for some enzyme. At all events, it seems desirable to bear this possibility in mind since it opens up the prospect of making synthetic peptides which may compete with bacterial enzymes for known metabolites. If we regard this suggestion as reasonable, research into the details of the mode of interaction of bacterial enzymes and their substrates acquires an added importance for, if we know the mechanism, we may be able to devise two types of inhibitor—one modelled on the substrate and designed to compete with it for the enzyme, the other modelled on the enzyme and designed to compete with it for the substrate.

No chemotherapeutic agent can be effective, either *in vitro* or *in vivo*, unless it can penetrate into the parasite. The ability to do this depends on a number of factors, relating chiefly to the ability to penetrate cell membranes and surfaces, which I have not attempted to discuss. Such factors are of very great importance and likely to complicate the task of designing a chemotherapeutic agent, e.g. by calling for the introduction of special groupings to assist permeation. There is no doubt that, both scientifically and practically, research on the permeability of bacterial cells is of similar importance to research on bacterial nutrition; knowledge of both types is essential for a truly rational approach to chemotherapy.

I may conclude this essay by outlining what I believe will be the most fruitful course for future chemotherapeutic studies. There are three main steps:

(1) The nutrition of the pathogenic organism in which we are interested must be studied and some nutritional weak spot found, i.e. some metabolite essential to the life of the organism.

(2) The inhibitory activities of one or more complete series of substitution products of this essential metabolite must be studied. Sooner or later we should be able to deduce the detailed chemistry of the combination between our chosen essential metabolite and its related bacterial enzyme.

(3) Knowing this we should then be able to devise other inhibitors which

may either combine with the enzyme in the same way as, but more firmly than, the metabolite or with the metabolite in the same way as the enzyme. It is at this stage that such questions as permeability and toxicity to the host must be investigated and allowed to modify the course of the research.

There will, of course, be short cuts (e.g. it would be foolish to ignore hints from established agents) and modifications (e.g. the use of isolated enzymes in the second stage) but this is the general pattern of a truly rational approach to chemotherapy in which conscious design completely replaces the old empiricism.

I am much indebted to Sir Paul Fildes, F.R.S., for introducing me to this field of work and to him and my other former colleagues in the Medical Research Council's Unit for Bacterial Chemistry for many stimulating discussions.

REFERENCES

ADAM, N. K. (1941). *The Physics and Chemistry of Surfaces*, 3rd ed. p. 250. Oxford.
ALBERT, A., RUBBO, S., GOLDACRE, R., DAVEY, M. & STONE, J. (1945). *Brit. J. Exp. Path.* **26**, 160.
ASTBURY, W. T. (1941). *Chem. & Ind.* **60**, 491.
ASTBURY, W. T. & BELL, F. O. (1941). *Nature, Lond.*, **147**, 696.
BATEMAN, J. B. (1945). In HÖBER, R., *Physical Chemistry of Cells and Tissues*, pp. 97, 145. Philadelphia and Toronto.
BRANCH, G. E. K. & CALVIN, M. (1941). *The Theory of Organic Chemistry*, p. 443. New York.
CURD, F. H. S., DAVEY, D. G. & ROSE, F. L. (1945). *Ann. Trop. Med. Parasit.* **39**, 157.
DAVIES, M. (1946). *Ann. Rep. Chem. Soc.* **43**, 5.
DITTMER, K., MELVILLE, D. B. & DU VIGNEAUD, V. (1944). *Science*, **99**, 203.
DITTMER, K. & DU VIGNEAUD, V. (1944). *Science*, **100**, 129.
DYER, H. M. (1938). *J. Biol. Chem.* **124**, 519.
EHRLICH, P. (1909). *Ber. dtsch. chem. Ges.* **42**, A, 17.
EHRLICH, P. & BERTHEIM, A. (1912). *Ber. dtsch. chem. Ges.* **45**, 756.
FILDES, P. (1940*a*). *Lancet*, **1**, 955.
FILDES, P. (1940*b*). *Brit. J. Exp. Path.* **21**, 67.
FILDES, P. & RYDON, H. N. (1947). *Brit. J. Exp. Path.* **28**, 211.
FOURNEAU, E., TRÉFOUEL, J. & VALLÉE, J. (1924). *Ann. Inst. Pasteur*, **38**, 81.
HARRIS, J. S. & KOHN, H. I. (1941). *J. Pharmacol.* **73**, 383.
HOPKINS, F. G., MORGAN, E. S. & LUTWAK-MANN, C. (1938). *Biochem. J.* **32**, 1829.
JOHNSON, O. H., GREEN, D. E. & PAULI, R. (1944). *J. Biol. Chem.* **153**, 37.
JORDAN-LLOYD, D. (1940). *Trans. Faraday Soc.* **36**, 886.
KUHN, R. (1942). *Chemie*, **55**, 1.
KUHN, R., WEYGAND, F. & MÖLLER, E. F. (1943). *Ber. dtsch. chem. Ges.* **76**, 1044.
LETTRÉ, H. (1937). *Angew. Chem.* **50**, 581.
LETTRÉ, H. (1943). *Ergebn. Enzymforsch.* **9**, 1.
LILLY, V. G. & LEONIAN, L. H. (1944). *Science*, **99**, 205.
LONDON, F. (1930). *Z. Phys.* **63**, 245.
LONDON, F. (1942). *J. Phys. Chem.* **46**, 305.
MCILWAIN, H. (1940). *Brit. J. Exp. Path.* **21**, 136.

McIlwain, H. (1941). *Brit. J. Exp. Path.* **22**, 148.

McIlwain, H. (1942). *Brit. J. Exp. Path.* **23**, 95.

Madinaveitia, J. (1946). *Biochem. J.* **40**, 373.

Martin, A. R. & Rose, F. L. (1945). *Biochem. J.* **39**, 91.

Michaelis, L. & Menten, M. L. (1913). *Biochem. Z.* **49**, 333.

Oesterlin, M. (1936). *Klin. Wschr.* **15**, 1719.

Pauling, L. (1940). *The Nature of the Chemical Bond*, 2nd ed., p. 265. Ithaca, N.Y.

Pauling, L. (1945). In Landsteiner, K., *The Specificity of Serological Reactions*, revised ed., p. 275. Harvard.

Pauling, L. (1947). Lecture to the XIth International Congress of Pure and Applied Chemistry, 17 July 1947.

Potter, V. R. & Elvehjem, C. A. (1937). *J. biol. Chem.* **117**, 341.

Quastel, J. H. & Wooldridge, W. R. (1928). *Biochem. J.* **22**, 689.

Robbins, W. J. (1941). *Proc. Nat. Acad. Sci., Wash.*, **27**, 419.

Roblin, R. O. (1946). *Chem. Rev.* **38**, 255.

Roblin, R. O., Lampen, J. O., English, J. P., Cole, Q. P. & Vaughan, J. R. (1945). *J. Amer. Chem. Soc.* **67**, 290.

Rubin, S. H., Drekter, L. & Moyer, E. H. (1945). *Proc. Soc. Exp. Biol., N.Y.*, **58**, 352.

Rydon, H. N. (1948). *Biochem. Soc. Symp.* **1**, 55.

Sidgwick, N. V. (1937). *The Organic Chemistry of Nitrogen*, 2nd ed. p. xvii. Oxford.

Snell, E. E. (1941*a*). *J. Biol. Chem.* **139**, 975.

Snell, E. E. (1941*b*). *J. Biol. Chem.* **141**, 121.

Tracy, A. H. & Elderfield, R. C. (1941). *J. Organ. Chem.* **6**, 54.

Voegtlin, C. (1925). *Physiol. Rev.* **5**, 63.

Witt, O. N. (1876). *Ber. dtsch. chem. Ges.* **9**, 522.

Woods, D. D. (1940). *Brit. J. Exp. Path.* **21**, 74.

Woolley, D. W. (1944*a*). *Proc. Soc. Exp. Biol., N.Y.*, **55**, 179.

Woolley, D. W. (1944*b*). *J. Biol. Chem.* **152**, 225.

Woolley, D. W. & White, A. G. C. (1943*a*). *J. Biol. Chem.* **149**, 285.

Woolley, D. W. & White, A. G. C. (1943*b*). *J. Exp. Med.* **78**, 489.

Wyss, O., Rubin, M. & Strandskov, F. B. (1943). *Proc. Soc. Exp. Biol., N.Y.*, **52**, 155.

ANTIBIOTICS DERIVED FROM *BACILLUS POLYMYXA*

By GEORGE BROWNLEE

The Wellcome Physiological Research Laboratories, Kent, England

I. INTRODUCTION

The antibiotic at first referred to as 'Aerosporin' and now known as polymyxin A was discovered at The Wellcome Physiological Research Laboratories, Kent, England, in the course of routine screening of possible antibiotics in 1945.

The productive strain of the organism at first identified as *Bacillus aerosporus*, Greer, is now recognized to be identical with *B. polymyxa*, a widely distributed bacterium whose natural habitat appears to be the soil.

Isolated by Ainsworth, Brown & Brownlee (1947) the antibiotic was shown to be chemotherapeutic and selectively active against Gram-negative pathogens. It attracted attention in the course of screening operations by reason of its high activity against *Haemophilus pertussis* and *Eberthella typhosa*. Subsequently its chemotherapeutic and pharmacological properties were described by Brownlee & Bushby (1948). They described its bactericidal nature and emphasized the difficulty with which resistant strains emerged. After the paper of Brownlee & Bushby had been lodged for publication the report on 'Polymyxin' by Stansly *et al.* (1947) became available. 'Polymyxin' as described is produced by another strain of the organism which produces 'Aerosporin' and appears to have an identical antibacterial spectrum; however, as is apparent from the report, it differs both in antibacterial efficiency and in certain pharmacological properties from polymyxin A. There is an even earlier report of the antibacterial nature of crude liquid cultures of *Bacillus polymyxa* presented by Benedict & Langlykke (1947) which anticipated by three days a description by Ainsworth *et al.* (1947), of the chemotherapeutic activity of relatively pure material. The preliminary clinical report of Swift (1948) indicated that the new antibiotic appeared to be effective in aborting early cases of pertussis and in modifying the course of the disease in later cases in which mixed infections were already present. Simultaneously with the development of the pharmacology and chemotherapy of the material described as 'Aerosporin' were published the studies on its isolation and purification (by Catch & Friedmann, 1948), and on its chemical nature by Catch & Tudor Jones (1948) and by Tudor Jones (1948*a*, *b*). It was early observed that even 'pure' samples of the antibiotic appeared to be contaminated by a nephrotoxic principle (Brownlee & Bushby, 1948). This problem was studied by Brownlee & Short (1948), and it was shown that the lesions were restricted to the distal portion of the renal convoluted tubules and their extent could be estimated histologically and also by observing the resultant proteinurea in rats or dogs. Efforts to eliminate the toxic factor took the form of purification, identification of the site of damage and the development of counter-measures, and strain selection to eliminate its formation. The third of these procedures proved to be the most productive and quickly provided a second antibiotic which differed chemically from both 'Polymyxin' and 'Aerosporin' and proved to be free from nephrotoxic factor both when examined experimentally and clinically. Evidence will be presented later in this Symposium to justify these comments. There was a second and not altogether unexpected outcome of strain selection. There was quickly built up knowledge of a number of antibiotics of essentially similar biological and chemical structure, pharmacological differences being noted from one to the other.

The recognition of this related series, of which 'polymyxin' was one, raised in acute form the question of nomenclature. The reinvestigation of the taxonomic derivation of *B. aerosporus* and *B. polymyxa* (Porter, McClesky & Levine, 1937) has shown the two to be identical and confirms the latter as the generic name. It is therefore appropriate that the generic name for the antibiotics shall be polymyxin and agreement in this sense has been recently brought about between the British and American workers involved. 'Aerosporin' has thus been renamed polymyxin A and the 'polymyxin' of Stansly *et al.*, polymyxin D, while the newly investigated antibiotics which are free from the nephrotoxic property are polymyxin B and polymyxin E. There is also described polymyxin C.

The chemical basis on which this classification rests has been given by Jones (1948*c*) and is reviewed in this present account.

The substance polymyxin A, formerly known as 'Aerosporin' was early shown to be a basic substance forming salts with mineral acids and insoluble precipitates with acidic dyestuffs. The observation of Catch & Friedmann (1948) that the purest preparations gave the ninhydrin reaction in solution, both before and after acid hydrolysis, gave reason to suppose that part of the molecule, at least, consisted of polypeptide. The application of the method of partition chromatography on paper (Consden, Gordon & Martin, 1944) to the products of the acid hydrolysis of polymyxin A preparations by Jones (1948*a*) led to the discovery of the presence of leucine and threonine and of a basic substance with some of the properties of an amino-acid, as major constituents. The basic substance was later identified with αγ-diaminobutyric acid and was isolated from the mixed amino-acids of the hydrolysate by precipitation with picric acid, as the DL-picrate (Catch & Jones, 1948). This acid has since been shown to be present in hydrolysates prepared under milder conditions as the L-form (Catch, Jones & Wilkinson, 1948). Receipt of a sample of polymyxin D, the 'polymyxin' of Stansly *et al.* (1947) kindly made available by Dr Carey of the Lederle Laboratories Division of the American Cyanamid Company, enabled Catch *et al.* (1948) to identify the same amino-acids in this antibiotic, with serine as an additional acid. The optical configuration of the leucine and threonine present in polymyxin A has been determined on the paper chromatogram by an enzymic method (Jones, 1948*b*). These amino-acids were shown to be D-leucine and L-threonine. Subsequent isolation of D-leucine and L-threonine from the hydrolytic products by Catch *et al.* (1948), has amply confirmed the result obtained by the micro-technique.

It was early noted that when adequate samples of relatively crude polymyxin A were hydrolysed by acid in sealed tubes, globules of an oily appearance separated from the acid solution and a rancid odour was present; evaporation of the acid on a steam-bath in an open dish left a crystalline residue only. The fatty material was at first discounted as an impurity, but was noted for further study. A private communication from Dr Malcolm of the Lederle Laboratories Division of the American Cyanamid Company informed us of the presence, in the polymyxin molecule, of a carboxylic acid similar to pelargonic acid. Examination of our purest preparations of polymyxin A showed that there was little ether-soluble material present in the intact substance but that, after acid hydrolysis, ether indeed extracted an acidic substance. Removal of the ether and fractional distillation *in vacuo* yielded an optically active mobile oil, having the properties and composition of a saturated fatty acid $C_9H_{18}O_2$, different from pelargonic acid but as yet unidentified, a description of which has been given by Catch *et al.* (1948).

II. PARTITION CHROMATOGRAPHY OF POLYMYXIN A AND POLYMYXIN D

The elegant method of Consden *et al.* for the partition chromatography on filter paper of amino-acids and lower peptides, was successfully used in the identification of the amino-acid components of polymyxin A. In this method, filter paper, the moisture content of which is maintained constant, becomes the support for the stationary aqueous phase and a mobile, usually largely immiscible, phase is caused to flow over the sheet. The mixture, applied to the paper in solution in a small drop of water or other volatile solvent, which is then removed, is thus subjected, during the flow of the mobile phase, to countless distributions between the two phases and the components separate as spots. The positions taken up on the paper are dependent mainly upon the distribution coefficients of the components of the mixture for the two phases employed.

The method has the twofold value of not only enabling the separation and identification of complex mixtures but of providing constants, mainly dependent upon the solubility relationships of substances being studied, with the consumption of a minimum amount of material.

Preparations of polymyxin A of adequate purity moved on paper as single, rather elongated, spots. However the reproducibility of the positions of the spots on the paper, relative to the solvent front, in different chromatograms, prepared under what appear to be identical conditions, is not as good as that of simple amino-acids.

A typical chromatogram (Jones, 1948*c*) is shown in Pl. 1, fig. 1. The solvent flow is upwards with the solvent front completely off the leading edge of the paper. It is seen that polymyxin A and polymyxin D moved on the paper at different rates. In the same figure, is the chromatogram derived from a mixture of the two antibiotics, using the same amounts as in sections *A* and *C*, and, within the limits of the method, this chromatogram is the superimposition of those obtained with the single antibiotics. There can be no doubt that the two antibiotics are different substances. These observations proved to be of particular value at the time of the receipt of the polymyxin D sample since this was known from its stated assay value to be at most 67·5% pure, based on the published figure for the purest material, and the amino-acid serine, additional to the amino-acids known to be present in polymyxin A, might otherwise have been considered to be a component of an impurity.

III. ANTIBIOTICS FROM FURTHER STRAINS OF *B. POLYMYXA*

The two antibiotics considered above, while exhibiting certain differences, were shown by Brownlee, Bushby & Short (1948) to have some biological properties in common, one of which was the important property of renal toxicity (Brownlee & Bushby, 1948). Examination of the products of numerous other strains has led to the recognition of other antibiotics, some characterized by their chromatographic behaviour as the intact material and others by more detailed analysis following hydrolysis. At the time of writing, five groups are known, including polymyxin D supplied by the American Cyanamid Company.

Pl. 1, fig. 2, is a reproduction of a chromatogram prepared from seven only of the strains tested. The products of *B. polymyxa* are identified here by the culture numbers of the selected strains and the polymyxin D chromatograms were derived from three separate samples from strain B–71 of the Lederle Laboratories. In this figure three separate groups are recognizable. Strain CN 1419 produces an antibiotic which moves fastest on the paper. The strains CN 1984, CN 2136, CN 114J, CN 2002 and CN 121 yield products which move at comparable speed and more slowly than does CN 1419. Polymyxin D moves at an intermediate rate. It was shown in Pl. 1, fig. 1, that the material derived from strain CN 1419 (section *D*) and polymyxin D (section F) indeed move at different rates and that when mixed (section *E*), a partial separation may be achieved. The distinction between the products of these strains is of importance and has been confirmed using other solvents. If chromatograms derived from intact material provided the only evidence for the chemical identification of these antibiotics, only three groups would be recognizable from the diagram. An analysis of the chromatogram, Pl. 2, fig. 1, of the hydrolysed antibiotics enable additional differences to be detected (Table 1).

IV. PARTITION CHROMATOGRAMS OF THE HYDROLYSED ANTIBIOTICS

There is given in Pl. 2, fig. 1, a chromatogram of the hydrolytic products of the antibiotics. The main amino-acid spots are readily identified as leucine, phenyl-alanine, threonine, serine and $\alpha\gamma$-diaminobutyric acid in that order downwards. The direction of flow of the solvent on the paper is from the line at the bottom in an upward direction; one dimensional partition chromatography sufficed to separate the small number of amino-acids present.

We now have the data for the classification of the products of the *B. polymyxa* group of organisms (Jones, 1948*c*). There now appear to be four distinct classes of antibiotics, the component amino-acids of which are shown in Table 1. The classes of antibiotic are then distinguished by the agreed generic name polymyxin followed by alphabetical suffixes, such as appear earlier in this paper. Table 1 provides the evidence upon which the present assignment of names is based; this evidence is thus confined, at present, to the qualitative amino-acid composition of the antibiotics. The basis of the classification may require modification at a later date, should evidence of quantitative differences distinguish new members of the group. In Table 1, the signs given under the headings of the amino-acids indicate spots of an intensity known by experience to represent a significant amount of the amino-acid. The spots given by phenyl-alanine in the reproduction of Pl. 2, fig. 1, are much fainter than those of the freshly prepared chromatogram; moreover, the colour of the spot following use of the acid solvent mentioned above is characteristic for phenyl-alanine. Other relatively faint

Table 1. *The qualitative amino-acid composition of polymyxin A, B, C and D*

Polymyxin	Culture no.	Leucine	Phenyl-alanine	Threonine	Serine	α-γ-diamino-butyric acid
A	1984 2002 121	+	−	+	−	+
B	1419	+	+	+	−	+
C	2136 114J	−	+	+	−	+
D Stansly *et al.* (1947)	B 71 (Lederle)	+	−	+	+	+

spots are present in the chromatogram, e.g. CN 2136 and CN 114J give products in which there is a faint spot in the leucine position; spots in the serine position appear in the chromatograms given by other products. The question whether these classes of polymyxin represent single antibiotics or mixtures of two or more is not answerable in our present state of knowledge. Most of the materials investigated in this comparative study are not more than 50% pure. The fainter spots could therefore be due to the presence, either of impurities devoid of antibacterial activity, or of small proportions of other antibiotics. To give an example, the faint spots for leucine in the polymyxin C chromatograms could be due to admixture with a small proportion of polymyxin A. That polymyxin B is not a mixture of polymyxins A and C, which could be one interpretation of its amino-acid composition, is shown, chemically, by the chromatograms of the intact materials (Pl. 1, fig. 2), and, biologically, by the absence of nephrotoxicity (Brownlee *et al.* 1948) even in preparations known to be impure. Should purification of the polymyxin C preparations yield homogeneous material still showing the presence of leucine in the smaller proportions indicated above, the molecular weight of this antibiotic would be rather greater than the present estimates based on polymyxin D and polymyxin A. What is not known at present is whether each class is a single antibiotic of settled structure or a mixture of several substances containing the same amino-acids. The available evidence based on the purification of polymyxin A (Catch *et al.* 1948) and on the present work appears to favour the concept that each strain of the organisms produces one main antibiotic, which, when isolated in a standard manner, appears substantially homogeneous.

V. THE FATTY ACID COMPONENT

As mentioned above, the fatty acid components of these preparations have been characterized but not yet identified, and the acids have been isolated from polymyxins A and B and D as the *p*-bromobenzylthiuronium salts. These show identical crystalline form, melting-points and mixed melting-points (Catch *et al.* 1948).

VI. *IN VITRO* ANTIBACTERIAL ACTIVITY

Throughout the biological comparisons which follow, the concentrations of polymyxin A are stated in terms of 'Polymyxin A hydrochloride Standard 1947' containing 10,000 'units'/mg. Material of this standard may, for all practical purposes, be considered to be the pure hydrochloride. Polymyxin B has not yet been obtained as a pure substance and its intrinsic potency is unknown. Preliminary studies indicate, however, that its potency will not differ materially from that of polymyxin A. Polymyxin B has therefore been expressed as 'units' of polymyxin A. The assay procedure is the short turbidometric method using *Esch. coli* (Brownlee & Bushby, 1948). Polymyxin D is referred to in terms of pure antibiotic of 2000 units/mg. (Stansly *et al.* 1947).

The minimal concentrations of polymyxin A, B and D which inhibited growth of large inocula of a range of organisms were determined by streaking 24 hr. old broth cultures of the organisms on nutrient agar incorporating 2% horse blood and containing twofold dilutions of the antibiotics. Repetitions using the same strains show slight variation in the minimum inhibitory concentration. A typical comparison is given in Table 2.

Table 2. *Comparative minimal concentrations of polymyxin A, B and D inhibiting large inocula*

Polymyxin A 10,000 units/mg. (Brownlee & Bushby, 1948), B assayed in terms of A: D 2000 units/mg. (Stansly *et al.* 1947).

Culture C.N.	Organism	Polymyxin A (μg./ml.)	Polymyxin B (μg./ml.)	Polymyxin D (μg./ml.)
512	*Eb. typhosa*	0·25	0·25	0·30
312	*Esch. coli*	0·25	0·25	0·30
740	*Br. bronchisepta*	0·125	0·125	0·15
247	*V. comma*	0·25	0·25	0·30
164	*Sal. schottmuelleri*	0·125	0·25	0·30
1520	*Sh. paradysenteriae*	0·016	0·03	0·04
214	*Br. abortus*	1·0	1·0	1·25
144	*A. aerogenes*	0·25	0·25	0·45
1833	*Past. Muriseptica*	0·25	0·25	0·45
200	*Ps. aeruginosa*	1·0	2·0	1·5
181	*Kleb. pneumoniae*	2·0	2·0	2·5
13	*St. viridans*	>4·0	>4·0	>4·0
491	*Staph. aureus*	>4·0	>4·0	>4·0
10	*St. pyogenes*	>4·0	>4·0	>4·0
329	*Pr. vulgaris*	>4·0	>4·0	>4·0
127	*H. pertussis**	0·25	0·25	0·15

* Bordet-Gengou.

No significant difference in the sensitivity of these organisms to the three antibiotics is detected by the method using large inocula.

VII. EFFECT OF THE SIZE OF THE INOCULUM

Polymyxin B is bactericidal like polymyxin A and D. Time, number of organisms, and concentration appear to bear a simple relation one to the other. The fact that the antibacterial action of the three antibiotics depends on the number of organisms present is illustrated in Table 3. This shows the effect of twofold dilutions of the antibiotics in nutrient broth, inoculated with 0·05 ml./5 ml. of a 24 hr. broth culture of *Eberthella typhosa* diluted to 10^0, 10^{-2}, 10^{-4} and 10^{-6}. The cultures were incubated at 37° C. for 18 hr. Polymyxin A and B are similarly influenced by the number of bacteria present, D is less so.

Table 3. *The antibacterial efficiency of polymyxin A, B and D varies with the number of organisms. This is less so with D*

Polymyxin units in terms of A	Inoc.	2·0	1·0	0·5	0·25	0·125	0·06	0·03
A	10^0	−	+	+	+	+	+	+
	10^{-1}	−	−	−	−	+	+	+
	10^{-4}	−	−	−	−	−	+	+
	10^{-5}	−	−	−	−	−	−	+
B	10^0	−	+	+	+	+	+	+
	10^{-1}	−	−	−	−	+	+	+
	10^{-4}	−	−	−	−	−	+	+
	10^{-5}	−	−	−	−	−	−	+
D	10^0	+	+	+	+	+	+	+
	10^{-1}	−	−	+	+	+	+	+
	10^{-4}	−	−	−	+	+	+	+
	10^{-5}	−	−	−	−	+	+	+

VIII. *IN VIVO* ANTIBACTERIAL ACTIVITY

The chemotherapeutic worth of polymyxin A, B and D was assessed in a side-by-side comparison, in groups of mice infected with Gram-negative pathogens.

Haemophilus pertussis. Groups of mice, infected with 0·05 ml. of a broth suspension containing 10^5 organisms (Kendrick *et al.* 1947) washed from a 48 hr. culture on Bordet-Gengou medium, and constituting about 10,000 average lethal doses, die consistently within 4–7 days. With such groups the relative chemotherapeutic effect of polymyxin A, B and D were compared, for a period of 14 days during which survivors were recorded every second day. A protocol of a typical experiment is reproduced in Table 4 in which the average survival time is calculated by adding together the number of days survived by each mouse and dividing the sum by the number of animals in the group. This table shows polymyxin A and B to be similar and to be more effective than polymyxin D.

Table 4. *Protective action of polymyxin A, B and D for mice infected with* 10,000 *lethal doses of* Haemophilus pertussis

Polymyxin A as 'pure' polymyxin A 10,000 units/mg. (Brownlee & Bushby, 1948), Polymyxin B in terms of A, and D as 'pure' polymyxin D 2000 units/mg. (Stansly *et al.* 1947).

	Dose (mg.)	No. of doses	No. of mice	No. of mice surviving on day				Average survival rate (days)	% survival
				2	6	8	14		
Polymyxin A	0·1	2	36	36	36	35	28	13·1	77·8
	0·05	6	66	66	66	65	60	13·6	91·0
	0·025	6	36	36	32	28	20	10·8	55·5
Polymyxin B	0·1	6	6	6	6	6	6	14·0	100·0
	0·05	6	11	11	11	10	9	12·9	81·8
	0·025	6	13	13	12	8	6	10·2	46·0
Polymyxin D	0·25	6	23	23	23	23	21	13·7	91·3
		2	21	21	21	19	12	12·3	57·0
	0·125	6	19	19	19	16	11	11·8	57·9
	0·1	2	14	14	14	11	3	9·6	21·4
	0·625	6	21	21	20	17	9	10·6	43·0
Lethal controls	—	—	48	48	21	3	0	4·9	0·0

Eberthella typhosa. Mice infected intraperitoneally with 0·5 ml. of a suspension of Rawlings strain, containing 1000 average lethal doses in 5% mucin, develop an overwhelming infection so that most die within 6 hr. A typical protocol of a comparison between the three antibiotics in these test animals is reproduced in Table 5. The antibiotics were administered immediately after infection, 6 hr. later and twice daily for the following 3 days with subcutaneous injections of varying doses of the antibiotics.

Polymyxin B and D show similar protection, and appear to be less efficient than polymyxin A.

Table 5. *The effect of polymyxin A, B and D in protecting mice infected with* 1000 *lethal doses of* Eberthella typhosa

Polymyxin A is 'pure', 10,000 units/mg. (Brownlee & Bushby, 1948), B is in terms of A, D is 'pure' 2000 units/mg. (Stansly *et al.* 1947).

	Dose (mg.)	No. of organisms	No. of mice	No. of mice surviving on day				Average survival rate (days)	% survival
				1	3	5	7		
Polymyxin A	0·1	5×10^7	18	18	18	18	18	7·0	100
	0·05	5×10^7	18	18	17	16	14	6·4	76·7
Polymyxin B	0·1	5×10^7	18	17	13	13	13	5·3	72·2
	0·05	5×10^7	18	16	16	13	12	5·5	66·7
Polymyxin D	0·2	5×10^7	12	12	12	11	11	6·9	91·7
	0·1	5×10^7	12	12	9	9	9	5·6	75·0
Lethal controls	—	5×10^7	18	1	—	—	—	0·0	0·0
	—	5×10^6	6	4	—	—	—	1·0	0·0
	—	5×10^4	6	3	3	3	3	3·5	50·0

Kleb. pneumoniae. Protection experiments were designed also with groups of mice infected intraperitoneally with 10,000 and 1000 lethal doses of *Kleb. pneumoniae* suspended in 0·5 ml. of 5% mucin. They were treated immediately, 6 hr. later, and twice daily for the following 3 days with subcutaneous injections of varying doses of polymyxin A, B and D. Protection against this strain of the organism is not high (Table 6), and it is evident that the higher infecting dose was overwhelming. Comparison at the lower infecting dose shows the relative efficiency of the three antibiotics to be close.

Table 6. *The effect of polymyxin A, B and D in protecting mice infected with* 10,000 *and* 1000 *lethal doses of* Kleb. pneumoniae

Polymyxin A, in terms of 'pure' substance, 10,000 units/mg. (Brownlee & Bushby, 1948), B in terms of A, D in terms of 'pure' substance, 2000 units/mg. (Stansly *et al.* 1947).

	Dose (mg.)	No. of organisms	No. of mice	No. of mice surviving on day				Average survival rate (days)	% survival
				1	3	5	7		
Polymyxin A	0·1	5×10^8	6	5	1	1	1	1·8	16·7
		5×10^7	6	5	2	2	1	3·8	16·7
	0·05	5×10^8	6	6	3	3	3	4·0	50·0
		5×10^7	6	6	3	3	3	4·0	50·0
Polymyxin B	0·1	5×10^8	6	1	—	—	—	0·0	0·0
		5×10^7	6	6	5	4	4	4·8	66·7
	0·05	5×10^8	6	2	—	—	—	0·0	0·0
		5×10^7	6	6	5	2	2	3·3	33·3
Polymyxin D	0·2	5×10^8	6	6	2	1	1	2·5	16·7
		5×10^7	6	6	2	2	2	3·3	33·3
	0·1	5×10^8	6	6	1	1	1	2·5	16·7
		5×10^7	6	6	3	3	2	2·5	33·3
Lethal controls	—	5×10^8	6	0	—	—	—	0·0	0·0
	—	5×10^7	6	0	—	—	—	0·0	0·0
	—	5×10^6	6	6	2	2	2	3·0	33·0
	—	5×10^4	6	6	4	4	4	5·0	66·7

IX. TOXICITY

Acute toxicity to mice. The acute toxicity to mice was studied after the intravenous injection of varying doses of polymyxin A, B and D into groups of ten animals, weighing 18–20 g. With lethal doses of polymyxin A and B, death occurs from respiratory failure in less than 2 min. With near-lethal doses there is vasoconstriction, muscular incoordination and respiratory distress; this stage is followed by clonic convulsions and then flaccid paralysis associated with respiratory embarrassment and cyanosis, but the animals recover within 10 min.

The symptoms seen with polymyxin D are similar. Differences are that recovery may be delayed for 30 min. or more, the paralysis is much less evident and deaths may sometimes be delayed as long as 30 min. The acute intravenous toxicities of several batches of polymyxin A and B and of three batches of polymyxin D have been determined. Inspection of the slopes relating probits corresponding to mortalities and the logarithms of the doses used for all three antibiotics show them to be of the same order. There is no evidence of more than one agent contributing to the death of the mice.

The mean intravenous L.D. 50 for polymyxin A corresponds to 6·9 mg./kg. of 'Polymyxin A Standard (1947)', with a range of 4·2–13·5, mean of estimates on twenty-one different samples.

The mean intravenous L.D. 50 for polymyxin B corresponds to 6·1 mg./kg. of 'Polymyxin A Standard (1947)', range 4·2–10·7, mean of estimates on twelve different samples.

The mean intravenous L.D. 50 for polymyxin D corresponds to 11·9 mg./kg., mean of estimates on three samples, with a range of 8·7–18·0.

Acute toxicity to mice—intraperitoneal route. The acute toxicity after injection by the intraperitoneal route was observed in groups of mice; the antibiotics were each dissolved in 0·4 ml. of distilled water. The toxic signs were essentially similar for all antibiotics and were evident 30 min. after injection, with weakness of hindlimbs, followed by shivering and loss of temperature control. Clonic convulsions developed in 15–30 min., with respiratory embarrassment; death from respiratory failure occurred after 2–4 hr. Recovery from near toxic doses occurs in 3–4 hr.

The mean intraperitoneal L.D. 50 for polymyxin A corresponds to 13·9 mg./kg. 'Polymyxin A Standard (1947)' with a range of 10·5–19·0 for a mean of five samples.

The mean intraperitoneal L.D. 50 for polymyxin B corresponds to 12·1 mg./kg. 'Polymyxin A Standard (1947)' with a range of 8·7–17·5, a mean of eleven samples.

The mean intraperitoneal L.D. 50 for polymyxin D corresponds to 24·5 mg./kg., a mean of two samples, range 24·0–25·0.

Acute toxicity to mice—subcutaneous route. The acute toxicity after subcutaneous injection of the three antibiotics was observed in groups of mice. The toxic signs were similar to those observed after administration by other routes, but were delayed for 30–60 min. Deaths occurred until 36 hr. and the estimates were made at 48 hr. For polymyxin A, the mean L.D. 50 corresponded to 87·5 mg./kg. with a range of 79·2–87·5 mg./kg., a mean of three samples.

For polymyxin B the mean L.D. 50 was 82·5 mg./kg. with a range of 70·5–92·4 mg./kg. for one sample.

For polymyxin D the mean L.D. 50 was 160 mg./kg. with a range of 112·5–220·0 mg. for two samples.

The manifold differences between the acute toxic doses for these antibiotics by the intravenous or intraperitoneal routes when compared with their toxicities by the subcutaneous route may be due to fixation of the antibiotics in local tissue.

Acute toxicity—intracisternal—rabbit. Groups of rabbits injected intracisternally with polymyxin A contained in 0·25 ml. isotonic buffer show respiratory depression and respiratory failure; other central effects appear to be absent.

The average lethal dose appears to be 0·6 mg./kg. with a range of 0·5–0·75, a mean of eight observations.

Vestibular function. The injection intracisternally in rabbits of sublethal doses of polymyxin A, in the range of 0·3–0·5 mg./kg., reveals no evidence of vestibular disfunction. A temporary, infrequent, rotatory nystagmus such as follows the injection of sterile saline is seen during the first hour. Thereafter, nystagmus after rotation in the lateral plane is normal during periods of 12 days observation. Animals similarly treated with 3 mg./kg. of pure streptomycin sulphate show, for periods of about 4 hr. severe respiratory depression, extensor tremors, stupor and inability to hold the head horizontal, although jerking attempts were made to do so. However, other signs of vestibular disfunction, such as rate and duration of nystagmus were unaltered.

Circulatory effects. In cats anaesthetized with pentobarbitone sodium traces of depressor substances are detected in impure batch preparations of polymyxin A and B. The amount of depressor substance, which is readily removed by further purification is of the order of 0·2–3 μg. equivalent of histamine per mg. of 50% pure antibiotic.

Kidney-damaging factor. Polymyxin A preparations were shown to contain a nephrotoxic principle (Brownlee & Bushby, 1948). Since nephrotoxic action and purity were not directly related from batch to batch it appeared possible that the principle was not necessarily also the antibiotic. Indeed, at the present time the position remains unproved, yet the purest material available to us produces substantial proteinuria in the experimental animal and in man. Meanwhile the observation that the closely allied polymyxin B is free from the nephrotoxic principle, although itself impure, offers an alternative explanation for the batch variability. The lesions produced by polymyxin A preparations when studied histologically in the rat and guinea-pig appear to be restricted to the distal portion of the convoluted tubules. Estimates of the resultant proteinuria, in rats and in dogs have enabled semi-quantitative assays to be made, and potential antagonists to be tested. It is convenient to relate the logarithm of the dose of polymyxin A administered and the mean weight of protein excreted in 72 hr. by a group of four rats (Wistar, inbred stock), expressed in terms of mg./100 gm. of rat. It is to be noted that the 'normal' rat excretes 0–12 mg. protein. This test, which is capable of refinement, has proved useful.

It is known that the nephrotoxic action of DL-serine is due to the unnatural isomer, D-serine (Artom, Fishman & Morehead, 1945), and the site of damage to be the distal portion of the convoluted tubules (Wachstein, 1947*a*). Further, Wachstein (1947*b*) found DL-methionine and glutathione, but not cysteine, thioglycollic acid or 2, 3-dithiopropanol to protect against D-serine. DL-alanine, glycine, DL-threonine, and glycollic, butyric and pyruvic acids gave some protection.

Brownlee & Short (1948) confirmed the effects of these substances in rats and dogs by estimating the degree of proteinuria, and have further found that substances which antagonize the nephrotoxic action of D-serine, also antagonize the nephrotoxic activity of preparations of polymyxin A. A number of the substances tested is listed in Table 7. Unless otherwise stated they were injected parenterally. The most effective protecting agent is DL-methionine, followed by S-methylcysteine, then animal protein hydrolysate. Animal protein hydrolysate is also effective orally and further protection may be devised prophylactically by the latter means. One substance rich in reactive-SH groups, 2:3-dithiopropanol, is ineffective and L-cysteine gives little protection. Glycine and DL-alanine which contain no sulphur are intermediate in their protective capacity.

Table 7. *The effect of various substances on the nephrotoxic factor of polymyxin A in rats*

Dose: 1 mg. (10,000 units) per 100 gm. rat.

Substance		Protein excreted in 72 hr. (mg./100 gm. rat)
Glycine	400 mg.	24·5
DL-Alanine	400 mg.	30·5
L-Cysteine	400 mg.	40·0
DL-Methionine	200 mg.	12·0
S-Methyl-L-Cysteine	200 mg.	15·1
2:3-Dithiopropanol	10 mg.	55·0
Protein hydrolysate horse of muscle:		
Parenterally	100 mg.	17·4
Orally	5 g.	17·0
In diet	50 mg. daily for 4 days	14·0
In diet	1 g. for 1 day	16·5
Polymyxin A	Alone	58·8, 57·5, 62·0
Controls	—	11·4, 12·3, 8·0

Protection by methionine is complete in the dog as is illustrated in Table 8, where it is shown that 10 mg. of methionine injected once will protect against the nephrotoxic action associated with 1 mg. of polymyxin A given four times daily for 3 days. Complete protection with methionine is not seen in man. It is an interesting observation that the tubular damage in man appears to be quite temporary. Thus while the stimulus is applied for 5 days in the form of doses of polymyxin A, the response is skew; it is at a maximum at about the fourth day. In other words the kidney has been able to mobilize its own repair devices during the period of exhibition of the damaging agent.

Table 8. *The protective effect of methionine against the nephrotoxic factor in dogs*

Dose: 0·8 mg. (8000 units) per kg. 3-hourly 4 times

Methionine (mg./kg.)	Max. conc. protein in urine in 96 hr. (g./l.)
20	0
10	0
5	0
2·5	0, 0·2
1·0	0·8, 1·4
0·0	0·7, 0·2, 0·3
Dose: 1·0 mg. (10,000 units) per kg. 3-hourly 4 times daily for 3 days	
10	0
5	0, 1·0
2·5	7·1
0·0	5·0, 2·0

Serine is absent from toxic preparations of polymyxin A, and the hydrolytic product of the antibiotic, and its individual amino-acid components leucine, threonine and $\alpha\gamma$-diaminobutyric acid are not themselves nephrotoxic.

It may be deduced from the observations that the toxic principle is a peptide with an unnatural amino-acid make-up, that the lesions arise during tubular-resorption, and that labile methyl groups may be involved in the process.

X. A COMPARISON OF KIDNEY-DAMAGING EFFECTS OF POLYMYXIN A, B AND D IN RATS

The total urinary protein excreted by groups of four Wistar rats in the 72 hr. after the subcutaneous injection of 1 mg./100 gm. of a number of batches of polymyxin A was compared with that produced by a standard batch 'A 1 P48', and a control group receiving saline.

There has been listed in Table 9 an index of the nephrotoxic effect of ten representative plant-scale batches of polymyxin A. Following these are five hydrochloride fractions obtained by the successive helianthate fractionations of one batch of the antibiotic. It is to be noted that there is no direct relation between impurity and nephrotoxic action. An alternative Reineckate fractionation, also given, shows no diminution in nephrotoxic activity with purification. Even although preparations 256 and 56 are to be considered only of 90% purity, it must be considered a possibility that the nephrotoxic property associated with polymyxin A is an intrinsic property.

When a comparison is made between polymyxin A, B and D it is observed that polymyxin B causes no significant change in proteinuria in the rat. Polymyxin A and D cause a marked increase in proteinuria.

Table 9. *Nephrotoxic principle associated with polymyxin A*

Polymyxin A

	Batch	Units/mg.	Wt. protein excreted (mg.)		
			Test	Standard	Control
First hydrochloride	2	5250	164	335	55
	12	3400	300	129	9
	23	5612	323	235	52
	24	4800	540	235	52
	36	6000	180	298	48
	37	4000	438	209	44
	85	5000	146	107	27
	104	6500	247	130	37
	105	5900	204	130	37
	106	5500	176	130	37
Hydrochloride prepared from successive helianthate fractions	256	8750	148	280	75
	257	5330	234	280	75
	258	8320	235	280	75
	259	6960	135	280	75
	260	2430	236	280	75
Hydrochloride prepared from successive Reineckate fractions	54	7940	214	238	49
	55	8260	116	238	49
	56	8500	139	238	49
	57	7460	234	238	49

Table 10. *A comparison between the nephrotoxic properties associated with polymyxin A, B and D in the rat*

Mean weight of protein excreted per rat, in a group of four, in 72 hr.

	Batch	Units/mg.	Wt. protein excreted (mg.)		
			Test	Standard	Control
Polymyxin A: hydrochloride prepared from successive Reineckate fractions	54	7940	214	238	49
	55	8260	116	238	49
	56	8500	139	238	49
	57	7460	234	238	49
Polymyxin B: first hydrochloride	93	1200	24	173	39
	108	2800	89	132	52
	109	2900	14	263	35
	120	1600	48	168	45
	121	2660	61	168	45
	123	4500	44	168	45
Polymyxin D	—	1000*	147	73	12
	—	1000	210	260	35

* Stansly *et al.* (1947), 2000 units/mg.

Renal toxicity in rabbits. Comparison between polymyxin A and B. Rabbits given 3-hourly subcutaneous injections of 3 mg., 30,000 units/kg., four times daily for 3 days show no significant increase in proteinuria in the case of those given polymyxin B (Table 11). Those treated with polymyxin A show a peak excretion of protein at 24–48 hr. A microscopical examination of the urine of rabbits given polymyxin B shows no casts and only an occasional renal cell, in contrast to those given polymyxin A which showed the presence of renal cells and casts at the 24 and 48 hr. periods, and renal cells at 48 and 72 hr.

Table 11. *Nephrotoxic action of polymyxin A and B in the rabbit treated with* 30,000 *units/kg.* 4 *times daily for* 3 *days*

	Rabbit	Batch	μ/mg.	Concentration of protein (g./l.) in urine after			
				0 hr.	24 hr.	48 hr.	72 hr.
Polymyxin A	1006	52	4340	0·18	0·3	1·2	1·8
	1010	85	5000	0	3·4	2·6	0·5
Polymyxin B	1004	83	4600	0·18	0·6	0	0·13
	1008	83	4600	0	0·2	0·02	0·08
	1005	87	3800	0·1	0·8	0·2	0·9
	1009	88	4500	0	0·14	0·27	0·24
	1007	93	1200	0	0	0·12	0·05
Controls	—	—	—	0	0	0·1	0·3
	—	—	—	0	0	0	0·08

Histological examination. Rabbits numbered 1006 and 1010 receiving polymyxin A showed the usual congestion and stripping of the epithelium of many tubules with disintegration of cells of Henle's loops.

Rabbits 1004, 1008, 1005, 1009 and 1007, given polymyxin B, were normal or with slight cloudy swelling of some secretory tubules.

XI. COMPARISON OF RENAL TOXICITY OF POLYMYXIN A AND B IN DOGS

A comparison of the nephrotoxic action of polymyxin A and B in dogs given 3-hourly subcutaneous injections of 1 mg./kg. four times a day for 3 days is shown in Table 12 in which typical results are given. This is a critical experiment in view of the known nephrotic tendency of the dog. As with the rat and rabbit, it is observed that polymyxin B differs from polymyxin A in causing no significant degree of damage. The peak excretion of protein is at 48 to 72 hr.

Microscopic examination of the urine of the treated dogs showed the presence of many renal cells but only occasional casts in animals treated with polymyxin A, and no, or only occasional, renal cells in the case of dogs treated with polymyxin B.

Table 12. *A comparison of the nephrotoxic action of polymyxin A and B in dogs*

1 mg. (10,000 units) 4 times daily for 3 days. Polymyxin A and B in dogs

	Dog	Batch	Units/mg.	Concentration of protein in urine (g./l.) after					
				0 hr.	24 hr.	48 hr.	72 hr.	96 hr.	120 hr.
Polymyxin A	8	37	4000	0	0·1	2·6	4·4	0·8	—
	11	52	4340	0	0·08	3·7	1·0	0·5	0·25
	18	85	5000	0	0·8	2·5	1·4	—	—
	14	85	5000	0	0·15	1·8	2·6	1·0	1·0
Polymyxin B	19	108	2800	0	0	0	0·07	—	—
	20	108	2800	0	0·20	0·3	0·3	—	—
	12	123	4500	0	0·56	0·25	0·2	0·3	0·4
	2	124	6000	0	0·12	0·14	0·7	0·5	0·7
	17	125	6000	0	0·5	0·8	0·8	—	0·6

Histological examination of kidneys. Dogs treated with polymyxin A show severe acute nephrosis. The secretory epithelium is seen in all stages of disintegration; cells are shrunken rather than swollen; large numbers of granular, hyaline and cellular casts are seen. There is no apparent glomerular damage apart from possibly slight swelling of tuft epithelium. (But this is seen in the 'normal' dog.) There is also a little active acute, and chronic, focal, interstitial nephritis.

In dogs treated with polymyxin B the only abnormalities observed in the histology of the kidney were a mild hyaline droplet degeneration of convoluted tubular epithelium and an occasional foreign body (round worm larvae) granulomata. Similar changes were observed in controls.

Dogs given subcutaneous injections of 1 mg./kg. of polymyxin A excrete large amounts of the antibiotic in the urine both during, and for several days, after treatment. The greater the kidney damage, the higher appears to be the concentration of antibiotic in the urine.

Similarly, the concentration of polymyxin B found in urine is small but is sufficient to sterilize Gram-negative organisms (Table 13).

Enzyme treatment. Neither the nephrotoxic principle nor the antibacterial activity of polymyxin A or D is affected by digestion with trypsin, pepsin, or papain.

Renal damage and urinary concentration in man. A comparison is made in Table 14 of the effects on urinary concentrations of both antibiotic and protein in man treated with polymyxin A and varying doses of methionine, and with polymyxin B.

As with the experimental animal, there appears to be a direct relation between the extent of renal damage and the concentration of antibiotic in the urine. Nevertheless, the concentrations in urine are adequate to inhibit Gram-negative pathogens.

Table 13. *Nephrotoxic action and renal excretion of polymyxin A and B*

'P' is the units in plasma. 'Uu.' is the units in urine. 'Up.' is the protein in urine. The damaged kidney (polymyxin A) appears to leak antibiotic.

	Dog	Polymyxin (μ/ml.) Protein (g./l.)	Day of treatment			Day after treatment		
			1	2	3	1	2	3
Polymyxin A	11	P.	8	8	32	—	—	—
		Uu.	7	106	213	106	53	26
		Up.	0	0·1	3·7	0·7	0·5	0·25
	18	P.	4	16	8	4	—	—
		Uu.	3	56	106	426	—	—
		Up.	0	0·75	2·5	1·4	—	—
	14	P.	24	3	6	—	—	—
		Uu.	3	3	213	106	2·6	7
		Up.	0	0·25	1·8	2·6	1·0	0·8
Polymyxin B	12	P.	8	4	4	—	—	—
		Uu.	3	3	3	3	3	3
		Up.	0	0	0·8	0·2	0	0
	8	P.	4	16	32	—	—	—
		Uu.	7	3	203	26	—	—
		Up.	0	0	0	0·4	0·5	0
	19	P.	4	8	4	4	—	—
		Uu.	3	13	26	13	—	—
		Up.	0	0	0	0·07	—	—

Table 14. *A comparison of the nephrotoxic activity of polymyxin A and B in man*

No kidney damage is seen with polymyxin B. As with the dog, the kidney damaged with polymyxin A appears to leak protein and antibiotic.

Polymyxin A and B in man

Case	Dose (mg./kg.)	Methionine	(*a*) Antibiotic units/ml. (*b*) Protein g./l.	Day of treatment					Day after treatment		
				1	2	3	4	5	1	2	3
			Polymyxin A								
C	1	×5	(*a*) units/ml.	—	—	250	380	73	<3	—	—
			(*b*) g./l.	—	—	1·0	0·9	0·5	0·2	—	—
E	1	×10	(*a*) units/ml.	3	20	133	266	533	426	106	26
			(*b*) g./l.	—	0·02	0·17	1·1	3·7	1·5	1·0	0·7
Mac.	1	×5	(*a*) units/ml.	6	13	7	13	106	326	106	26
			(*b*) g./l.	—	—	0·2	0·2	1·4	4·4	0·7	0·3
K	1	×10	(*a*) units/ml.	7	26	53	26	213	326	213	26
			(*b*) g./l.	—	—	0·1	0·8	0·6	1·4	0·5	0·1
Bur.	1	×5	(*a*) units/ml.	—	3	3	7	106	213	106	13
			(*b*) g./l.	—	—	—	Tr.	2·5	1·7	2·0	Tr.
			Polymyxin B								
B	0·5	—	(*a*) units/ml.	3	7	<3	<3	<3	3	<3	<3
			(*b*) g./l.	Tr.	—	Tr.	—	—	Tr.	—	—
W	0·2	—	(*a*) units/ml.	3	3	<3	<3	<3	<3	<3	<3
			(*b*) g./l.	—	—	—	Tr.	Tr.	Tr.	—	—

Polymyxin B: side-effects. Two apparently related side-effects have been observed after the administration of polymyxin B in experimental animals and in man. They are a mild pyrexia, reversible with antipyretic drugs and a mild local reaction which is of delayed onset. Both have proved to be of minor importance and have not proved contra-indications to the use of the drug. The preparations in use are impure and it is not yet possible to implicate drug or impurity.

XII. SUMMARY

1. 'Aerosporin' and 'Polymyxin' are two representative members of a series of antibiotics produced by strains of *Bacillus polymyxa*.

2. The application of paper chromatography to the product of different strains of *B. polymyxa* has revealed the multiple nature of the antibiotics and provides a means of classification. They are now called polymyxin A, B, C, D, E, and so on.

3. All polymyxins at present known contain threonine, $\alpha\gamma$-diamino-butyric acid and an optically active, fatty acid of empirical formula $C_9H_{18}O_2$. Polymyxin A has D-leucine, and polymyxin C phenyl-alanine in addition. Polymyxin D has leucine and serine in addition.

4. Pure strains appear to produce single antibiotics.

5. Of the series at present known, polymyxin A, B and D have been examined systematically and are shown to be powerful antibacterial substances selectively active against Gram-negative pathogens. They are chemotherapeutic.

6. In experimental infections with *Haemophilus pertussis*, polymyxin A is superior to polymyxin B, and both are superior to polymyxin D. In the very acute infections with *Eberthella typhosa* and *Kleb. pneumoniae*, polymyxin A is more efficient than polymyxin B and D.

7. Estimates of toxicity to mice (L.D. 50) by intraperitoneal, intravenous or subcutaneous routes show polymyxin D to have one-half to one-third the acute toxicity of polymyxin A or B.

8. Polymyxin A and D cause gross renal damage in the rat, guinea-pig, rabbit and dog. Polymyxin A causes proteinuria in man.

9. The proteinuria may be partly prevented in the rat, and totally in the dog, by the simultaneous administration of methionine parenterally or orally, or by protein hydrolysate orally. Polymyxin B produces slight proteinuria under experimental conditions but no histological changes in the kidney. It does not cause proteinuria in man.

10. Polymyxin A and B appear in urine. The concentration is greater with A than B and in the former case is proportional to the damage to tubules.

11. Two side effects observed in experimental animals with samples of impure polymyxin B, a mild pyrexia reversible with antipyretic drugs, and a local reaction have not proved contra-indications to the use of this antibiotic in man.

REFERENCES

AINSWORTH, G. C., BROWN, A. M. & BROWNLEE, G. (1947). *Nature, Lond.*, **160**, 263.

ARTOM, C., FISHMAN, W. H. & MOREHEAD, R. P. (1945). *Proc. Soc. Exp. Biol., N.Y.*, **60**, 284.

BENEDICT, R. G. & LANGLYKKE, A. F. (1947). *J. Bact.* **54**, 24.

BROWNLEE, G. & BUSHBY, S. R. M. (1948). *Lancet*, no. 254, p. 127.

BROWNLEE, G., BUSHBY, S. R. M. & SHORT, E. I. (1948). *Ann. N.Y. Acad. Sci.* (in the Press).

BROWNLEE, G. & SHORT, E. I. (1948). *Biochem. J.* **42**, liii.

CATCH, J. R. & FRIEDMANN, R. (1948). *Biochem. J.* **42** (in the Press).

CATCH, J. R. & JONES, T. S. G. (1948). *Biochem. J.* **42** (in the Press).

CATCH, J. R., JONES, T. S. G. & WILKINSON, S. (1948). *Ann. N.Y. Acad. Sci.* (in the Press).

CONSDEN, R., GORDON, A. H. & MARTIN, A. J. P. (1944). *Biochem. J.* **38**, 224.

JONES, T. S. G. (1948*a*). *Biochem. J.* **42**, xxxv.

JONES, T. S. G. (1948*b*). *Biochem. J.* (in the Press).

JONES, T. S. G. (1948*c*). *Ann. N.Y. Acad. Sci.* (in the Press).

KENDRICK, P. L., ELDERING, G., DIXON & MISNER, J. (1947). *Amer. J. Publ. Hlth*, **37**, 803.

PORTER, R., MCCLESKY, C. S. & LEVINE, M. (1937). *J. Bact.* **33**, 163.

STANSLY, P. G., SHEPHERD, R. G. & WHITE, H. S. (1947). *Bull. Johns Hopk.* **81**, 43.

SWIFT, P. N. (1948). *Lancet*, no. 254, p. 133.

WACHSTEIN, M. (1947*a*). *Arch. Path.* **43**, 503.

WACHSTEIN, M. (1947*b*). *Arch. Path.* **43**, 515.

EXPLANATION OF PLATES

PLATE 1

Fig. 1. A paper partition chromatogram (Whatman 4) of intact polymyxins in butanol-acetic acid mixture, run for 17 hours at 24° C. *A*, is polymyxin A, *C*, is polymyxin D and *B* is a mixture of both which has separated. *D*, is polymyxin B, *F*, is polymyxin D, and *E* their mixture. This chromatogram provides evidence for the existence of three distinct substances.

Fig. 2. A paper partition chromatogram (Whatman 4) of intact polymyxins run for 17·5 hours at 22° C. Representatives of three groups are identified. That derived from CN 1419 is polymyxin B, that from B 71 is D and that in the third group is A and C; they are identified by their component amino-acids (Plate 2).

PLATE 2

Fig. 1. A paper partition chromatogram (Whatman 4) of the acid hydrolysis of seven strains of *B. polymyxa*. The chromatogram was run for 17·5 hours at 22° C. with butanol-acetic acid mixture. Four distinct polymyxins are shown and are classified in Table 1.

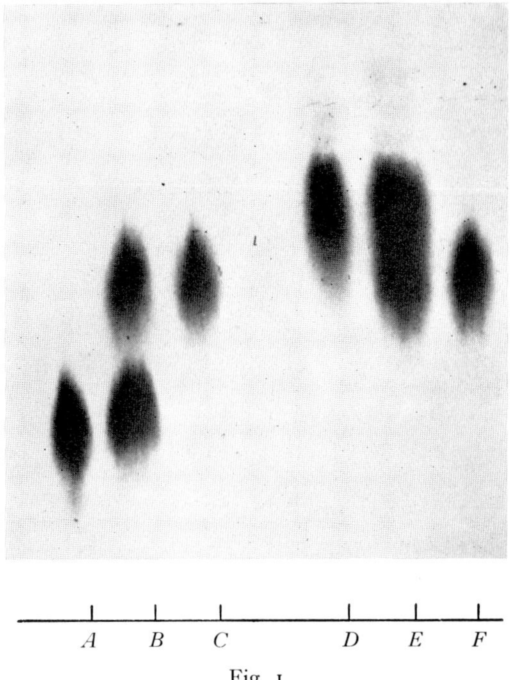

Fig. 1.

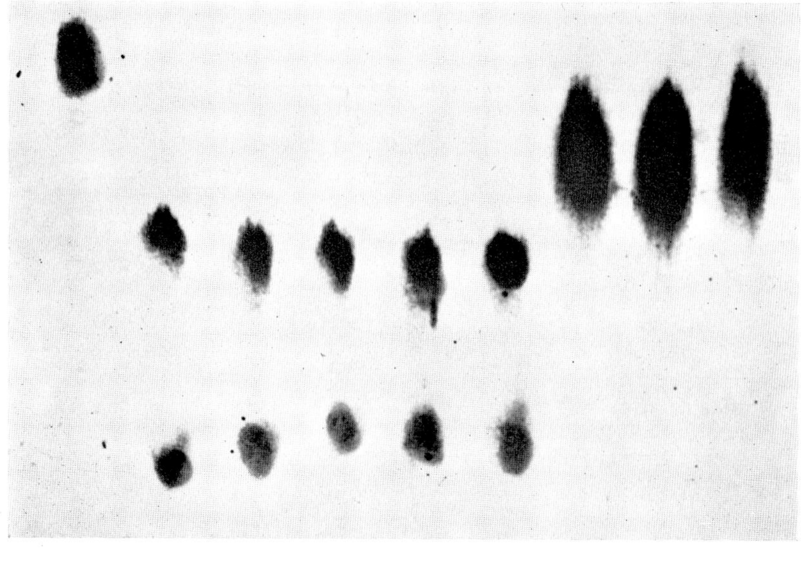

Fig. 2.

For explanation see p. 100

PLATE 2

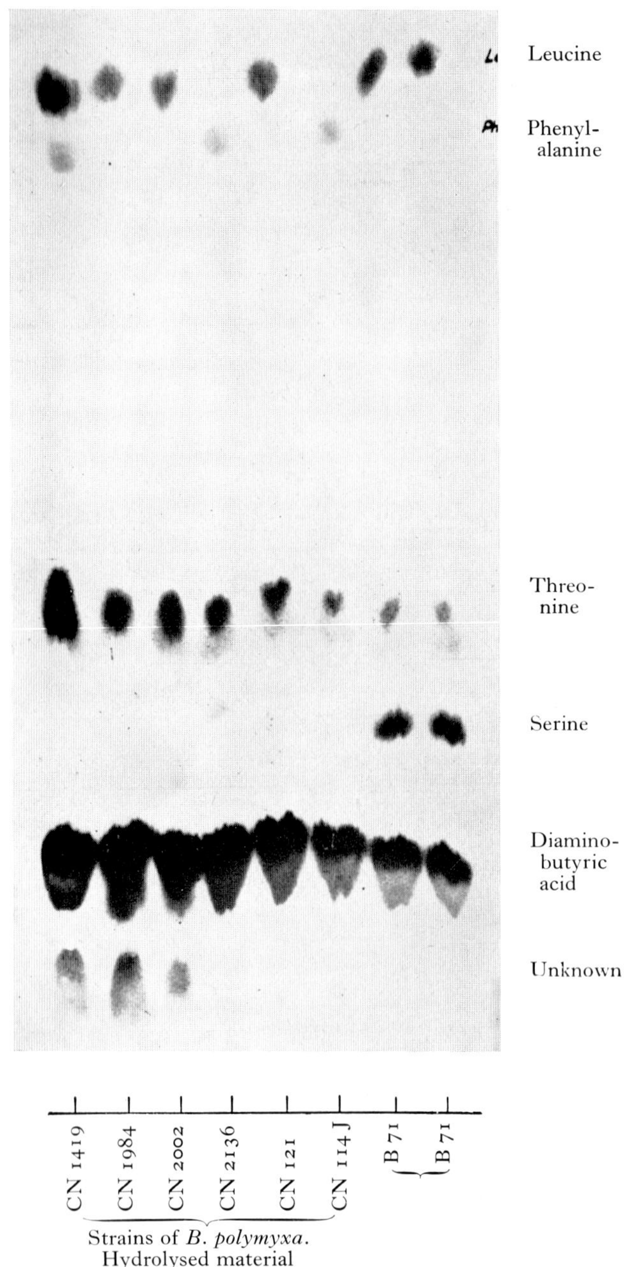

Fig. 1.

For explanation see p. 100

THE RESINS OF HOPS AS ANTIBIOTICS

By L. R. BISHOP

From the Laboratory of the Stag Brewery, London, S.W. 1

The striking success of the antibiotics derived from fungi and bacteria calls attention to the resinous constituents of hops.

Before A.D. 1500 beer was made in this country, much as it had been made for some 10,000 years before, by allowing hot water to act on germinated cereals so that the amylase converts the starch to a sugary and dextrinous solution which is boiled and then fermented by yeast. During the boiling various herbs were often added, but in A.D. 1510–1520 the use of hops was introduced from Holland and in time the hopped 'beer' replaced the unhopped 'ales'. Gradually it was realized that the value of hops lay, not only in the odour of the essential oils, but in the preservative or antibiotic action, which allows yeast to grow freely, but, even in very small amounts, retards or prevents the growth of many of the potential spoilage bacteria, and this preservative action of hops appears to have been recognized at least as early as their introduction to this country.

This exceptional behaviour fits in with the rather isolated botanical position of the genus *Humulus* which has been placed in the cohort Urticales with no closely related genera. It is distantly related to the Canabacae, Moracae and Urticaceae. Hops are climbing plants with separate male and female plants, and it is the female flowers which are dried, compressed and used in brewing. The flower is a cone of greenish bracts which, when fertilized, each contain seeds at the base. On the surface of the bracts are small yellowish glands which can readily be rubbed off. These contain the resinous matters and some essential oils.

In the course of time it became clear that the preservative action was associated with the resins. This was shown by Hayduck in 1885. The resins were differentiated into the soft resins which showed the preservative action and the hard resins which comprise oxidation and polymerization products. Both contribute to the bittering taste of hops. Later the soft resins were differentiated further in seeking the source of the preservative action. In 1885 Hayduck separated the resins into α, β and γ fractions. In 1891 Bungener & Lintner isolated the α acid from the α fraction. This is now known as humulon, and a β acid was isolated from the β fraction and is now known as lupulon. The formulae suggested for them in 1925 by Wieland are given on p. 102.

These formulae are, however, not finally established and may need some modification.

Both these compounds were found to have strong preservative action. The stronger, lupulon, was shown by Walker and colleagues to inhibit acid production by *Lactobacillus bulgaricus* at a concentration of approximately 0·5 to 1 part per million. Humulon is about one-third to one-quarter as effective as lupulon. Only some 15% of the antiseptic activity of the hops used actually reaches the wort and beer, but this concentration is effective against most beer spoilage bacteria and, in conjunction with the relatively low pH of beer, is entirely sufficient to prevent the danger of infection by water-borne pathogenic bacteria.

```
                                 CO
                               /    \
(CH3)2C:CH.CH2.HC                    C.CO.CH2.CH(CH3)2
                 |                   ||
                OC                   C.OH
                   \               /
                          C
                   /               \
                HO                   CH:CH.CH(CH3)2
```

Humulon

```
                                 CO
                               /    \
(CH3)2C:CH.CH2.CH                    C.CO.CH2.CH(CH3)2
                 |                   ||
                OC                   C.OH
                   \               /
                          C
                   /               \
(CH3)2CH.CH:CH                       CH:CH.CH(CH3)2
```

Lupulon

The annual production of hops in this country is about 10,000 tons weighed in the dried state containing about 12% moisture. About 5% of this is humulon and 1% lupulon, which gives an annual production of about 500 tons of humulon and 100 tons of lupulon. At present the needs of the brewing industry are greater than the supply.

The humulon can be quantitatively estimated by precipitation of its lead salt from methyl alcohol, while the β soft resins (including the lupulon) can be estimated by difference as the remainder of the soft resins. The quantities found vary considerably with the season, with the variety and the length and conditions of storage. Breeding, carried on in this country for many years by Prof. E. S. Salmon, has produced new varieties with higher contents of antiseptic resins combined with desirable flavour from essential oils.

The many apparent contradictions in the preservative action were gradually sorted out, and it was found that the effect was bacteriostatic rather than truly antiseptic and that the chief effect lay in prolonging the initial lag phase in development. It gradually became clear early in this century that the magnitude of the effect depended to a great extent on the pH of the medium, the antiseptic action being at a maximum on the acid side of neutrality—say between pH 5 and 6—while on the alkaline side the inhibition rapidly disappeared for some organisms but remained for others.

In addition, some organisms were found to be highly sensitive to hop resin while others were unaffected. This was rationalized in 1937 when Shimwell showed that hop resins were effective only against Gram-positive bacteria, most of which are inhibited by minute amounts of hop resin. However, with one Gram-positive lactic acid-producing species, *Lactobacillus pastorianus*, rather larger amounts are necessary, and this bacterium is liable to cause some spoilage.

The resins are only slightly soluble in water and are sensitive to hydrolytic and oxidative change. Nevertheless, they are highly active, fairly stable in acid solution and readily obtainable. In addition, it can be taken for granted that they are entirely non-injurious to human beings, at least when taken orally. They constitute the first antibiotics to be utilized and recognized, but do not appear to have been considered by those engaged in the modern study of antibiotics, and the purpose of this note is to draw attention to the somewhat interesting features of the hop resins.

REFERENCES

Introduction of Hops

'Turkey, carp, hops, pickerill and beer
Come into England all in a year.' (1511).

(J. Bannister, *Synopsis of Husbandry*, 1799.)

'Hops were first brought from Flanders into England, Anno 1524 in the 15th year of K. Henry the 8th, before which Alchoof, Wormwood etc., was generally used for the Preservation of Drink.'

(Laurence, *New System of Agriculture*, 1726.)

Preservative Action

'The continuance of the drink is alwaie determined after the quantitie of the hops, so that being well hopped it lasteth longer.'

(Harrison, *Description of England*, 1577.)

WALKER, T. K. and co-workers. *J. Inst. Brew.* 1923 onwards (Summary). *J. Inst. Brew.* 1935, p. 193.

SHIMWELL, J. L. *J. Inst. Brew.* 1937, p. 111.

Isolation of Resin Constituents

HAYDUCK, F. *Wschr. Brau.* 1885, **3**, 268.

WOELLMER, W. *Ber. dtsch. chem. Ges.* 1916, p. 780.

BEYAERT, M. & CORNAUD, P. *Congrès Int. des Ind. de Fermentation, Ghent*, 1947, p. 236.

Constitution of Resin Constituents

WIELAND, H. *Ber. dtsch. chem. Ges.* 1925, **58**, 102.
WOELLMER, W. *Ber. dtsch. chem. Ges.* 1925, **58**, 672.

Estimation of Antiseptic Resins

Biological Estimation.

BROWN, A. J. & CLUBB, D. *J. Inst. Brew.* 1913, p. 261.
CHAPMAN, A. C. *J. Inst. Brew.* 1925, p. 13.
WALKER, T. K., HASTINGS, J. J. H & FARRAR, E. J. *J. Inst. Brew.* 1931, p. 512.

Chemical Estimation.

SILLER, R. *Z. Untersuch. Nahr. -u. Genussm.* 1909, **18**, 241.
FORD, J. & TAIT, A. *J. Inst. Brew.* 1924, p. 426 and 1932, p. 351.
WOELLMER, W. *Allg. Brau. -u. HopfZtg,* 1930, 1531.
GOVAERT, F. & VERZELE, M. *Congr. Int. Ind. Fermentation, Ghent,* 1947, p. 279.

Effects of Storage and Drying Conditions

BURGESS, A. H. *J. Inst. Brew.* 1920 onwards (Summary). *J. Inst. Brew.* 1940, p. 5.

Breeding and Selection of Varieties

SALMON, E. S. *J. Inst. Brew.* 1908 onwards (Summary). *J. Inst. Brew.* 1943, p. 29.

SYSTEMIC INSECTICIDAL PROPERTIES INDUCED IN PLANTS BY TREATMENT WITH FLUORINE AND PHOSPHORUS COMPOUNDS

By HUBERT MARTIN

Long Ashton Research Station, University of Bristol

The rich success of chemotherapy in medicine has inevitably encouraged the application of the principle to plant pathology, but, until recently, with no great success. Failure has been generally attributed to the absence in the plant of a circulatory system analogous to the blood stream of animals, for it was often found that the injected chemical remained localized near the point of injection. The rules governing the distribution of an injected substance in woody plants are now well known (see e.g. Roach, 1939), but a second obstacle, the requirements of selective toxicity, remains. Brooks & Storey (1923) concluded, after many attempts to cure silver leaf in plums by injection methods, that there was little hope of finding a substance toxic to the pathogen *Stereum purpureum* yet harmless to the host plant. In the more successful tests, such as those involving the use of a hydrolysable cyanide, Moore & Ruggles (1915) found that the hydrocyanic acid so produced was confined to the older tracheae of the tree, and suggested that phytotoxic action would follow the penetration of the gas into the protoplasm-containing cells.

The need for selective toxicity is reduced if the pest or the disease is associated with those parts of the plant which are devoid of protoplasm. Vascular diseases, in which the invading pathogen induces the formation of toxins or causes gum formation with a resultant interference in water movement and the translocation of nutrients within the plant, would therefore seem amenable to chemotherapeutic treatment. Zentmyer, Horsfall & Wallace (1946) have explored this possibility and have reported some success in the use of 8-hydroxyquinoline derivatives and of a dithiocarbamate in the protection of young elms against Dutch elm disease.

Alternatively, the difficulties of ensuring an effective distribution of the chemotherapeutic chemical may be side-stepped by using compounds analogous to those already present and translocated as metabolites within the plants. This method arose almost by accident from the observation of Hurd-Karrer & Poos (1936) that wheat growing on seleniferous soils was not attacked by aphides. The watering of growing plants with dilute solutions of sodium selenate has since been found effective in keeping the plants free not only from aphis but also from red spider (see e.g. Speyer, 1941). The practical applications of the method are severely limited because the plant is rendered poisonous to man and to stock. Indeed, it would be unwise to recommend the method even for flower and seed production until ways are devised for overcoming the ill consequences of the selenium remaining in the soil.

The insidious nature of selenium is shown by a recent Long Ashton experiment (Bennett & Martin, 1948) in which wheat sown in 10 in. flower pots was watered at the flowering stage with various amounts of a 0·1% solution of anhydrous sodium selenate. The grain was harvested at maturity and was subsequently infested with adult grain weevils (*Calandra granaria*). Four weeks after infestation the adult weevils were removed and counted, the grain then being stored for a further 4 weeks when the number of adult weevils present was again determined. The results are given in Table 1, from which it will be seen that the selenate-treated wheat has not been directly toxic to the weevils but, at the higher doses, has inhibited their reproduction. The evidence is not conclusive, for the grain from the plants receiving the heavier amounts of selenate was much inferior in size and quality, but the result strongly suggests that the cause of the absence of eggs and larvae from the treated wheat is the selenate treatment.

Table 1. *Infestation of selenium-treated wheat by* Calandra granaria

Treatment 0·1 % Na_2SeO_4		Infestation		
Vol.	No. of applications	No. *Calandra* added	At 4 weeks no. dead	At 8 weeks no. adult
10	1	56	1	429
50	1	50	2	124
50	2	50	1	75
50	3	61	5	0
100	1	72	0	139
Nil	Nil	44	4	344

The ability of selenium compounds to function as systemic insecticides is probably to be associated with the chemical similarities between selenium and sulphur which are adjacent members in the same group of the Periodic Table. The relationships between the uptake by plants of sulphur and selenium compounds from soil were studied by Hurd-Karrer (1937), who showed that the sulphur/selenium ratio of the plant tissue increased with the sulphur content of the seleniferous soil. It may then be assumed that selenium is absorbed and translocated in selenium-tolerant plants in the same way as sulphur. Moreover, the first symptoms of selenium poisoning in animals are associated with the sulphur-containing keratin. The failure of *Calandra* to infect the selenium-treated wheat suggests that certain sulphur-containing proteins, co-enzymes or vitamins are uniquely concerned in the reproduction process which becomes deranged if the analogous selenium compounds are present.

The absorption and translocation within the plant of a foreign molecule is less easy to visualize and, in view of previous failures in the chemotherapeutic treatment of plants, it was hard to accept without test German claims of the discovery of certain compounds which when watered on the growing plants rendered those plants systemically insecticidal. It would seem that the routine examination carried out by Kükenthal at the Biologische Institut of I.G. Leverhusen and Elberfeld of various synthetic compounds as insecticides included such a test, and that the property was at last discovered in certain phosphorus and fluorine compounds synthesized by Gerhard Schrader (1947). Shaw and I (1947) were told by Kükenthal and Schrader of their work when we visited them in 1946, but we were unable to secure samples of the compounds. But, fortunately, somewhat similar compounds had been under investigation by McCombie & Saunders at Cambridge, and a sample of one of the compounds used by Schrader, bis-(2-fluoroethoxy)-methane, $CH_2(OCH_2CH_2F)_2$, was kindly supplied by Dr Saunders.

The first trial made at Long Ashton with this sample was on groundsel plants potted in 3 in. pots and watered with 10 ml. of a 1% aqueous solution of the compound. The following day nearly fully grown larvae of the Cinnabar moth (*Callimorpha jacobaeae* L.) were placed on the plants, and on those in untreated pots placed alongside. On the latter control plants the larvae fed normally and in 16 hr. had eaten nearly all the leaf laminae. On the treated plants, however, there was little feeding, and most of the larvae had dropped to the soil and were giving a peculiar side-to-side body twitching with little co-ordinated movement. Larvae placed on other plants 18 hr. after watering with 10 ml. of a 1·0% solution of the compound had all fallen from the plants within 24 hr., and, although some were showing the twitching movement mentioned above, most were moribund.

Having confirmed the existence of the systemic effect, further trials have been made for the purpose of determining the order of the amount required to produce the effect, first against aphides and secondly against leaf-eating caterpillars.

Bean aphis. 8 in. flower pots holding five or six tick bean plants about 2 ft. high were used, the pots being watered with various amounts of 1% aqueous solution of the fluoroethoxymethane. After treatment the plants, which were growing in an unheated glasshouse, were infested with *Aphis fabae* by placing a leaf or sprig of bean heavily infested with the aphid in the axil of a leaf about half-way up the plant. As the infested sprig withered, the aphides moved to the growing plant and usually wandered to the growing point where an infestation built up. The results are assembled in Table 2, which includes the degree of infestation of the plants on 12 July 1947, 5 days after attempted infestation.

Table 2. *Infestation of treated beans by* Aphis fabae

Treatment 1 % $CH_2(OC_2H_4F)_2$ per pot	Date of treatment	Degree of infestation
100 ml. +50 ml. +50 ml.	30. vi. 47 2. vii. 47 8. vii. 47	Nil (three pots)
100 ml. +50 ml.	30. vi. 47 3. vii. 47	Nil on one pot, almost nil on second and very slight on third
100 ml.	30. vi. 47	Nil on one pot, other two almost free
10 ml. +10 ml. +10 ml.	30. vi. 47 3. vii. 47 8. vii. 47	Very slight infestation on all plants
10 ml. +10 ml.	30. vi. 47 3. vii. 47	Two pots practically free and one slightly infested
10 ml.	30. vi. 47	Slight to moderate infestation on three pots a number of dead aphides on one
Nil	—	Slight infestation on two pots, moderate on third and heavy on fourth, but infestation rapidly thinning

Table 3. *Infestation of treated beans by* Aphis fabae

Treatment 1 % $CH_2(OC_2H_4F)_2$ per pot	Date of treatment	Degree of infestation
100 ml. +50 ml.	30. vi. 47 3. vii. 47	Slight with many dead aphides on upper surfaces of leaves
100 ml.	30. vi. 47	Only a few aphides feeding, no young aphides but many dead
10 ml. +10 ml.	30. vi. 47 3. vii. 47	Slight infestation
Nil	—	Infestation building up with many young aphides present

In a parallel experiment, recorded in Table 3, the transfers of aphid-infested material were made on 5 July 1947, and the degree of infestation recorded 4 days later.

In other experiments bean plants already infested with *A. fabae* were watered with even smaller amounts, namely, 5 and 10 ml. of 1% fluoroethoxymethane. When examined 3 days later the aphid population on the treated plants was obviously less than on the control plant, and dead aphides of all stages of growth were found on the upper surfaces of the lower leaves of the treated plants.

Although environmental conditions did not favour the rapid development of heavy aphid infestations, the systemic effect of the fluoroethoxymethane was clearly evident even at such low dosages as 0·05 ml. per 8 in. pot.

Lepidopterous larvae. 10 in. pots each containing a cauliflower plant, just beginning to show curd formation, were watered with various amounts of the fluoroethoxymethane. The plants were left outdoors to be naturally infested. The untreated controls became attacked by larva of *Pionea forficalis*, *Pieris brassicae* and *P. rapae*, *Plutella maculipennis* and *M. brassicae*, but all the treated plants, including those receiving only 0·1 g. of the insecticide, were clearly less damaged than the untreated.

Similar plants brought into the glasshouse and treated with 50 ml. of 1% fluoroethoxymethane on 29 July 1947 remained free from damage by *Pieris brassicae* and *P. rapae* until 21 August 1947, though the adjacent untreated plants were heavily infested. Eggs of both species were freely laid on the treated plants and, although egg-hatching was not apparently affected, the larvae died presumably after eating a trace of the leaf tissue. Plants treated with only 10 ml. of the 1% solution remained almost free from attack for a fortnight. An attempt was made to determine the degree of toxicity of leaves of the treated plants to cabbage-moth larvae, but the caterpillars refused to eat disks cut from the treated leaves.

In laboratory tests designed primarily to establish that the insecticidal effect was not a fumigant action, the petioles of cabbage leaves were inserted in 3×1 in. specimen tubes containing water or a 1% aqueous solution of the fluoroethoxymethane, the specimen tubes then being placed in a beaker containing the insecticide solution or water respectively so that each leaf was over both insecticide and water. Young larvae of *P. brassicae* were then transferred to the leaves, and only those leaves with their petioles in water were eaten by the caterpillars.

The results of these experiments amply confirm Schrader's claims of systemic insecticidal properties for the fluoroethoxymethane and, in this series of compounds, he found a greater activity with the doubled molecule $CH_2(OCH_2.CH_2OCH_2.CH_2F)_2$. But the property was first observed with bis-dimethylamino-fluorophosphine oxide, $[(CH_3)_2N]_2POF$, and, later, with the fluorine-free pyrophosphate, $[(CH_3)_2N]_2PO.O.PO[N(CH_3)_2]_2$. Schrader concluded that, in the organic phosphates, the N—P linkage is associated with the systemic insecticidal effect, for the alkyl phosphates, although including many potent insecticides, are inactive systemically. Moreover the P=O group is requisite, for the corresponding thiophosphates are inactive systemically though some are highly effective as contact insecticides.

Clearly the poisonous properties which these systemic insecticides impart to the treated plants is a severe limitation to their practical utilization, though the non-fluorine phosphorus derivatives may be of use in seed production or in the raising of virus-free stocks. But their academic interest needs no emphasis, and their use, particularly if made from the radioactive P^{32}, will greatly facilitate the study of molecular transportation in plants.

REFERENCES

Bennett, S. H. & Martin, H. (1948). *Ann. Rep. Long Ashton Res. Sta. for* 1947, p. 147.

Brooks, F. T. & Storey, H. H. (1923). *J. Pomol.* **3**, 117.

Hurd-Karrer, A. M. (1937). *J. Agric. Res.* **54**, 601.

Hurd-Karrer, A. M. & Poos, F. W. (1936). *Science*, **84**, 252.

Martin, H. & Shaw, H. (1947). *Brit. Intell. Obj. Sub-comm.*, Final Report 1095.

Moore, W. & Ruggles, A. G. (1915). *Science*, **42**, 33.

Roach, W. A. (1939). *Ann. Bot., Lond.*, **3**, 155.

Schrader, G. (1947). *Brit. Intell. Obj. Sub-comm.*, Final Report 714.

Speyer, E. R. (1941). *Rep. Exp. Res. Sta. Cheshunt for* 1940, p. 49.

Zentmyer, G. A., Horsfall, J. G. & Wallace, P. P. (1946). *Bull. Conn. Agric. Exp. Sta.* no. 498.

SURFACE ACTIVITY AND PERMEABILITY AS FACTORS IN DRUG ACTION

By A. R. TRIM, Department of Biochemistry
and A. E. ALEXANDER, Department of Colloid Science,
University of Cambridge

I. INTRODUCTION

The past two decades have been marked by a number of brilliant practical achievements in chemotherapy which were based mainly upon an empirical approach. At the same time, and undoubtedly stimulated by these practical achievements, theories of drug action have advanced and have been integrated. So far these theoretical developments have not led to the production of any new drugs of outstanding properties, but have had an important general influence upon research and day-to-day routine in the field of chemotherapy.

The earliest theories of drug action were developed by Claude Bernard (1875), who focused attention upon the effect of drugs on intracellular colloids—a physico-chemical approach—and by Brown & Fraser (1868), and later Ehrlich, who were more concerned with specific interactions between chemical groups in the drug and receptor groups in the organism—a strictly chemical approach. Present views on the subject have carried over much of the essentials of these theories and their subsequent developments. These essentials have now been clarified, integrated and enriched by material drawn from wide fields of both theoretical and practical work.

The major theories have emphasized the physico-chemical aspects of drug action, and there has been a tendency to regard a particular type of physico-chemical interaction as the explanation of the observed biological response. Indeed, it has been easy to demonstrate a striking parallel between various physical properties of some drugs and their biological effects. Thus it is well known that Overton & Meyer (1899) and their school laid supreme importance upon the solution of drugs in cell lipoids in their explanation of narcosis, whereas Traube (1904) considered that capillary activity, i.e. adsorption on cell surfaces, is paramount. Höber (1907), Lillie (1909) and Winterstein (1915) developed Traube's view further and focused attention upon the changes of cell permeability which result from the adsorption of drugs upon their surfaces. On the other hand, Bancroft & Richter (1931) developed Claude Bernard's theory in terms of changes in the colloidal state of protoplasm which can be induced by drugs.

Coupled with Ehrlich's theories these ideas carried the germ of the principal concepts of most modern theories. At the time when they were developed the classical theories of drug action tended to exaggerate the role of selected aspects of drug action and created a conflict of opinion too great to be justified by the experimental evidence which could be brought to support any one of them. Important achievements of the recent period have been the integration of older views and great improvements in the experimental approach. It is more clearly understood now that the action of a drug is usually a complex process involving a number of stages in a definite sequence, the events at each stage being dependent upon those in the previous one. The same intrinsic characteristics of a drug molecule may be revealed in one or more ways at all stages: in the medium in which the drug and biological system are brought together, at the interface between the medium and the organism, and in transport through any boundary membranes, through the tissues or cell contents to the final stage of interaction with cell receptors. This is the most common sequence of events, although it is possible that in some systems the biological receptors are to be found upon the surface, so that the response is determined only by reactions in the medium and at the interface between the medium and the organism.

Since a particular physico-chemical quality may endow a molecule with a variety of possibilities for interaction with biological systems, it is usually impossible to infer the mode of action of the drug from any simple correlation between biological action and a particular physical property. The well-known homologous series effect serves to illustrate this point. It frequently happens that the biological response to an homologous series of drugs bears an exponential relationship to the number of carbon atoms in the hydrocarbon chain, as, for example, in the bactericidal action of the series of *n*-alkyl phenols. The logarithm of the equitoxic concentrations of the serial members, or the logarithm of the rate of killing of equimolar solutions, bears a straight-line relationship to the number of carbon atoms in the alkyl side-chain. As was pointed out by many workers and developed particularly by Ferguson (1939), many physical properties such as surface activity and fat solubility show precisely the same exponential relationship to the number of carbon atoms in the alkyl chain. Fig. 1 (from Ferguson, 1939) illustrates this point. Accordingly, a correlation of toxicity or other biological effects in an homologous series with any one of these properties means that the slowest physical process in the overall effect depends on a partition property of the molecule. The particular one involved cannot be specified, and various partition effects may be involved in the complete process of drug action.

Since a drug usually acts on an organism in a succession of stages, it is possible for interactions depending on the same physical characteristics to produce conflicting contributions to the total action. In the simplest example the properties of a drug which enable it to act upon an organism may also make it liable to react with components of the medium and

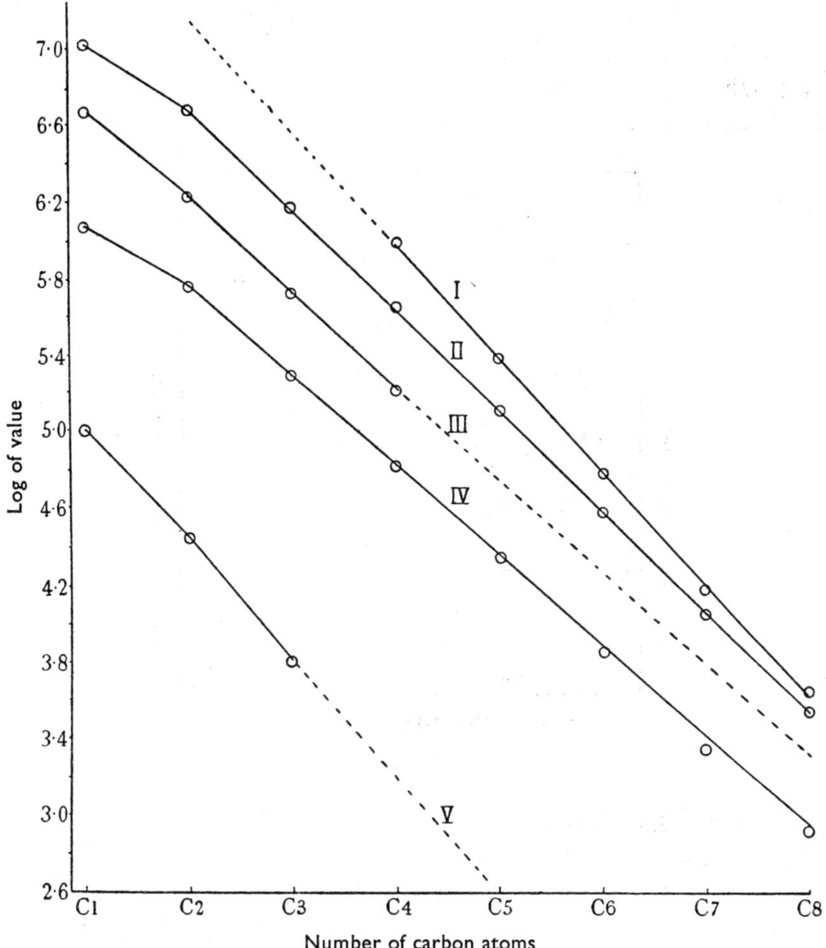

Fig. 1. Properties of normal primary alcohols. I, solubility (mol. $\times 10^{-6}$/l.); II, toxic concentration for *B. typhosus* (mol. $\times 10^{-6}$/l.); III, concentrations reducing S.T. of water to 50 dynes/cm. (mol. $\times 10^{-6}$/l.); IV, vapour pressure at 25° (mm. $\times 10^{4}$); V, partition coefficient between water and cotton-seed oil ($\times 10^{3}$).

become neutralized, as often happens with colloidal electrolytes like the synthetic detergents. A more subtle kind of 'opposed' activity is shown in the biological effects of combinations of soaps and phenols (Alexander & Trim, 1946; Alexander & Tomlinson, 1948). Considering a fixed concentration of phenol the principal points can be enumerated as follows (see Fig. 2):

(1) On increasing the soap concentration from zero the biological activity, as measured by the penetration into *Ascaris* or the rate of killing of bacteria, rises from its initial value *A* to a maximum at *B*.

(2) Over this range of soap concentrations the interfacial tension of the mixture falls to a minimum at *E*.

(3) With increase of soap concentration beyond that corresponding to *B* (or *E*) the biological activity falls enormously.

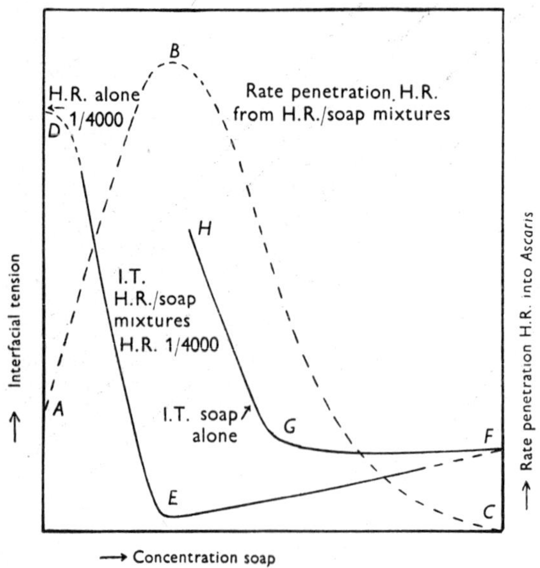

Fig. 2.

The following explanation will account for these observations (for further details see p. 128). The structural features of soaps make them highly surface active and also cause them to form colloidal aggregates, termed micelles, above a certain critical soap concentration. In the type of system described here, addition of soap in low concentration increases the interfacial activity of the mixture, so that the soap/phenol complex is adsorbed on to the surface of the organism and makes it more permeable to the drug. Thus at this stage of increasing soap concentration the biological activity of the drug component is increased. However, beyond a certain soap concentration micelles begin to form, and, by taking up the phenol from solution, reduce the amount available for interaction with the organism. The biological activity thus falls to very low values if sufficient soap is added.

It must be reiterated that the mode of action of a particular drug in a particular biological system can only be determined by experiment. Nevertheless, a consideration of the physical characteristics of a drug can be a useful guide to its mode of action. For this purpose substances may be divided into four main categories (see also § II):

(*a*) Ionized substances of low surface activity and low lipoid solubility, e.g. thiocyanates, penicillin and alkaloids such as nicotine.

(*b*) Ionized substances which are very surface active and have a low lipoid solubility, such as the natural and synthetic soaps and the bile salts.

(*c*) Unionized compounds of high surface activity and high lipoid solubility, e.g. phenols, cresols and alcohols.

(*d*) Unionized compounds of low surface activity and high lipoid solubility such as chloroform.

The different types of compound will tend to interact in particular ways with all biological systems, for example, highly surface-active compounds may be expected to be involved more in interactions of the Traube type, whereas substances like chloroform will act by accumulation in lipoid components of the organism.

It will be convenient to divide the process of interaction of the drug with the organism into four main stages:

(*a*) Interaction of the drug with the environment.

(*b*) Interaction of the drug with the boundary between the organism and the environment, i.e. adsorption phenomena.

(*c*) The passage of the drug through and from boundary layers of the organism, i.e. absorption and diffusion.

(*d*) Intracellular action of the drug.

Consideration of the last stage does not lie within the province of this paper, although the physical characteristics involved at earlier stages will also operate at this stage. It includes specific effects such as inhibition of enzyme systems (reversibly or irreversibly), as with sulphonamides and triphenylmethane dyes, specific action on —SH groups, which are active groups in the functioning of a number of intracellular enzymes, reactions with metal complexes in enzymes as with azide and H_2S, and the general denaturation of proteins and the splitting of protein-non-protein complexes. Another mode of action, operative with the very surface-active materials such as soaps, is disruption of the surface layers, the ultimate action being lysis and escape of vital cell constituents.

We are concerned here with the role of adsorption and diffusion processes as they affect the passage of drugs into an organism. It is already clear that there are many ways in which a drug may be made more or less effective at the stage of its action where these processes predominate. They fall into two categories: (*a*) interactions of the drug and other substances before they reach the site of adsorption, (*b*) interactions at the site of adsorption. In addition, although little is known about them, interactions of the drug and the organism at later stages may also influence the biological response.

As will be seen, much of the biological data in the present paper is drawn from studies upon *Ascaris* and bacteria, not only because we have personally worked upon such systems, but also because the necessary quantitative data are more extensive. Reviews of the bacteriological work, with more detailed references than is possible here, are given by Suter (1941), Dubos (1942), Daniels (1943), Rahn (1945) and Valko (1946). In the case of acridine derivatives a very exhaustive study has been carried out by Albert, Rubbo and their co-workers (see Albert *et al.* (1945) and earlier papers).

II. CLASSIFICATION OF DRUGS

As pointed out in the Introduction, it seems certain that different drugs may act in different ways to produce a given biological response. For the purpose of the present approach, it is convenient to divide drugs into four principal groups according to certain physico-chemical similarities, rather than according to their chemical nature. Like all such classifications, however, there is no hard-and-fast line between these groups, and in some cases a particular compound may fall into one or another of these classes, depending upon the conditions. For example, organic acids and bases of low molecular weight would fall into group (*a*) if ionized, but into group (*c*) if present in the unionized form.

These four principal groups are as follows:

(*a*) *Ionized substances of low surface activity and low lipoid solubility.* The extreme case would be simple inorganic salts possessing biological activity (e.g. cyanides, thiocyanates). More usually this group would include low-molecular weight organic salts, acids and bases (when ionized), such as nicotine, many acridine and quinoline derivatives, penicillin and certain dyes.

(*b*) *Ionized substances very surface active but of low lipoid solubility.* The principal members of this group are the natural and synthetic soaps, but it also includes many dyes, certain bile salts, certain phospholipoids such as lecithin, and some ionized drugs containing long paraffin chains (e.g. some arsonium and phosphonium salts). These types of compounds are termed 'colloidal electrolytes', combining an ionic character with the formation of colloidal aggregates ('micelles') in the higher concentration ranges.

(*c*) *Unionized compounds of high surface activity and high lipoid solubility.* Most of the polar (but not ionized) organic compounds would fall into this group. Important members are the phenols and cresols, together with their alkyl and chlorinated derivatives, the alkyl resorcinols, the alcohols, etc.

(*d*) *Unionized compounds of low surface activity but of high lipoid solubility*. Principal members are the hydrocarbons, chloroform, ethylene dichloride and other chlorinated hydrocarbons.

III. THE INFLUENCE OF THE METHOD OF APPLICATION OF THE DRUG AND THE BIOLOGICAL ENVIRONMENT

The common methods of application can be grouped as follows:

(I) The drug is applied in an *aqueous* medium generally as a solution, less commonly in suspended form or as an emulsion, either of the neat drug or of the drug in an organic solvent.

(II) The drug is applied dissolved in an organic solvent.

(III) The drug is applied in neat form, which may be a solid, liquid or vapour.

Of these we propose to restrict ourselves almost entirely to method (I), and to cases where the drug is in true solution, not emulsified or suspended in any way. A great deal of biological data refer to such conditions, and much more is known about the behaviour of aqueous systems than of the non-aqueous type. A little will, however, be said about methods (II) and (III) in § V below.

The biological environment will include the drug, the carrier medium (water throughout the present discussion unless otherwise stated), and other constituents present either naturally or as deliberate additives. These other constituents, even if quite biologically 'inert', may nevertheless profoundly modify the action of drugs, as will be discussed in detail in §§ IV and V. The principal variables in the biological environment, other than the drug and carrier medium, may for convenience be subdivided into: (i) pH, (ii) simple inorganic salts, (iii) other constituents (natural or added). These will be briefly discussed below.

(i) pH

Change of pH will always affect the charge on any biological surfaces in contact with the aqueous phase, and in addition will alter the degree of dissociation of drugs containing ionizable groups, such as acids, amines, alkaloids such as nicotine, acridines, quinolines and many antimalarial compounds, penicillin, sulphonamides, etc.

The latter effect will influence the biological response, since as a rule the uptake of the two forms (e.g. unionized and ionized for a primary dissociation) will not be the same. The relative amounts of the various forms can readily be calculated if the dissociation constants are known, and by working over a suitable pH range the biological activity of each entity may be determined (see p. 125).

The former effect will only be of importance with ionized drugs (see p. 125). Changes in surface charge can be obtained from measurements of the electrokinetic potential which is readily measurable in some systems (e.g. with bacteria from their electrophoretic mobility).

(ii) *Simple inorganic salts*

The concentration and nature of simple salts present in the medium can markedly influence the adsorption of ionized drugs; unionized substances are much less affected. In the case of colloidal electrolytes (soaps, dyes, bile salts, etc.), addition of simple salts may cause aggregation to micelles, and this can bring about radical changes in biological response. All these questions are discussed in more detail in § IV, since they are usually related to changes in adsorption.

(iii) *Other constituents* (*natural or added*)

Many tests of drug activity are carried out in the presence of substances other than simple inorganic salts and the drug. These constituents may either be of natural occurrence, such as bile salts, carbohydrates, proteins, etc., with drugs acting *in vivo*, or may have been deliberately added, such as glucose, peptones, agar, etc., in much bacteriological work. As we shall see later, the colloidal constituents may modify drug activity considerably, particularly with ionized drugs, leading as a rule to varying degrees of detoxication.

IV. ADSORPTION

Unless the drug enters through capillary pores the first step in all biological systems is an uptake by *adsorption* upon the bounding surfaces or by *absorption* (solution) in the extreme boundary layer. (Entry through capillaries seems unlikely in most cases, and there is definite evidence against it in some systems, e.g. *Ascaris* with soap/hexyl resorcinol mixtures.) With ionized substances and with unionized molecules of high surface activity (types (*a*), (*b*) and (*c*) of § II) the *ad*sorbed film will be monomolecular in thickness, the concentration of molecules in the monolayer being greater with the more surface-active compounds such as soaps and long-chain phenols. With substances not normally reckoned to be capillary active, e.g. hydrocarbons, CCl_4 and other members of type (*d*), it is more accurate to talk of *ab*sorption in the boundary layers, rather than an *ad*sorption on the surface, since it is unlikely that even the initial amounts will be confined to an adsorbed monolayer.

In biological systems, we are frequently dealing with non-equilibrium conditions, the drug, if it is soluble in the membrane, tending to diffuse away from the surface layer. Therefore the system may involve a *rate* of adsorption upon the surface, and/or a *rate* of desorption from the surface into the interior of the membrane. The former is unlikely to be of importance save with compounds of extremely low water solubility and so will not be considered here; the latter is treated further in § V below.

Here we will consider the factors leading to adsorption of types (*a*), (*b*) and (*c*), paying particular attention to the very surface-active substances (types (*b*) and (*c*)) where the concentration by adsorption will be most pronounced. Type (*c*) differs from types (*a*) and (*b*) chiefly in that electrical (Coulomb) forces are not of such importance. As pointed out earlier, the biological activity of the most extreme members of this series—the soaps—arises undoubtedly from their power of concentrating at the biological surface, leading to an alteration in permeability and sometimes to lysis with the escape of the cell contents.

As indicated in § I there have been numerous attempts to correlate surface activity with biological activity. The former has generally been obtained from surface-tension measurements (i.e. at the air/water interface), less frequently from interfacial tension against an oil such as medicinal paraffin. Of these the latter is undoubtedly the closer approximation to most biological surfaces, but it cannot be too strongly emphasized that the only fundamentally satisfactory system is the biological surface itself. This very frequently presents great difficulty, although in a few cases it has been possible, particularly with bacteria where the use of thick suspensions gives a suitably large adsorbing surface (McMullen & Alexander, 1948). The chief divergence between adsorption at a biological and an air/water or oil/water interface is likely to arise when charge effects are of importance, particularly when the biological surface and the drug are oppositely charged. As will be seen below, there are many systems in which a reasonable agreement between surface and biological activities is found, the apparent discrepancies arising almost entirely in systems where the drug and biological surface are oppositely charged.

Factors in adsorption

The principal physico-chemical factors which influence adsorption are outlined below together with, in each case, biological examples where this particular factor appears to be the most important.

(i) *Concentration*

With all the drugs considered here the surface concentration increases with the bulk concentration, as formulated by the Gibbs adsorption isotherm. The surface activity thus reaches its maximum with a saturated solution of the drug. The production of supersaturated solutions or of metastable forms (which will possess a higher solubility) is possible in some cases, particularly if a comparatively insoluble drug is dispersed by solution in water-soluble organic solvents such as alcohol or chloroform. This may explain certain increases in activity produced by such means, although it must be borne in mind that other factors, particularly rates of solution and diffusion, may enter in these heterogeneous drug systems. In the case of soaps and some other colloidal electrolytes micelle formation sets an effective limit to surface activity at concentrations which may be considerably less than the saturation value (see also (vi) below).

As the surface activity, measured by the lowering of surface or interfacial tension, increases, so in many cases and in a general way does the biological activity. As pointed out above, the best parallelism is shown with uncharged drugs and with ionized drugs where the organic ion of the drug and the biological surface are similarly charged. The apparent discrepancies when the drug ion and biological surface are oppositely charged arise from the electrical (Coulomb) forces involved, and are discussed further in (v) below.

The influence of fourteen surface-active compounds (alcohols, anionic and neutral soaps) upon the growth of acid-fast bacteria showed a general parallelism with the lowering of surface tension, complete inhibition being obtained below about 30 dynes/cm. regardless of the chemical nature of the depressant (Alexander & Soltys, 1946). Media containing depressants showed an increase in surface tension during growth, showing that these compounds were taken up by the bacteria, presumably by adsorption.

Similarly, the work of Höber & Höber (1942) upon the lysis of cells, particularly red blood cells, has shown that with surface-active anionic soaps a close parallelism exists between lytic activity and lowering surface-tension.

A recent study of the bacteriostatic activity of bile salts and related compounds (Stacey & Webb, 1947) also shows a reasonable agreement, a lowering of *c.* 5 dynes/cm. below the value for the growth medium being sufficient to inhibit growth. With certain other compounds this parallelism was not observed, but inspection of their formulae reveals that they would exist as cations at the pH used, whereas the bile salts would be in the form of anions. In such cases, as previously mentioned, the measurement of surface tension would not be expected to give a true indication of their adsorption on the biological surface (see (v) below).

In addition to the examples mentioned above, the correlation of biological and surface activity with unionized drugs is shown by the rates of penetration of hexyl resorcinol into *Ascaris*, and of phenols into bacteria, both alone and in the presence of soaps (Alexander & Trim, 1946; Alexander & Tomlinson, 1948). Further examples are given below in the consideration of homologous series and molecular structure.

(ii) *Molecular weight (homologous series)*

For a given type of molecule (e.g. the *n*-alkyl phenols) the surface activity increases with the molecular weight, i.e. with the number of $—CH_2$ groups added to the side-chain. For such homologous series the well-known rule of Traube states that the bulk concentration required to produce a given lowering of surface tension decreases by a constant factor (*c.* 3) for each $—CH_2$ group added. (Alternatively, as the chain length increases arithmetically the surface activity increases logarithmically.) If the biological activity parallels the surface activity as measured by the lowering of surface tension, then Traube's rule should apply to homologous series of drugs. Many examples are known where this is the case, although many apparent exceptions have been reported.

A considerable number of the apparent exceptions can be explained by the diminishing solubility with increasing chain length, as discussed below and on p. 112; others are due to the basic assumptions of Traube's rule not being valid (see Alexander, 1948). Of these assumptions the most important is that the amount of adsorption is very small, which may be far from the truth for biological systems not very sensitive to the particular drug series. The limitation set by solubility arises from the fact that most tests of drug action are associated with dynamic rather than with equilibrium conditions, and with compounds of very low solubility the rate of solution and/or diffusion to the biological surface may become the limiting factor in drug access.

It is not proposed here to list all the homologous drug series which behave in accordance with Traube's rule, for this reference may be made to other reviews (e.g. Ferguson, 1939; Clark, 1937), but rather to consider a few apparent anomalies.

The 4-*n*-alkyl resorcinols, for example, have been extensively studied as germicides and anthelmintics. With some organisms such as *Staphylococcus aureus*, the Traube relationship is obeyed up to the *n*-nonyl member, but with others, such as *B. typhosus*, an apparent optimum occurs at *n*-hexyl. These data were derived from Rideal-Walker coefficients and accordingly refer to dynamic conditions. One explanation for the difference is that the latter organism has a lower permeability, so that the surface concentration required to reach a given rate of penetration is not attainable with the higher members owing to solubility limitations. This seems reasonable in view of the generally greater resistance of Gram-negative organisms.

In the case of anthelmintic activity an optimum at hexyl resorcinol has also been reported (see Fig. 3). Measurements of the rate of penetration into *Ascaris* have shown, however, that provided the comparison is made at concentrations low enough to avoid solubility troubles, the *n*-heptyl member is certainly more active than the *n*-hexyl (Trim, 1948, and Fig. 3).

The activity of homologous series of enzyme inhibitors often shows a maximum when comparison is made at a fixed molar concentration. For example, with amine oxidase and mono-amidines at 10^{-3} M concentration this optimum occurs with ten C atoms in the chain (Blaschko & Duthie, 1945). In this case and in many others solubility appears to set the limit,

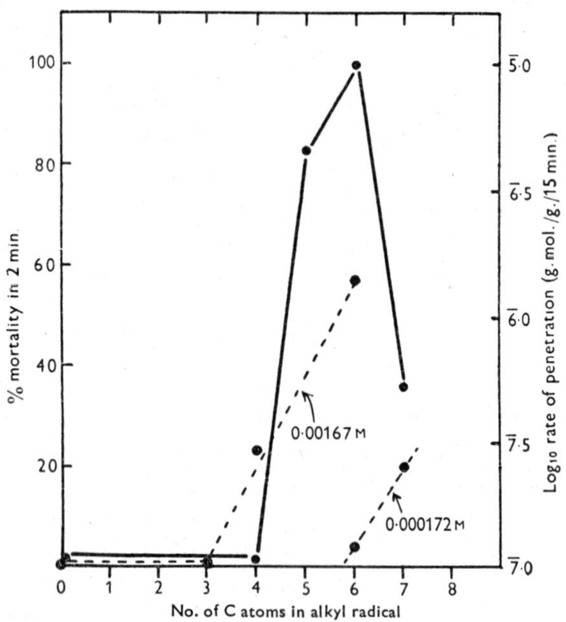

Fig. 3. ●——● percentage mortality of *Ascaris lumbricoides* var. *suis* after immersion for 2 min. in 1/1000 solution of 4-*n*-alkyl resorcinols (after Lamson *et al.* 1935). ---●---●--- relation between $\log_{10}$ of rate of penetration of 4-*n*-alkyl resorcinols into *Ascaris* and length of alkyl radical. Concentration of drug as indicated, temperature 37°.

for the surface activity as measured by interfacial tension against nujol shows a maximum at about the same member (Alexander, unpublished).

The question of apparent 'optima' in homologous series is considered from a somewhat different angle in § V.

(iii) *Structure of the molecule*

Even with a fixed concentration and molecular weight the adsorption of any particular isomer is also influenced by its molecular structure, for example, whether it consists of rings, straight chains or branched chains. The question has been most studied in the case of fatty acids and synthetic soaps. With these ionized compounds the branched-chain isomers are less surface active at low concentrations than the straight-chain compound, but there is frequently a reversal at high concentrations. The reason for this is the greater difficulty of the branched-chain isomers in forming micelles

(Hartley, 1941), which, as previously mentioned, set a very effective limit to surface activity (see also (vi) below).

It is well known that differences in biological activity are often found between the various isomers in any given case, although data upon comparative adsorption or surface activity seem very limited. In the case of the alkyl resorcinols the bactericidal activity (from the Rideal-Walker coefficient) of an iso-compound appears to be less than its straight-chain isomer, and this would also be the anticipated order of surface activity.

Adams and his co-workers, who prepared a large number of synthetic organic acids for testing against acid-fast bacteria, found that the branched-chain compounds were more active than their straight-chain isomers, which may be due to differences in surface activity for the reasons mentioned above. Biological activity was usually optimal with molecules having 15–18 carbon atoms, and with the lower members (but not the higher ones) this paralleled the surface-tension lowering power. The apparent divergence of the higher members may be due to effects arising from low solubility, as in some of the examples previously discussed.

It should also be mentioned that with drugs containing ionizable groups the dissociation constant of these groups will rarely be the same in the different isomers. This will then influence the overall activity (see § III), since most comparisons are done at a fixed pH (see Albert *et al.* (1945) and also (iv) and (v) below).

(iv) *Charge effects*

Charge effects (due to Coulomb forces) arise with all ionized drugs, owing to the mutual interaction between the organic ions (adsorbed and in solution) and the charges on the biological surface. The influence of the biological surface is considered separately in the next section.

The concentration, by adsorption at a surface, of soaps and other ionized drugs, gives rise to a mutual repulsion between the similarly charged ions, and this tends to oppose further adsorption. Like all ionic effects it can be reduced by addition of simple salts (provided these do not precipitate the drug chemically nor salt it out physically), which thus permits further adsorption to take place. The effect is determined primarily by the concentration and valence of the ion with the opposite charge to that on the drug ion. Thus, in the case of anionic soaps $CaCl_2$ is more effective than NaCl, and Na_2SO_4 more than NaCl with the cationic type.

Addition of salts frequently leads to precipitation of soaps, particularly if the salt contains a di- or trivalent ion oppositely charged to the organic ion, the surface activity first increasing, then reaching a maximum with the onset of precipitation, and thereafter declining. Such precipitation is not always obvious to the eye, owing to the dispersing action of the soap, but it can be followed by surface-tension measurements.

The antibacterial activity of soaps is found to be modified by inorganic salts in the way predicted from the changes in surface activity. Thus addition of amounts insufficient to bring about precipitation increases the bactericidal activity in all cases; with cationic soaps small amounts of Na_2SO_4 are very effective (Hill & Hunter, 1946), and with anionic soaps $CaCl_2$ is much more effective than NaCl (Agar & Alexander, unpublished).

Addition of organic salts also influences the adsorption of ionized drugs. If the organic ions are similarly charged the effect of low concentrations is little different from that produced by an equivalent amount of an inorganic salt; in relatively high concentration competitive surface displacement may become of importance. If, on the other hand, they are oppositely charged, then very considerable increases in surface activity can be brought about by *very small* amounts of additive, and in addition the saturation solubility and the critical concentration for micelles (for aggregating systems) are both markedly lowered. A number of cases of synergism in bacterial systems can probably be explained with the above concepts; for example, penicillin with cationic dyes such as gentian-violet and brilliant-green (George & Pandalai, 1946), sulphonamides with synthetic soaps (Gershenfeld & Sagin, 1946) and with certain dyes (Thatcher, 1945), and *N*-*N'*-dichloro-azo-dicarbonamidine with sodium tetradecyl sulphate (Petroff & Schain, 1940).

If sufficient capillary-active salt is added to bring about precipitation of the drug, then the surface activity rises to a maximum at the onset of precipitation and thereafter falls with further additions. The surface activity may in some cases be considerably reduced by this means, leading to a detoxication. Quantitative measurements of the antibacterial activity of a number of such precipitating systems (e.g. cationic soaps and dyes with anionic soaps) have shown good correlation with the variation of surface activity (Tomlinson & Alexander, 1948).

The repulsive forces between adsorbed organic ions can also be reduced by adding polar organic molecules, such as alcohols, phenols, amines. These are believed to act by forming 'interfacial complexes' with the organic ion, probably by hydrogen-bond formation. This, by permitting increased adsorption, leads to an increased surface activity and with biologically active substances like phenols to a concomitant increase in their activity. The effect is shown particularly clearly with soap/phenol systems. Fig. 4*a* shows the rate of penetration of hexyl resorcinol into *Ascaris* in the presence of the synthetic soap CTAB (Alexander & Trim, 1946). Up to the maximum, i.e. in the premicellar region (see (vi) below), the correlation between rate and surface-tension depression is quite obvious. The bactericidal activity of Aerosol M.A./0·5% phenol mixtures towards *Bact. coli*, plotted in Fig. 4*b*, shows precisely the same effect (Alexander & Tomlinson, 1948), as does the system cetyl pyridinium bromide/0·5% phenol against the same organism (Agar & Alexander, unpublished).

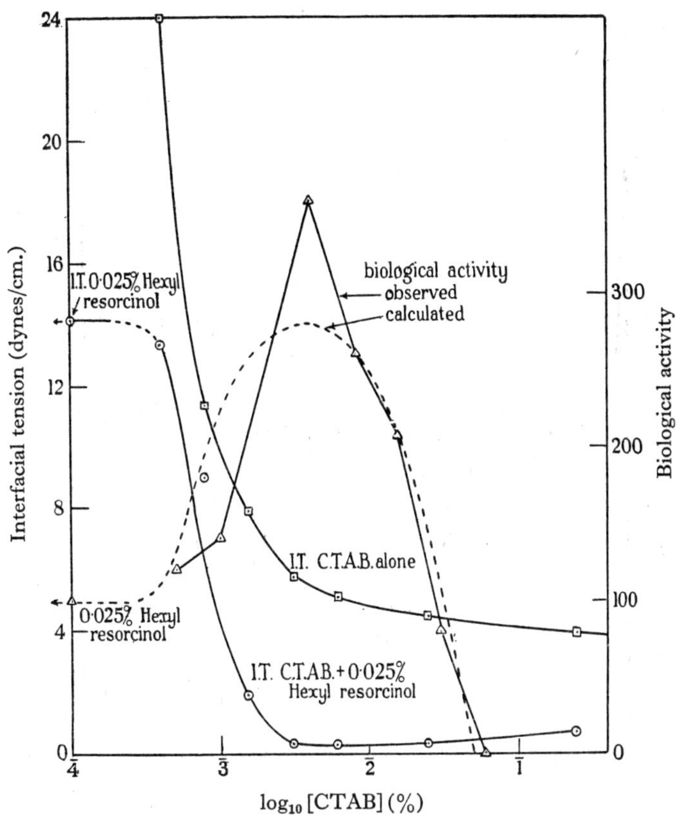

Fig. 4*a*. CTAB/0·025% hexyl resorcinol mixtures in 0·9% NaCl. Interfacial tension against nujol and biological activity measured by the penetration of hexyl resorcinol as percentage of control (0·025% hexyl resorcinol alone).

With drugs containing ionizable groups, the amounts in the unionized and various ionized forms depend upon the pH and the dissociation constants of these groups, as pointed out earlier in § III. Ionization will make the drug more water-soluble and thus at a given bulk concentration decrease its tendency to adsorb at surfaces, *unless the surface and organic ion are oppositely charged*. Thus, since most biological surfaces are negatively charged, drugs giving an organic anion should be more biologically active in the unionized form. This appears to be generally so, as shown by the organic acids and by substituted phenols, which become much less bactericidal when the pH is sufficiently high to cause ionization. With drugs giving an organic cation, ionization may increase or diminish the adsorption, depending upon the conditions, particularly the structure of the molecule, the nature of the biological surface (see below), and the salt concentration, since salts reduce all ionic effects.

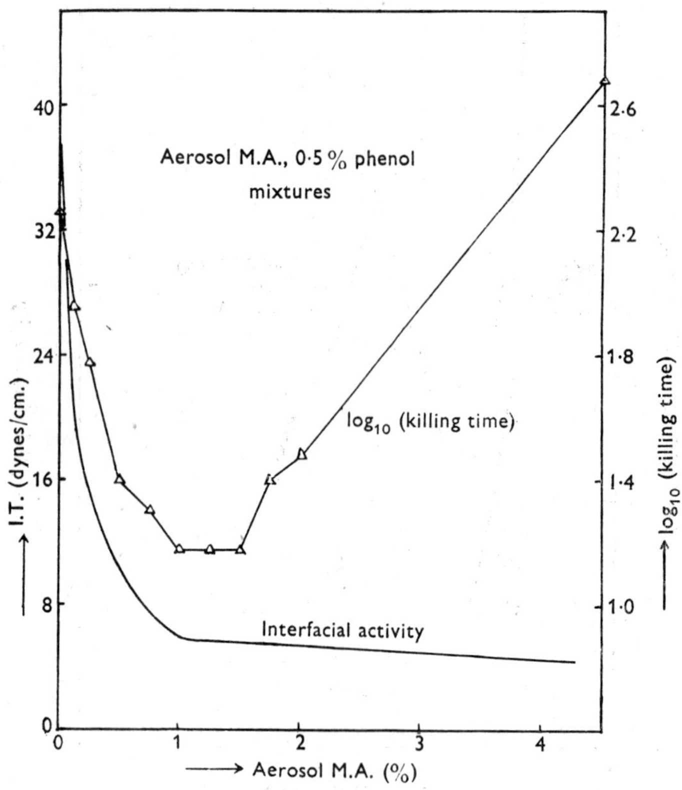

Fig. 4*b*. Aerosol M.A./0·5 % phenol mixtures. Interfacial tension against nujol and biological activity measured by time for complete kill of *Bact. coli*.

The effect of pH upon the penetration of nicotine through the cuticle of *Ascaris* in the presence of physiological saline (Trim, 1948) is shown in Fig. 5. The close parallelism between the degree of dissociation (full curve) and the experimental points shows that the doubly ionized molecule (for in this case we go from M^+ to M^{++} with decreasing pH) has relatively little permeating power, due undoubtedly to its greater hydrophilic nature and therefore decreased adsorption.

(v) *Effect of adsorbing surface*

It has been shown experimentally (McMullen & Alexander, 1948) that both the physical and chemical nature of the adsorbing surface influence the adsorption of soaps, and this can undoubtedly be generalized for all types of adsorbate. At low concentrations of soap solid surfaces adsorb much less strongly than fluid ones, but at high concentrations (above the micelle point) there is little difference between them. This approach has been used to explore the nature of the bacterial envelope, which appears from its adsorptive behaviour to be in a solid or plastic solid, rather than in a fluid, state.

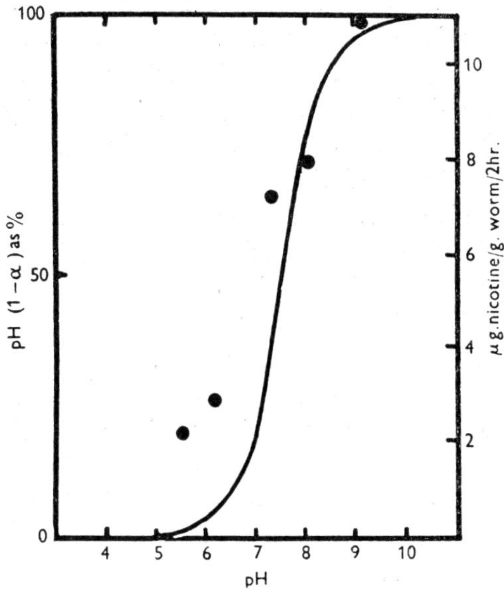

Fig. 5. The relation between the rate of penetration of *l*-nicotine into *Ascaris lumbricoides* var. *suis* from a 0·03 % solution and the dissociation of its methyl pyrrolidine group.

With ionized drugs the adsorption is also dependent upon the charge carried by the surface, as previously mentioned. Biological membranes are usually believed to be lipoprotein in nature (see p. 135), and their surfaces are thus likely to have a high density of ionized groups such as carboxyl, amino, phosphate, etc. These ionized groups, being immobile and frequently of high surface density (unlike the ions adsorbed at air/water and oil/water interfaces which are mobile and normally present only in relatively small numbers), can greatly modify the adsorption of ionized drugs. If the drug ion and membrane are oppositely charged adsorption is facilitated, if similarly charged adsorption is greatly diminished, although not necessarily to zero, since the hydrocarbon portion of the molecule also plays an important part (see (ii) and (iii) above). Addition of simple salts reduces these Coulombic interactions; with similarly charged drug ion and membrane this will increase the adsorption (see also (iv) above); when they are oppositely charged salts will tend to reduce the initial amounts of adsorption, but when the surface charge has been reversed by adsorption of drug ions salts will again facilitate adsorption. It is clear then that due regard must be paid to the physico-chemical factors involved in any given system.

The importance of the charge on the biological surface is shown particularly well with bacteria as test organism and synthetic soaps as drugs. Synthetic soaps are particularly suitable, since many of them, such as the long-chain quaternary ammonium salts, alkyl sulphates, etc., are unaffected by pH over a wide range (i.e. they remain fully ionized). Cationic soaps, dyes and other organic cations are well known to be much more bactericidal than their anionic analogues, and their toxicity increases as the pH is raised. With the anionic soaps, dyes, etc., the toxicity increases with decreasing pH (for substances where the ionization remains constant). Bacteria are normally negatively charged at neutral pH, the charge decreasing as the pH is lowered. The Coulomb forces thus tend to facilitate the adsorption of cations but to diminish that of anions. The differences between these types and their variation with pH are thus readily explained without invoking unionized molecules, which many bacteriologists have done, and which is impossible in many cases.

The sensitivity to anionic soaps also depends to some extent upon the bacterial type, Gram-negative organisms being relatively more resistant than Gram-positive. This difference cannot be ascribed to effects arising from their surface charges (which are in turn an indirect expression of the substances constituting the bacterial envelope), for as the early work of Stearn & Stearn showed, the Gram-negative organisms have the higher isoelectric point and are thus less strongly charged at neutral pH. Other things being equal surface-active anions will thus be taken up less by the Gram-positive organisms.

As previously mentioned (p. 123) many apparent discrepancies between the biological activity of cationic drugs and their adsorptive power as measured by surface tension can be resolved by paying due regard to the charges on the biological surface.

The above considerations should of course apply to the uptake of all ionized compounds, whether they happen to be drugs or essential metabolites, as with lysine and glutamic acid in the case of certain bacteria (e.g. Gale, 1947). In the case of metabolites, however, the quantities involved will, as a rule, be much greater than with drugs, and this will modify the picture.

(vi) *Effects arising from micelle formation*

As previously mentioned the substances termed colloidal electrolytes (soaps, many dyes, bile salts, lecithin, etc.) show spontaneous formation of colloidal aggregates, termed micelles, when the concentration exceeds a certain rather definite limit. (This limit is usually termed the 'critical concentration for micelle formation'.) Aggregation, which often sets in at concentrations well below the saturation solubility, resembles in some ways the separation of a new phase, so that the activity of the solution, and hence the surface activity, thereafter increases but little with further additions of soap. (This is shown, for example, in Fig. 4.)

Micelle formation can be facilitated (i.e. it occurs at a lower concentration) by several means:

(*a*) by lowering the temperature,

(*b*) by increasing the chain length of the soap,

(*c*) by addition of simple salts (provided these do not lead to precipitation) (see also p. 123),

(*d*) by addition of capillary-active salts, the effect being very marked if the two organic ions (soap and additive) are oppositely charged (e.g. $C_2H_5COO'Na^+$ and $C_{16}H_{33}\overset{+}{N}(CH_3)_3\ Br'$),

(*e*) by addition of many organic compounds, particularly those partially soluble in water and containing —OH and $—NH_2$ groups (e.g. phenols, alcohols, amines).

Such changes, by leading to the formation of micelles, can bring about marked effects in the biological behaviour of the system.

The micelle consists of aggregated hydrocarbon chains, in a fluid state, with an outer 'covering' of ionized groups in contact with the aqueous phase. In the case of the C_{16} straight-chain soaps its diameter is *c.* 40 Å., and it contains about fifty molecules. Owing to its high charge the micelle binds very strongly a considerable proportion of its gegenions (e.g. Na^+ in sodium soaps), often 75% or more, thus reducing its apparent 'valency' considerably. Owing to their peculiar structure micelles interact very strongly with, and thus indirectly influence the biological activity of, a wide variety of substances, ranging from water-insoluble hydrocarbons at the one extreme, through partially soluble organic molecules like phenol, to the very hydrophilic simple inorganic ions.

When small amounts of polyvalent ions of opposite charge are present (e.g. Ca^{++} with anionic soaps, or SO_4'' with those of the cationic type), these are strongly adsorbed by the micelles in preference to the univalent gegenions. This may conceivably influence biological behaviour in some circumstances, for example, by removing traces of toxic ions such as Cu^{++}.

With capillary-active salts, marked interaction with micelles would only occur with oppositely charged organic ions, this being linked with their ability to promote micelle formation. Provided the micelles are in considerable excess and are not themselves biologically active this can be a means for detoxicating certain ionized drugs, e.g. acridines and other organic cations by bile salts. (The reason for stipulating an excess of micelles is that, with comparable amounts, the increased surface activity of the mixture may lead to an opposite effect, as discussed on p. 130.)

Micelles exert a powerful 'solubilizing' action upon organic molecules insoluble or only slightly soluble in water. Phenols, for example, are strongly taken up by micelles which, if present in considerable number, may lead to the solution becoming largely depleted of the phenol. This

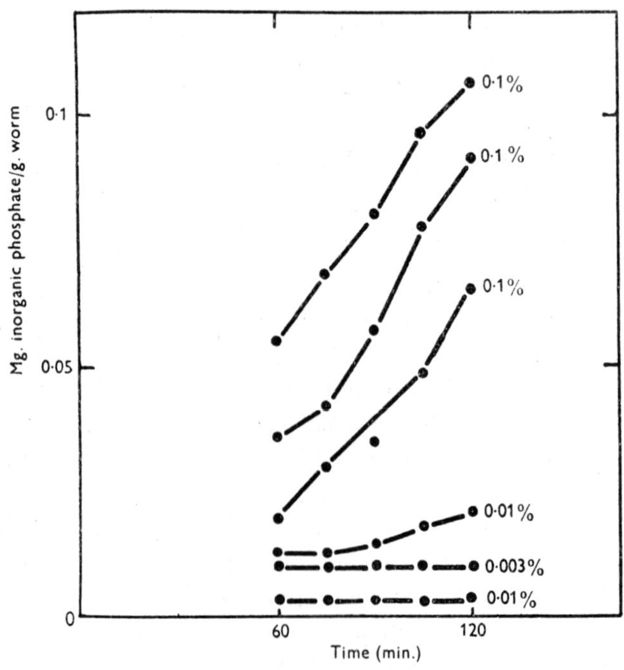

Fig. 6. Rate of leakage of inorganic phosphate from *Ascaris* immersed in CTAB solutions of concentrations indicated.

depletion results in the biological activity of the phenol diminishing, as has been shown by the anthelmintic action of soap/hexyl resorcinol mixtures, and by the bactericidal activity of phenol in the presence of anionic soaps. Fig. 4*a* shows the effect of the synthetic soap CTAB upon the rate of penetration of hexyl resorcinol (0·025%) into *Ascaris*. The optimum corresponds to the onset of micelle formation (shown by the break in the interfacial tension curve), and with higher soap concentrations the rate falls progressively to zero. The depletion, and hence the fall in activity, can actually be calculated in this case from surface chemical considerations, in good agreement with experiment (see Fig. 4*a*). Fig. 4*b* shows the same phenomenon as obtained with *Bact. coli* as test organism and Aerosol M.A./0·5% phenol mixtures, using a modified Rideal-Walker technique. This work has been able to explain the very conflicting results obtained by earlier workers concerning the bactericidal activity of soap/phenol mixtures.

Solubilization provides a mechanism for transporting sparingly soluble substances at a rate which may be enormously greater than the usual process of solution and diffusion. If, as seems not uncommon, the biological response is determined by a *rate* of adsorption or penetration, then such conditions may lead to an apparent increase in the biological activity of sparingly soluble substances.

Solubilization may also influence biological activity in a way which does not appear to have been previously pointed out. It is found that soaps have little influence upon the rate of phosphate leak from *Ascaris* until micelles are present in the solution (see Fig. 6, where the critical micellar concentration is *c.* 0·01 %). The sudden increase in permeability just at this point can reasonably be ascribed to the micelles actually 'deterging' some of the lipoid constituents of the boundary membrane, and thus removing the principal barrier to ionic diffusion.

(vii) *Effect of other constituents of the medium*

For the purpose of discussion, substances, other than the drug and carrier medium, can be grouped as follows:

(*a*) Simple inorganic salts.
(*b*) Organic (capillary active) salts.
(*c*) Unionized organic substances of low molecular weight.
(*d*) Polymers.

As previously pointed out these may be present naturally or may have been added deliberately.

(*a*) *Simple inorganic salts.* The surface activity of drugs which are ionized is much more susceptible to the nature and concentration of simple inorganic salts than of those which are not. With unionized but polar molecules such as phenols, addition of salt increases the surface activity and lipoid partition coefficients, but the effects are marked only at high salt concentrations. Nevertheless, this may be one of the factors responsible for the increased germicidal powers of phenols produced by added salts.

With ionized drugs addition of simple salts may bring about a number of changes, depending upon the conditions and the nature of the drug; it may increase the adsorption, bring about micelle formation (with colloidal electrolytes), or cause precipitation. These changes and biological examples are discussed in detail earlier in this paper.

(*b*) *Capillary-active salts.* Examples sometimes present in biological tests are the bile salts, soaps, dyes and certain phospholipoids such as lecithin and kephalin. These are frequently micelle-forming compounds and have thus to be considered in the pre- and post-micellar ranges as shown on pp. 129 and 130 respectively.

Turning now to the influence of capillary-active salts upon unionized drugs some general account was presented in the introductory section, and their interaction when micelles are present was detailed on p. 129. Compounds with some water solubility, and therefore containing polar groups, usually increase the surface activity of the mixture, and in aggregating systems lower the critical concentration for micelles (p. 130). As the surface activity increases so, in a general way, does the biological activity. This is shown particularly clearly by the effect of soaps upon the anthelmintic activity of hexyl resorcinol (see Fig. 4*a*), and upon the bactericidal action of phenols (Fig. 4*b*).

Non-polar compounds (e.g. paraffins and chlorinated hydrocarbons) show little interaction with soaps unless micelles are present (when they dissolve to some extent in the micelles).

(*c*) *Unionized organic substances of low molecular weight.* Substances of low surface activity and high water solubility, such as glycerol, glucose and other sugars, exert little influence upon the adsorption of drugs, whether ionized or not. More surface-active substances, such as the lower alcohols, are themselves usually biologically active, so that it is not easy to disentangle their effect when mixed with other drugs. An interesting example of such a system comes from the study of the growth-rate curves of bacteria in alcohol/phenol mixtures of sublethal strength (Tatton & Dagley, private communication). Fig. 7 shows the effect upon the mean generation time (T = time in presence of alcohol/phenol mixture, T_0 = time in presence of phenol alone) of increasing amounts of ethyl alcohol when added to 0·074% phenol. The maximum appears to arise from a competitive displacement at the bacterial surface, the alcohol being intrinsically much less toxic than the phenol although not devoid of activity.

(*d*) *Polymers.* The principal substances to be discussed here are natural polymers such as agar, gelatine, proteins, peptones, etc., materials such as 'broth' which contain proteins or their degradation products, and a few synthetic materials such as polyvinyl alcohol and neutral (unionized) detergents (e.g. those based upon polyethylene oxides and polyglycerides). Such materials are usually regarded as being either biologically 'inert' or as facilitating growth. Certain of them, such as agar and 'broth', are commonly present in tests of antibacterial activity (e.g. the 'plate' method for penicillin assay).

Even when themselves quite biologically inert, polymers can nevertheless profoundly modify the effect of substances which do possess biological activity. The most probable ways of doing so would be either by diminishing the rate of diffusion to the biological surface or by adsorption of the active compound upon the polymeric material. The former will, of course, only enter under conditions where the slowest step is the rate of diffusion to the biological surface.

Polymers are well known for their power of forming viscous solutions and gels. The rate of diffusion through gels decreases very rapidly as the size of the diffusing molecule approaches the average pore diameter of the gel,

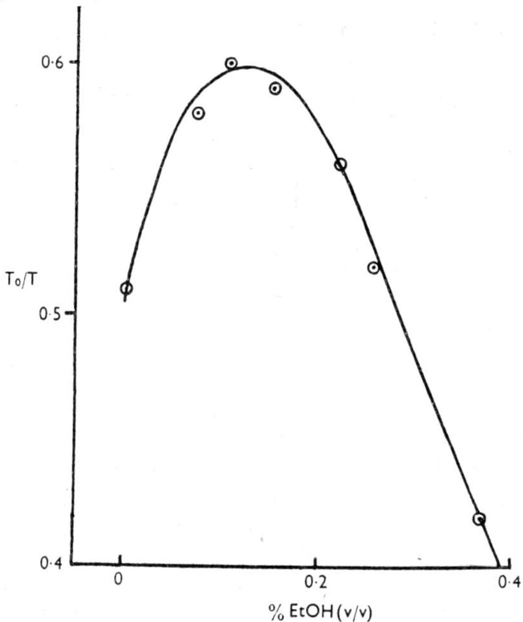

Fig. 7. The effect of ethyl alcohol on the antibacterial activity of phenol (by kind permission of J. O'G. Tatton & S. Dagley).

which in the case of 5% gelatine, for example, is about 50Å. With a C_{16} paraffin-chain soap the diameters of the single molecule and micelle in solution would be about 10 and 40Å. respectively. It is therefore clear that soaps, if present chiefly as micelles (as is frequently the case), will diffuse extremely slowly through most gel structures.

A further factor influencing the diffusion of ionized drugs through gels arises from the charge carried by the gel framework. If this is positive, as with gelatine below its isoelectric point (due largely to $—NH_3^+$ groups), the gel is only readily permeable to anions, if negatively charged only to cations. This preferential ionic permeability, like all electrical phenomena arising from the Coulombic forces between ions, is reduced by addition of simple inorganic salts.

The other means by which polymers may influence the biological activity of drugs is by combination, either by a definite chemical reaction or by physical adsorption on their surfaces. In the former process the polar groups of the polymer will be of special importance. For example, anionic drugs (negatively charged organic ion) would combine strongly with cationic groups (e.g. $—NH_3^+$) in gelatine, and cationic drugs with anionic groups (e.g. —COO′ in gelatine, $—SO_4'$ in agar). Attachments to unionized polar groups could also take place, but in general this would only be at all marked with such groups as —OH or $—NH_2$ which are capable of forming hydrogen bonds. Such combinations lead to a reduction in the biological activity of the drug, to changes in the gel-forming capacity of the mixture, and in some cases to mutual precipitation. A few biological examples may be briefly mentioned. The bactericidal activity of anionic soaps is reduced by polyvinyl alcohol ($—CH_2—\underset{OH}{\underset{|}{CH}}—CH_2—\underset{OH}{\underset{|}{CH}}—$, etc.) (Alexander & Tomlinson, unpublished), the attachment in this case probably being through a hydrogen bond between —OH and soap anion, and by neutral detergents (Agar & Alexander, unpublished); agar similarly reduces the germicidal activity of cationic soaps (Quisno, Gibley & Foster, 1946), a finding readily explicable by combination with anionic groups, particularly $—SO_4''$ and COO′, known to be present in this polymer. The last example concerns the growth of acid-fast bacteria in the presence of surface-active substances (alcohols, neutral and anionic soaps) (Alexander & Soltys, 1946). In some cases addition of 2% agar permitted growth in a previously inhibiting solution. This appeared to be due to the diminished surface activity of the mixture presumably by adsorption of depressant upon the agar molecules, since the surface tension was thereby increased.

With unionized drugs adsorption upon the polymer surfaces would also lead to detoxication, although the effects would, as a rule, be less marked than in the examples mentioned above.

V. TRANSPORT ACROSS BIOLOGICAL MEMBRANES

The best method of measuring the rate of drug transport is direct chemical determination of the amount passing into the organism, but so far this method has not been widely applied owing to experimental difficulties. Rates have usually been assessed from data on plasmolysis and biological responses such as 'death' and loss of irritability. There are, however, data available which are sufficiently well defined for the present discussion. For example, extensive direct measurements of the permeability of plant cells to many organic molecules have been made by Collander & Barlund (e.g. Collander, 1937). From these results Collander concluded that the rates of penetration were determined primarily by the lipoid solubility of the molecule and its size. The rates of penetration of anthelmintics and related substances into *Ascaris lumbricoides* var. *suis* have been followed quantitatively by direct chemical measurements in the work of Trim (1944, 1948), some of whose results are shown in Table 1.

Table 1. *Permeabilities of* Ascaris *cuticle*

Substance	Mol. wt.	B, olive oil/water partition coefficient	Permeability* $P \times 10^{16}$	$P/B \times 10^{18}$	$PM^{\frac{1}{2}}/B \times 10^{18}$
Resorcinol	110	2·6	0·47	18	270†
Propyl resorcinol	152	70	0·66	0·94	16·5
Butyl resorcinol	166	210	2·0	0·95	17·4
Hexyl resorcinol	194	1800	9·8	0·55	10·9
Heptyl resorcinol	208	5700	24·7	0·43	8·8
Thymol	150	(700)	5·0	0·71	12·4
Carvacrol	150	700	6·6	0·94	16·4
α-Naphthol	144	700	6·6	0·94	16·0
Chloroform	119	(380)	13·2	3·47	37·5
l-Nicotine++	158	8	0·008	0·1	1·78
l-Nicotine+	159	42	0·04	0·1	1·68
In presence of soaps (maximal rates):					
Hexyl resorcinol			*c.* 37·3		
Nicotine+			*c.* 0·6		
Chloroform			*c.* 13·2		

* Permeability P defined as g.mol. passing through 1 μ^2 of the membrane per sec. per g.mol. concentration difference across the membrane.
† Values of M are all double the simple formula weight, except in the case of chloroform

In the absence of a large body of accurate quantitative data it is not possible to make well-founded generalizations as to the prevailing modes of transport. However, our present knowledge of the chemical structure and micro-morphology of cell membranes, capsules, the integuments of Metazoa, etc., shows that there is a very great variety among natural membranes. As far as general selective permeability properties are concerned, these membranes will tend to behave as if they were more or less homogeneous, and as if they were more or less lipoid in composition. Nevertheless, it is reasonable, on the other hand, to expect them to show subtle variations which might endow them with high selectivity towards certain individual drugs or even classes of drugs, in which case a thorough investigation of permeability in relation to their detailed structure and chemical composition is desirable, together with the development of the complementary study of simple artificial membranes.

Further sources of variability and specificity in the selective permeability of the membranes of living organisms are the developmental changes in the organism. They change in many ways according to the age of the organism.

They are often, as in the case of bacteria, involved in the exchange of both nutrients and excretory products between the medium and the interior of the organism. These changes can have great effects upon selective permeability.

Many natural membranes do not appear to present a physically homogeneous barrier to the medium. Many lipoid membranes, although differing in detail, are believed to consist essentially of a thin lipoid layer (possibly no more than 50–100 Å. thick) held within a protein framework. The chief hindrance to the entry of most drugs is then thought to arise from this lipoid layer. In general, it would appear that the lipoid layer is in a solid rather than a fluid state. Evidence for such conclusions comes from several lines of investigation, such as the study of the water permeability of insect cuticles (Wigglesworth, 1945; Beament, 1945), the adsorption of soaps on bacteria (McMullen & Alexander, 1948), and the general behaviour of biological systems viewed in relation to the permeability of synthetic lipoid membranes.

Up to the present three principal modes of penetration through membranes have been postulated:

(*a*) by diffusion through water-filled pores,
(*b*) by interfacial diffusion,
(*c*) by solution in, and diffusion through, lipoid.

We shall confine ourselves here to a discussion of the cases of *Ascaris* and bacteria, for which mechanism (*b*) appears improbable (for evidence see Alexander & Trim, 1946) and mechanism (*a*) less important than (*c*).

The question of diffusion through a lipoid layer will be examined below in some detail, and its predictions compared with experimental findings. In general the process of drug penetration can be subdivided into three stages: movement from the aqueous phase across the interface into the adjacent lipoid phase (the process which we term *ab*sorption), diffusion through the lipoid, and removal from the far surface of the lipoid (desorption). The overall rate will be set by the slowest of these processes, and if we consider only *initial* rates of penetration, by either the first or the second (i.e. by absorption or by diffusion through the lipoid).

(i) *Penetration into the membrane*

Of the drugs considered here those of types (*a*), (*b*) and (*c*) (e.g. phenols, nicotine) are likely to experience much more difficulty than those of type (*d*) (e.g. simple and chlorinated hydrocarbons) in penetrating a lipoid layer from their aqueous solutions. The reason is that, in the case of polar molecules, the transfer involves a complex dehydration and aggregation process which does not occur with the non-polar molecules of type (*d*). Although the importance of dehydration has been mentioned before (e.g. Danielli, 1941), no attention appears to have been drawn to the aggregation factor. Aggregation of polar molecules in lipoid systems is well established, and to pass from an adsorbed monolayer at the interface to an aggregated unit inside a lipoid phase is known to be a very slow, hindered process. If, therefore, the slow process is that from the aqueous solution to the lipoid layer, we should anticipate marked differences between the rates with polar and non-polar drugs.

The study of the effect of dilute soap solutions upon drug penetration provides one definite means of indicating whether absorption is actually the slowest step involved. Soaps would be unlikely to modify greatly the actual diffusion through the lipoid layer since they are relatively insoluble in lipoids (see also (ii) below), but in view of their surface activity might well influence the rate of attainment of equilibrium between the aqueous and lipoid phases (i.e. the absorption process). Experimentally it is found that soaps can accelerate the penetration in some systems (e.g. phenol into bacteria; hexyl resorcinol and nicotine into *Ascaris*), but have no effect upon others (e.g. chloroform-penetrating *Ascaris*). These data, although very limited, suggest that in the case of *Ascaris* it is only the polar molecules which are accelerated by soaps, indicating that the *absorption* process is much slower in such cases than for non-polar molecules such as chloroform, as was anticipated above. It may be of significance that the penetration of the ionized (and hence very polar) nicotine molecule can be accelerated to a much greater degree (fifteen times as compared with about four) than the less polar molecule hexyl resorcinol.

The most likely way for soaps to facilitate the absorption of polar molecules is by an alteration in the surface structure of the lipoid layer, for example, by breaking some of the bonds linking it to the protein framework. On this view the acceleration should run parallel to the surface activity of the soap solution, as found experimentally in the case of *Ascaris*. With soaps in the pre-micellar region the modified lipoid layer might well retain its continuity and thus still present a relatively impermeable barrier to oil-insoluble compounds (e.g. soaps and inorganic ions), but when micelles are present solubilization would permit removal of lipoid and ultimately render the membrane freely permeable to all substances, such as inorganic ions from the interior (p. 131).

(ii) *Diffusion through the lipoid layer*

If the slowest step in drug penetration is diffusion across a lipoid layer, then the *initial* rate of transport is given by

$$\frac{dQ}{dt} = -DA\frac{dc}{dx} = DA\frac{C_{\text{lipoid}}}{x},$$

where dQ is the amount diffusing in time dt, A the area and x the thickness of the membrane, D the diffusion coefficient, and $-dc/dx$ or C_{lipoid}/x the concentration gradient. Hence

$$P = \frac{dQ}{dt}\frac{x}{A} = DC_{\text{lipoid}},$$

where P may be termed the permeability.

The partition coefficient $B = C_{\text{lipoid}}/C_{\text{aq.}}$ and hence $P = DBC_{\text{aq.}}$, since by the above postulate the surface lipoid layer is in equilibrium with the aqueous layer of concentration $C_{\text{aq.}}$. Hence for a given biological system and drug concentration ($C_{\text{aq.}}$) the *initial* rates of permeation should be proportional to DB, i.e. to the product of the diffusion coefficient and the partition coefficient.

It should be realized that D refers to the drug in the lipoid layer, and allowance must be made if the drug is aggregated, as will certainly occur with all polar molecules (i.e. types (*a*), (*b*) and (*c*)). To assume that $D \propto 1/M^{\frac{1}{2}}$, where M is the molecular weight in an aqueous medium, as is often done, is bound to introduce errors, particularly when comparing polar with non-polar drugs. The true value of M is that in the lipoid phase, and could be obtained from partition measurements. If it is not known a reasonable guess would be complete dimerization for polar molecules and the usual monomeric form for simple and chlorinated hydrocarbons. Since, however, there appears to be no theoretical reason for assuming $D \propto 1/M^{\frac{1}{2}}$ (the Einstein equation, if applicable, would give $D \propto (\rho/M)^{\frac{1}{3}}$, where ρ is the density), and since the changes in D are usually very much less than those in B, it seems better to assume D constant and to see if P is proportional to B. A reasonable constancy over a range of polar and non-polar compounds would then indicate that the slowest step is that of diffusion through the lipoid layer.

Table 1 shows some data for the permeability of *Ascaris* to a number of drugs (Trim, 1948). As will be seen the ratio P/B is practically constant except for $CHCl_3$ and resorcinol, which are both much higher than the rest.* This would at first sight indicate that lipoid diffusion is the rate-determining step and accordingly that the lipoid layer is always in equilibrium with the aqueous phase. However, as discussed above, consideration of soap effects leads to the conclusion that it is chloroform, rather than the other group, which may have to be regarded as the 'normal' case; with the other drugs (all polar molecules) the energy barrier at the interface, which

* The data for resorcinol are not to be taken at their face value, for we believe that the small depletion from the external medium is due to adsorption upon the outer layers of protein rather than to a true penetration inside. This would introduce little error with the higher members and would be in accord with the absence of toxicity of the lower members (see Fig. 3).

determines the rate of absorption, is comparable to that involved in diffusion through the lipoid. The difference between the two cases, and the way in which soaps are believed to influence the rate, can be seen by reference to Fig. 8. Fig. 8*a* represents the case for non-polar molecules such as hydrocarbons and chloroform, Fig. 8*b* that for polar organic molecules such as hexyl resorcinol.

Considering the latter case, in the absence of soaps the concentration just inside the lipoid (i.e. C_{lipoid}), which determines the rate of diffusion, is, owing to the energy barrier between the adsorbed and absorbed layers, less than if these layers were in true equilibrium. Addition of very surface-active compounds like soaps would certainly reduce the concentration of

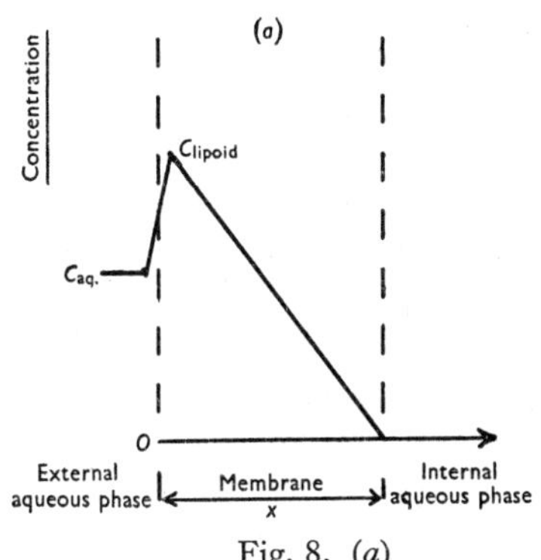

Fig. 8. (*a*)

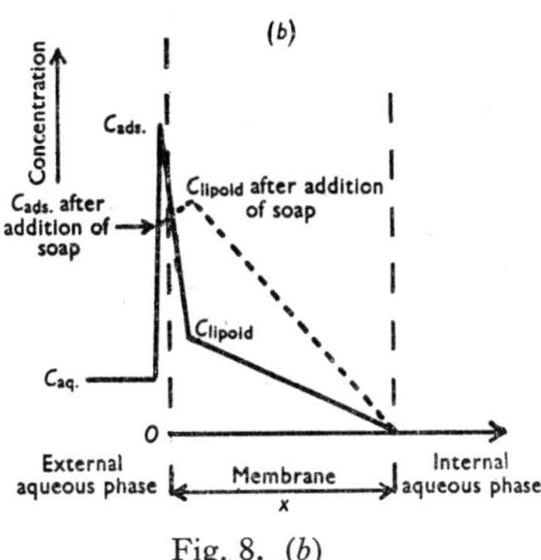

Fig. 8. (*b*)

adsorbed drug, but if at the same time the energy barrier hindering absorption were lowered sufficiently the *net* result might well be an increase in C_{lipoid} and thus an increase in rate of penetration. Such a postulated behaviour of soaps is not unreasonable in the light of their known effects upon adsorption and desorption processes at interfaces.

The close link between surface activity and permeability, in the case of polar molecules, is also apparent from Fig. 8*b*. In an homologous series, for example, the dehydration and aggregation processes involved in the jump from the *ad*sorbed to the *ab*sorbed layer would be expected to depend primarily on the nature of the polar group and only secondarily on the chain length. If these processes are hindered ones, as seems quite likely, then C_{lipoid} will be approximately proportional to C_{surface}, and in the extreme case the *overall* rate will be determined by surface activity rather than by lipoid solubility. This can be shown by applying the steady-state principle, since the gain by absorption must equal the loss by diffusion.

The rate of absorption will be approximately $k'AC_{\text{surface}}$, where k' is approximately constant for an homologous series by the argument given above. The rate of loss by diffusion is equal to $DA\frac{C_{\text{lipoid}}}{x}$. Hence $C_{\text{lipoid}} \propto C_{\text{surface}}$, approximately. The extent to which k' is likely to vary with the nature of the polar group is difficult to predict in the absence of experimental data. If the variations are not large then a general parallelism between surface activity and permeability would be found. This shows once again how, in the case of polar molecules, the surface activity and lipoid solubility aspects may be complementary rather than mutually exclusive, their relative importance depending upon the particular drug/biological system.

Homologous series

If the limiting step is diffusion through a lipoid layer then for two homologues with n and $(n-1)$ carbon atoms, compared as is usual at a fixed aqueous concentration, we have

$$P_n/P_{n-1} \simeq B_n/B_{n-1},$$

since D will be approximately the same for both. As a rule B_n/B_{n-1} = a constant, often *c.* 3, so that $P_n \simeq 3P_{n-1}$, as found experimentally in many systems (e.g. the higher resorcinols in Table 1). It can be seen, however, that this relation can only be expected to hold up to the saturation point of the higher homologue (this being the less soluble). Comparisons carried out at higher concentrations are certainly the reason for a large number of apparent 'optima' in homologous series. For example, hexyl resorcinol has frequently been quoted as having optimal activity, but comparison at low enough concentrations shows that the higher members penetrate more rapidly (see, for example, Table 1 and Fig. 3).

The permeability of drugs dissolved in organic solvents and of neat drugs

As shown above the permeability P is proportional to DB when lipoid diffusion is the limiting step. When water, the carrier medium previously considered, is replaced by an organic solvent (X), changes in both D and B are likely. B now becomes concentration in lipoid/concentration in liquid X, and for organic drugs is likely to be *reduced* as compared with the aqueous system. On the other hand, organic solvents are much more soluble in the lipoid membrane, and thus, by swelling it, may increase D considerably; the latter effect is likely to be especially pronounced if the lipoid membrane is normally relatively impermeable owing to a highly orientated or rigid structure. It thus becomes possible for a non-toxic solvent (carrier), by increasing D much more than decreasing B, to accelerate the penetration of a dissolved drug. Such considerations would explain many cases of synergism, for example, the effect of kerosene (swelling but non-toxic) in increasing the toxicity of alcohols (non-swelling but toxic) to certain types of insects (Hurst, 1943). It should also be pointed out here that where the drug concentration is not insignificant in comparison with the carrier, the mixed system may have quite markedly different swelling powers from either constituent.

The permeability of neat drugs (when liquid) can clearly be treated on similar lines to those above. The effect upon D is likely to be of most importance, so that permeability will be largely determined by the swelling action of the drug upon the lipoid membrane. Since this will be determined by the chemical and physical properties of the drug and by the nature of the lipoid, it becomes very difficult to make any predictions.

VI. CONCLUSION

We realize only too well the shortcomings of the present attempt at elucidating the physico-chemical factors involved in drug penetration. However, certain aspects, particularly those concerned with the effect of additives in the external medium, seem reasonably clear. Many of the difficulties arise from the lack of *quantitative* rates of penetration and from our limited knowledge of biological membranes. A detailed permeability study of synthetic lipoid layers of known structure is also a pressing need, for until simpler systems have been fully analysed the behaviour of biological membranes must necessarily remain obscure.

REFERENCES

Albert, A., Rubbo, S. D., Goldacre, R. J., Davey, M. E. & Stone, J. D. (1945). *Brit. J. Exp. Path.* **26**, 160.

Alexander, A. E. (1948). Faraday Discussion, *Surface Chemistry* (in the Press).

Alexander, A. E. & Soltys, M. A. (1946). *J. Path. Bact.* **58**, 37.

Alexander, A. E. & Tomlinson, A. J. H. (1948). Faraday Discussion, *Surface Chemistry* (in the Press).

Alexander, A. E. & Trim, A. R. (1946). *Proc. Roy. Soc.* B, **133**, 220.

Bancroft, W. D. & Richter, G. H. (1931). *J. Phys. Chem.* **35**, 215.

Beament, J. W. L. (1945). *J. Exp. Biol.* **21**, 115.

Bernard, C. (1875). *Leçons sur les anésthesiques et sur l'asphyxie.* Paris: Baillière et fils.

Blaschko, H. & Duthie, R. (1945). *Biochem. J.* **39**, 347.

Brown, C. & Fraser (1868–9). *Trans. Roy. Soc. Edinb.* **25**, 151.

Clark, A. J. (1937). *Trans. Faraday Soc.* **33**, 1057.

Collander, R. (1937). *Trans. Faraday Soc.* **33**, 985.

Danielli, J. F. (1941). *Trans. Faraday Soc.* **37**, 121.

Daniels, T. C. (1943). *Ann. Rev. Biochem.* **12**, 447.

Dubos, R. J. (1942). *Ann. Rev. Biochem.* **11**, 659.

Ferguson, J. (1939). *Proc. Roy. Soc.* B, **127**, 387.

Gale, E. F. (1947). *J. Gen. Microbiol.* **1**, 53.

GEORGE, M. & PANDALAI, K. M. (1946). *J. Sci. Industr. Res. India*, **58**, 79.
GERSHENFELD, L. & SAGIN, J. F. (1946). *Amer. J. Pharm.* **118**, 228.
HARTLEY, G. S. (1941). *Trans. Faraday Soc.* **37**, 130.
HILL, J. A. & HUNTER, C. L. F. (1946). *Nature, Lond.*, **158**, 585.
HÖBER, R. (1907). *Pflüg. Arch. ges. Physiol.* **120**, 492.
HÖBER, R. & HÖBER, J. (1942). *J. Gen. Physiol.* **25**, 705.
HURST, H. (1943). *Trans. Faraday Soc.* **39**, 390.
LAMSON, P. D., BROWN, H. W. & WARD, C. B. (1935). *J. Pharmacol.* **53**, 198.
LILLIE, R. S. (1909). *Amer. J. Physiol.* **24**, 14.
McMULLEN, A. I. & ALEXANDER, A. E. (1948). Faraday Discussion, *Surface Chemistry* (in the Press).
OVERTON, E. & MEYER, H. (1899). *Arch. exp. Path. Pharmak.* **42**, 109.
PETROFF, S. A. & SCHAIN, P. (1940). *Quart. Bull. Sea View Hosp.* p. 372.
QUISNO, R., GIBBY, I. W. & FOSTER, M. J. (1946). *J. Amer. Pharm. Ass.* (Sci. ed.), **35**, 317.
RAHN, O. (1945). *Bact. Rev.* **9**, 1.
STACEY, M. & WEBB, M. (1947). *Proc. Roy. Soc.* B, **134**, 523, 538.
SUTER, C. M. (1941). *Chem. Rev.* **28**, 269.
THATCHER, F. S. (1945). *Science*, **102**, 122.
TOMLINSON, A. J. H. & ALEXANDER, A. E. (1948). Faraday Discussion, *Surface Chemistry* (in the Press).
TRAUBE, I. (1904). *Pflüg. Arch. ges. Physiol.* **105**, 541.
TRIM, A. R. (1944). *Parasitology*, **35**, 209.
TRIM, A. R. (1948). (In the Press.)
VALKO, E. I. (1946). *Ann. N.Y. Acad. Sci.* **46**, 451.
WIGGLESWORTH, V. B. (1945). *J. Exp. Biol.* **21**, 97.
WINTERSTEIN, H. (1915). *Biochem. Z.* **70**, 130.
WINTERSTEIN, H. (1915). *Biochem. Z.* **75**, 71.

THE PERMEABILITY OF INSECT CUTICLE

By J. E. WEBB

Department of Natural History, University of Aberdeen

I. INTRODUCTION

Insects are killed when an insecticide interferes with some vital centre of the body such as the nervous system or the respiratory system. Before this can occur, however, the insecticide must penetrate the body by one or more of three routes. First, the insecticide may be eaten and absorption take place through the walls of the gut. Insecticides entering in this manner are referred to as stomach poisons irrespective of whether entry takes place through the cuticle-lined fore-gut or hind-gut, or through the mid-gut, which may or may not be protected by a peritrophic membrane; secondly, entry may take place through the tracheal system which is completely lined with chitin; or, thirdly, the insecticide may pass directly through the external cuticle covering the general surface of the body. Insecticides entering by means of the second or third routes are commonly known as respiratory or contact poisons respectively. As might be expected, the use of stomach poisons is almost entirely restricted to insects with biting mouthparts as opposed to those whose mouthparts are adapted for sucking liquid food. Similarly, respiratory poisons, other than fumigants, are limited in their use by the type of spiracle through which they must pass before entry into the tracheal system. The atrium of the spiracle in an insect serves to limit the amount of dust passing into the tracheal system with inspired air (see Webb, 1945*a*, *b*), and some spiracles are better adapted to carry out this function than others. It is not surprising, therefore, that some insects are very susceptible to insecticidal dusts entering by this route, whereas others show little or no susceptibility. Now with the possible exception of poisons penetrating the wall of the mid-gut, all insecticides, whether classed as 'stomach', 'respiratory' or 'contact', must pass through a layer of cuticle before they can reach those centres of the body upon which they exert their lethal effect. Yet it is well known that insecticides such as derris dust may act far more quickly as respiratory poisons than as contact poisons (Webb, 1945*a*). This may be due partly to the thinness of the internal cuticle of the gut and tracheae in comparison with the external cuticle of the body surface, but evidently not entirely, for some insects with excessively delicate external cuticles are well able to withstand the effects of certain contact insecticides such as rotenone. It seems almost certain then that the relative impermeability to insecticides of external as compared with internal cuticle is due to the main difference between these cuticles, namely, that external cuticle is endowed with waterproof qualities in its outer layers, which are lacking in an internal cuticle. This requires no further proof than the fact that insects with abraded cuticle or with the outer layer removed by detergents show increased susceptibility to insecticides (Wigglesworth, 1944, 1945), and the realization that, by their very functions in respiration and water absorption, the cuticle lining both the tracheae and the gut must be freely permeable to water and cannot, therefore, be waterproof. Now it has been shown, in particular by Ramsay (1935), Wigglesworth (1945) and Beament (1945), that the waterproof property of insect cuticle resides in the outermost layers of the epicuticle, and is due to a layer of oil, or more usually wax, molecules of which are highly orientated at right angles to the surface of the cuticle. It is, in fact, this orientation of the wax molecules, enabling them to be tightly packed together, that is responsible for the greater part of the impermeability of cuticle to water since, if the orientation is destroyed by heating, by the solvent action of some vapours of organic liquids such as chloroform or by detergents, the rate at which the insect loses water increases enormously. It is perhaps surprising that insecticides, which, almost without exception, are predominantly fat-soluble, should be more greatly hindered in their penetration into the insect body by cuticle covered by oil or wax than by one in which such a layer is absent. The examination of the problem of penetration through the outer layers of the cuticle is, therefore, of fundamental importance to the study of insecticide efficiency and is the chief concern of this paper.

II. THE STRUCTURE OF INSECT CUTICLE

Before passing to considerations of penetration of insecticides through insect cuticle, it will not be out of place to consider some of the structural peculiarities of cuticle in so far as they have a bearing on the problem. The bulk of the cuticle in an insect comprises two layers, an inner endocuticle and a central exocuticle. The endocuticle is a soft laminated structure composed of a chitin-protein complex while the exocuticle is a hard, apparently homogeneous or structureless layer of chitin associated with sclerotin. Both these layers are permeable to and are saturated with aqueous body fluid. The outermost layer of the cuticle lying beyond the exocuticle is the epicuticle, a complex structure which, according to Wigglesworth (1947), consists of an inner 'cuticulin layer' probably consisting of polymerized lipoproteins tanned by quinones, followed by a layer rich in polyphenols upon which the innermost wax molecules are adsorbed and orientated to form the 'wax layer' (see Fig. 1). In some, but not all insects, the 'wax layer' is covered by an outermost 'cement layer' (Wigglesworth, 1947; Webb, 1947). In *Rhodnius* and in the biting louse, *Eomenacanthus stramineus* (see Webb, 1947), at least, pore canals pass from the epidermal cells underlying the cuticle and penetrate the inner layers of the epicuticle or even as far as the wax layer and are filled with cytoplasmic contents. Dennell (1946), however, working with the larval cuticle of *Sarcophaga falculata*, found that only the outer zone of the endocuticle is penetrated by pore canals which stop short at the epicuticle, and that the inner region of the endocuticle secreted later in larval life contains no pore canals (there is no exocuticle in *S. falculata*). Furthermore, although the

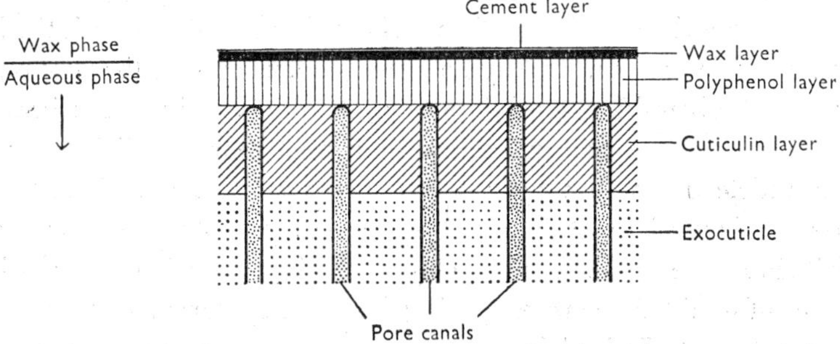

Fig. 1. Diagram showing the structure of the epicuticle in *Rhodnius prolixus* (compiled from Wigglesworth, 1947).

pore canals at first contain cytoplasmic filaments, these are later replaced by chitin. There appears, therefore, to be considerable variation in the structure of insect cuticle which might well have an important bearing on the problem of permeability. For instance, the presence of a 'cement layer', which is certainly absent in some insects, would be expected to impede penetration, and the existence of pore canals with cytoplasmic contents in which, in all probability, streaming would take place should facilitate transport of insecticidal material through the bulk of the cuticle once penetration of the outer layers had been effected. Furthermore, differences in the wax layer may also be important. It would be expected that the soft grease melting at just above 30° C., with which the cuticle of the cockroach is covered, would not provide so effective a barrier against the entry of insecticides as, for instance, the hard waxes melting at temperatures between 50 and 70° C. which coat the cuticles of *Rhodnius*, *Tenebrio* larvae or a *Pieris* pupa (Wigglesworth, 1945; Beament, 1945). The degree of natural abrasion of the cuticle, too, should be important, since insecticides are able to pass more easily through abraded areas of cuticle than through those in which the protective wax layer remains intact, hence, all other factors being equal, insects living in soil where the cuticle is continually and extensively abraded should show greater susceptibility to contact poisons than those in environments above ground. Clearly the variation in insect cuticle structure alone is sufficient to be responsible for most of the differences in susceptibility which even allied insects may show to any one contact insecticide.

III. THE CUTICLE AS A TWO-PHASE SYSTEM

If the wax on the cuticle of an insect is removed by abrasion with a hard dust such as carborundum or alumina, or by solvent action, as can be demonstrated by the dark staining of the exposed underlying polyphenol layer with ammoniacal silver hydroxide (Wigglesworth, 1945), then water evaporates from the abraded surface and continues to do so until the insect is completely desiccated. It may be concluded, therefore, that not only do the aqueous body fluids extend throughout the body and must be in contact with the base of the wax layer, but also they must be free to diffuse throughout the cuticle up to that layer. It may be assumed, therefore, that the cuticle of an insect is essentially a two-phase system of wax and water with a definite interface between the two layers. This conception does not imply that paths of lipoid elements extending from the epidermal cells to the wax layer do not exist. Indeed it seems certain that they are present at least at some time during an insect's life. For instance, where the pore canals contain cytoplasmic contents then a lipoid-containing plasma membrane will extend at least as far as the cuticulin layer of the epicuticle if not as far as the base of the wax. Wigglesworth (1945, 1947) has shown that, even in the adult, repair of damaged wax layer takes place through the bulk of the cuticle and not as an outpouring of wax from dermal glands. Thus it must be presupposed that channels are present throughout the cuticle along which such hydrophobe substances could pass. Such a simple concept of insect cuticle as a wax/water system with the water phase penetrated by strands of lipoid or lipophile elements may not be strictly accurate but is of great assistance in the understanding of the phenomena involved in the passage of insecticides through the cuticle.

IV. PHASE RELATIONSHIPS

It is a well-known physical phenomenon that in a system of two or more phases, such as a layer of oil upon water, a substance dissolved in one of the phases will diffuse into the other, the concentration in the second phase rising until it is in equilibrium with that in the first. The point at which equilibrium is attained depends on the solubility of the dissolved substance in each phase and is reached when the ratio between the concentration of dissolved substance and the concentration at saturation point is the same in both phases. In other words, equilibrium is established when the percentage saturation of dissolved substance is equal in both phases. The ratio between the solubility of the dissolved substance in each phase, therefore, is its partition coefficient between those phases, and is a measure of the degree to which the dissolved substance will be distributed between the phases.

Ferguson (1939) pointed out that, wherever substances are divided between two or more phases, equilibrium is governed by the same thermodynamic equations whatever the phases or the substances distributed and, conversely, that where equilibrium is established the thermodynamic activity or chemical potential or percentage saturation (these terms may be considered synonymous for present purposes) must be the same in all phases partaking in the equilibrium.

Ferguson maintained that '...in very many cases of physiological action, an equilibrium must exist between the concentration of the active substance in the circumambient medium or in the body fluids of the organism and its concentration in the particular cells or structures directly affected by it. It follows, therefore, that in those cases the measured equitoxic concentrations must be markedly influenced by the phase distribution relationship of the various toxic substances. In other words, the measured toxic concentration, though usually regarded as an index of toxicity, is in reality a function of the intrinsic toxicity of the substance and of its distribution equilibrium.' He then showed that, if instead of measured toxic concentrations, which for any one organism may vary very widely from substance to substance, the chemical potentials of the substances are substituted, the toxic indices so determined for that organism lie within a relatively narrow range of values. Furthermore, the actual chemical potential of a toxic substance at the seat of action is known, since, if true equilibrium is established, this is the same as the chemical potential in the external phase.

In support of this, Ferguson submits calculations based on a variety of published data. For instance, he plots the number of carbon atoms of the alcohols methyl-octyl first against the concentrations needed to reduce the surface tension of water to 50 dynes/cm., secondly against the vapour pressure at 25°, and thirdly against the partition coefficient between water and cotton-seed oil, showing that all these lines are approximately parallel and that the thermodynamic postulate is true (see Fig. 2). Next he plots the concentrations required for equal toxicity against *B. typhosus* and obtains a line approximately parallel to the others, thus showing that toxicity here is correlated with some common denominator of the three physical properties, presumably thermodynamic activity. In some of the data given by Ferguson, although the activities calculated for corresponding concentrations, that is, for equitoxic doses, lie within a relatively narrow range, they are not quite constant. This variation he ascribed to the effect of chemical constitution.

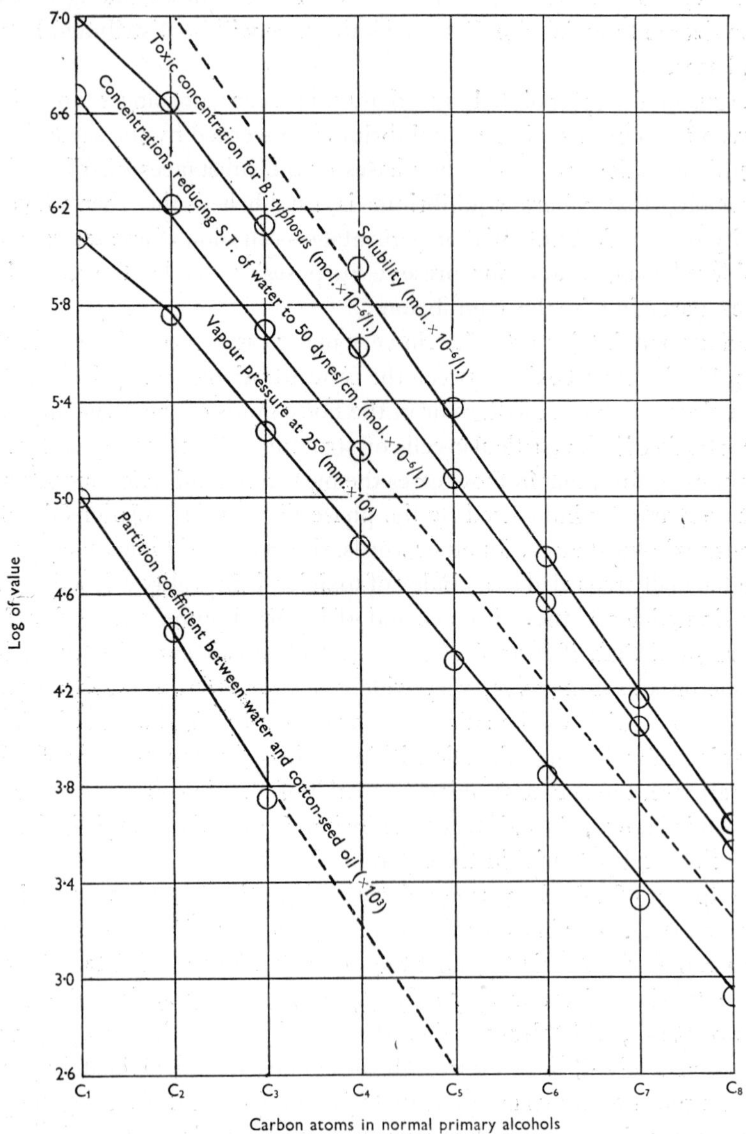

Fig. 2. Graph showing the properties of normal primary alcohols (after Ferguson).

An important modification of the Ferguson theory was made by Burtt (1945) working on the mode of action of sheep dips against the sheep tick, *Ixodes ricinus*. Burtt tested the effect of suspending ticks in baths of toxic solutions using chemical potentials rather than actual concentrations to compare their toxicities. He found that, if an estimate of toxicity is to be made, it is essential to know not only the concentration of the toxic agent in the solution but also its solubility, equitoxic solutions of any one toxic substance in different solvents being those in which the chemical potential or percentage saturation and not the actual concentration was the same. He pointed out, however, that if the percentage saturation were the only factor involved then all saturated solutions of a substance should be equally toxic even if the solvent were one in which that substance is almost insoluble. Obviously this cannot be true, since insoluble substances are without effect. Using the analogy that chemical potential may be compared with electric potential, he suggested that the missing factor may be compared with capacity. When the tick is suspended in a liquid containing dissolved toxic substance, some of the solute passes into its body and causes a lowering of concentration and hence of potential in the solution surrounding the tick. The degree to which chemical potential will fall will depend on the amount of toxic substance initially present in solution. Thus, where a solvent is used in which the toxic substance is sparingly soluble, the amount diffusing into the tick would greatly reduce the chemical potential of that substance in the liquid in the immediate neighbourhood of the tick's body. On the other hand, with a solvent in which that substance is very soluble and in which the initial chemical potential is the same, the same amount of dissolved substance diffusing into the tick would reduce the potential only slightly. Thus with high solubility of toxic substance in a solvent, although the initial chemical potential of a saturated solution is the same as a saturated solution of that substance in a solvent in which it has low solubility, the ability to maintain that potential against loss due to diffusion is proportionately greater. It is desirable then that solvents should be used in which the toxic substance is neither so soluble that an excessive amount is required to produce high chemical potential nor yet comparatively insoluble when loss due to diffusion into the tick lowers potential to the point where the toxic substance is no longer effective.

The concept postulated by Ferguson and Burtt is theoretically sound, and in certain cases at least is the principal factor governing the intensity of toxic effect. It undoubtedly correlates the major factors governing distribution of dissolved substance between two or more immiscible phases. Clearly then phase distribution relationships, although not the cause of toxicity, are so intimately bound up with some toxic effects, particularly in the case of those substances exerting their physiological effect by a physical mechanism rather than a chemical reaction (Ferguson gives criteria whereby these two types of toxic action may be recognized), that knowledge of this problem must be an integral part of a better understanding of the processes involved in cuticle permeability.

V. THE EFFECTS OF SOLVENTS ON PERMEABILITY

Using the sheep ked, *Melophagus ovinus*, as a test insect, and applying to the cuticle insecticidal dusts containing 0·25 % diphenylamine as the toxic agent, Webb & Green (1945) found that the addition of 1 % of certain organic liquids to the powder greatly reduced the time taken for the ked to die as judged by the cessation of heart beat. Furthermore, this increase in the rate of action of the insecticide was an example of synergy, for the effect was far greater than either that of the insecticide or the solvent applied separately or the sum of their effects when applied together. The ability of the solvent to induce such an increase in rate of action was presumably due to an enhanced rate of penetration of the insecticide through the cuticle and was termed the 'carrier efficiency' of the solvent. The carrier efficiency of a number of arbitrarily chosen organic liquids was measured and the results are given in Table 1 and in Fig. 3. It was found that whereas liquids

Table 1. In vitro *experiments with diphenylamine*

Solvent	Time of death of keds in hours		Time in hours at which keds became immobile after treatment with solvent control
	Solvent+ insecticide	Solvent control	
o-Cresol	1·5	10	1·0
m-Cresol	1·5	9	1·0
Xylenol	1·5	9	1·0
Benzyl alcohol	2·5	21	2·0
p-Cresol	2·5	9	1·0
Octyl alcohol	4	30	2·0
4-Methylcyclohexanol	5	30	2·0
Quinoline	6	27	2·0
Cyclohexanone	8	26	2·0
Diacetone alcohol	13	—	—
Cyclohexanol	17	21	2·0
Acetophenone	22	28	1·0
Benzonitrile	23	—	6·0
Aniline	24	—	1·0
Carbitol	24	26+	20
Dimethylaniline	27	—	1·0
Methyl benzoate	27	27	2·0
Castor oil	26+	—	—
Anisole	27+	—	—
o-Dichlorobenzene	21	19	3·0

N.B. Keds dusted with 0·25 % diphenylamine die in 25–30 hr.

such as the cresols, xylenol and benzyl alcohol showed high carrier efficiency, others such as quinoline, aniline and castor oil showed medium, low or no carrier efficiency respectively. It was thought at first that part of the effect might be due to the greater area of contact with the cuticle obtaining when the diphenylamine is dissolved in a solvent as compared with its particulate condition in those powders to which no solvent had been added. Clearly

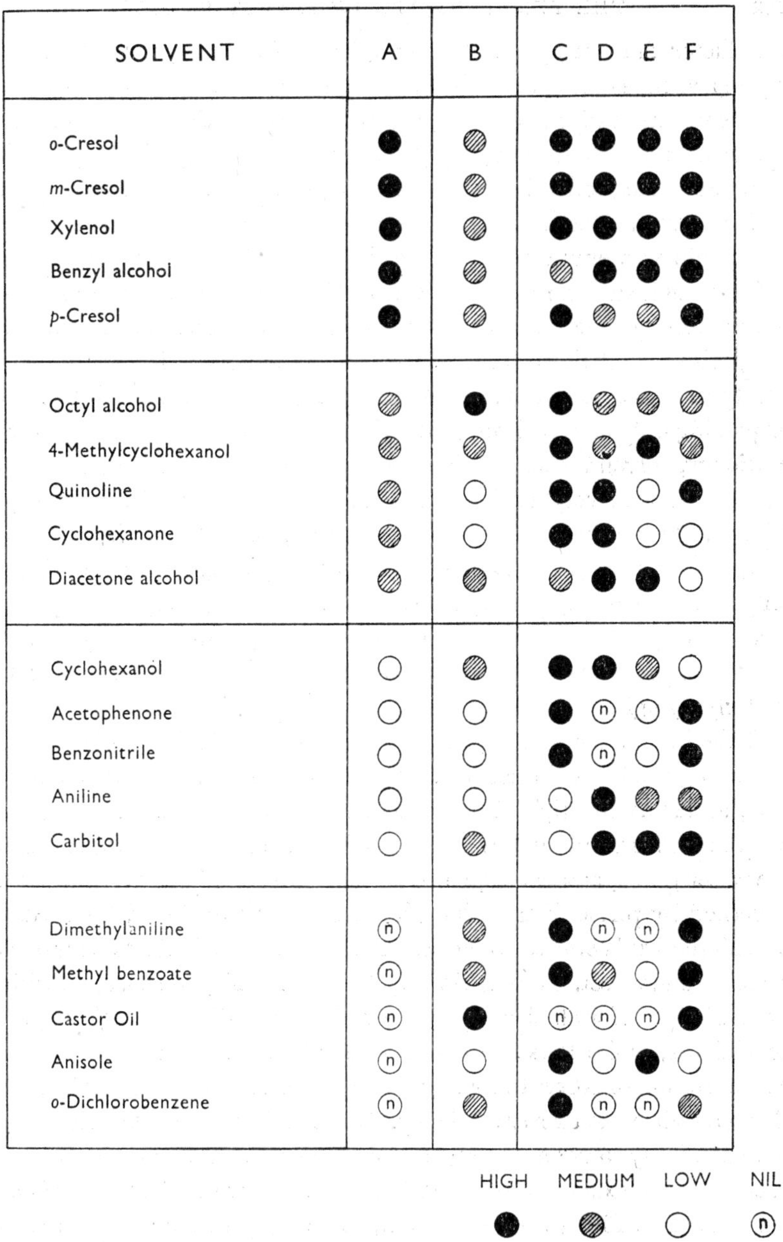

SOLVENT	A	B	C	D	E	F
o-Cresol	●	◍	●	●	●	●
m-Cresol	●	◍	●	●	●	●
Xylenol	●	◍	●	●	●	●
Benzyl alcohol	●	◍	◍	●	●	●
p-Cresol	●	◍	●	◍	◍	●
Octyl alcohol	◍	●	●	◍	◍	◍
4-Methylcyclohexanol	◍	◍	●	◍	●	◍
Quinoline	◍	○	●	●	○	●
Cyclohexanone	◍	○	●	●	○	○
Diacetone alcohol	◍	◍	◍	●	●	○
Cyclohexanol	○	◍	●	●	◍	○
Acetophenone	○	○	●	ⓝ	○	●
Benzonitrile	○	○	●	ⓝ	○	●
Aniline	○	○	○	●	◍	◍
Carbitol	○	◍	○	●	●	●
Dimethylaniline	ⓝ	◍	●	ⓝ	ⓝ	●
Methyl benzoate	ⓝ	◍	●	◍	○	●
Castor Oil	ⓝ	●	ⓝ	ⓝ	ⓝ	●
Anisole	ⓝ	○	●	○	●	○
o-Dichlorobenzene	ⓝ	◍	●	ⓝ	ⓝ	◍

		HIGH ●	MEDIUM ◍	LOW ○	NIL ⓝ
A	Carrier efficiency (time of death of keds) in hours	< 2·5	2·5–15	15–25	25–30
B	Initial percentage saturation with diphenylamine	> 50	50–10	< 10	
C	Rate of penetration through beeswax in minutes	< 30	30–60	> 60	∞
D	Partition coefficient between beeswax and water	> 0·005	0·005–1	< 0·001	nil
E	Solubility (mg./100 c.c.) of diphenylamine in solution of the solvent in water	> 6	6–4	< 4	nil
F	Boiling point °C. (volatility)	> 190	190–170	< 170	nil

Fig. 3. Diagram illustrating the physical properties essential in liquids showing carrier efficiency.

this factor did not operate to any appreciable extent, for the addition of liquids such as dimethylaniline, methyl benzoate and anisole, which are good solvents of diphenylamine, did not reduce the time of kill of the insects below that obtained with 0·25% particulate diphenylamine (see Fig. 3). The next factor considered was the ability of the added solvent to penetrate an insect wax. Here a series of artificial membranes of beeswax supported on muslin was employed, and the time taken for each of the solvents to pass through such a membrane was assumed to be a measure of its ability to pass through the wax on the surface of insect cuticle. The results showed that, whereas all those solvents with high carrier efficiency possessed the ability to pass through the wax membrane quickly and those unable to pass or passing only slowly through the membrane possessed little or no carrier efficiency, solvents such as dimethylaniline, anisole and *o*-dichlorobenzene penetrated wax very rapidly and still showed no carrier efficiency (see Fig. 3). From this we may conclude, first, that the ability to penetrate wax, though essential to a solvent showing high carrier efficiency, is not the only factor involved, and secondly, that, however diphenylamine penetrates insect cuticle, it does not do so preferentially through any strands of lipophylic elements which may extend from the outer wax layer to the underlying epidermal cells.

If the insecticide diphenylamine does not pass via lipophilic elements in traversing the cuticle then it must pass through the bulk of the cuticle at least as far as the tips of the pore canals where these have cytoplasmic contents. This would involve passage of insecticide from the wax phase into the water phase, and consequently its concentration in each phase would be governed by phase distribution phenomena. Accordingly, the next factor considered by Webb & Green was the partition coefficients of the various solvents under test. Here again they found that all those solvents with high or medium carrier efficiency possessed high or medium water/wax partition coefficients, while those with a low partition coefficient, as measured by concentration of solvent in water over concentration of solvent in beeswax, showed little or no carrier efficiency. Nevertheless, some solvents possessing both an ability to penetrate wax and a high or medium partition coefficient, such as cyclohexanol and methyl benzoate, still showed little or no carrier efficiency (see Fig. 3). Clearly, then, there are properties other than these two which are essential in a solvent of high carrier efficiency.

Had it been merely a question of the permeability of insect cuticle to the solvent then it would have been expected that the partition coefficient between wax and water would govern the rate of penetration, but here it was the rate of penetration of insecticide dissolved in the solvent and not that of the solvent alone which was measured by the period of time taken for the insect to die. It was argued then that the missing factor or factors would be expected to apply to the partition of the insecticide between the wax phase and the water phase and the influence on this of the dissolved solvent. Now diphenylamine was known to be very soluble in all the solvents used, at least to the extent that 0·25% diphenylamine would be completely in solution in 1% of any of the solvents at 30° C., the temperature at which the experiments were carried out. Thus as diphenylamine is predominantly oil-soluble it would be expected to be equally soluble in the mixture of wax and solvent which would form the oil phase on top of the cuticle after the dust has been applied. The solubility of diphenylamine in water, on the other hand, is very low, being only 5·5 mg./100 c.c. at 30° C. The partition coefficient of diphenylamine between the wax plus solvent and water phases, therefore, would also be very low, so low in fact that it is surprising that the evidence suggests that this insecticide normally penetrates the bulk of the aqueous part of the cuticle in quantities sufficiently large to prove toxic to the insect. It seemed probable, therefore, that the effect of solvents with high carrier efficiency lay in their ability to increase the solubility of diphenylamine in the aqueous phase, thereby raising its partition coefficient and proportionately increasing the rate of passage into that phase. In view of this supposition, solutions of each solvent in distilled water were made at 80, 60, 40 and 20% saturation, and the solubility of diphenylamine in these solutions was measured for comparison with its solubility in water. It was then found that, whereas in some cases the presence of the solvent in the water made little or no difference to the solubility of diphenylamine, in others, notably those with high carrier efficiency, the solubility of the insecticide was almost doubled, while in yet a further group, including methyl benzoate and cyclohexanone, solubility was depressed to a figure almost one-third of that in water (see Fig. 3). Here, then, was a means whereby the presence of a solvent might considerably alter the partition coefficient of diphenylamine between wax + solvent and water. In comparing this effect with the data already obtained on penetration through beeswax and the partition coefficient for each solvent tested, it was found that, with the exception of diacetone alcohol and cyclohexanol, the factors examined accounted for the degree of carrier efficiency exhibited by each solvent, those with low or no carrier efficiency being deficient in some respect while those with high carrier efficiency were positive for each factor. The apparently anomalous position of diacetone alcohol and cyclohexanol with lower carrier efficiencies than might have been expected was accounted for by the comparatively high volatility of those solvents (see Fig. 3). This was shown to be the case by carrying out tests in the vapours of the solvents, and by reducing evaporation of solvent from the powder coating the insect, thus raising their carrier efficiency to a figure in keeping with the remainder of the series. A solvent with high volatility evaporating from the surface of the cuticle would lose carrier efficiency once its volume fell below that necessary to maintain the insecticide in solution.

From this analysis of the problem it was possible to outline the physical properties which should be possessed by any liquid with high carrier efficiency. They are as follows:

(1) The insecticide must be soluble in the liquid.

(2) The liquid must penetrate the wax layer of the epicuticle comparatively rapidly.

(3) The liquid must possess a high partition coefficient between water and wax, i.e. the fraction, solubility in water/solubility in wax, should be greater than 0·005.

(4) The insecticide should be appreciably more soluble in a solution of the liquid in water than in water alone.

(5) Where the liquid is applied to the insect under conditions favouring rapid evaporation (e.g. spread in small quantities upon a dust as against use in bulk solutions), the liquid should be comparatively non-volatile.

VI. PENETRATION BY THREE-DIMENSIONAL DIFFUSION

Webb & Green suggested a mechanism whereby liquids with the above physical properties might increase the rate of penetration of insecticides through cuticle. It was not clear then, however, and is not so now, whether the passage of the insecticide through the aqueous phase of the cuticle involved diffusion across the entire width of the exocuticle and endocuticle before contact was made with living cells, or whether only the very narrow region from the base of the wax layer to the tips of the pore canals in the cuticulin layer of the epicuticle was involved. In the latter case transport of insecticide across the greater part of the cuticle might well be greatly facilitated by streaming of cytoplasm within the pore canals. Perhaps both conditions obtain in different insects according to the presence or absence of cytoplasm in the pore canals. In either case the mechanism suggested remains essentially the same. First, the solvent with dissolved insecticide either dissolves the epicuticular wax or, disorientating the molecules, passes through the greatly increased intermolecular spaces to the interface between the wax layer and the underlying aqueous phase. Here the solvent, by virtue of its solubility in water (high partition coefficient), passes on into the aqueous phase. The immediate result of this is to reduce the volume of solvent in the wax-solvent phase, thereby increasing the percentage saturation of the insecticide in that phase. Thus, at whatever figure that percentage saturation should stand initially, there would always be a tendency for this figure to increase, provided that the insecticide is less soluble in water than the solvent. With increasing percentage saturation of the insecticide in the wax-solvent phase the rate of passage of the insecticide across the interface will also increase. Further, as the presence of solvent in the aqueous phase raises the solubility of the insecticide in that phase, the partition coefficient of the insecticide will also rise causing a still further increase in rate of passage across the interface. The process, therefore, is analogous to an injection of the insecticide through the wax-water interface. There is every reason to suppose that the percentage saturation of solvent in the aqueous phase in the immediate neighbourhood of the interface rises to a comparatively high figure because the water permeating the cuticle is free from convection currents tending to disturb the equilibrium.

If this interpretation of the facts is correct, then the mechanism involved whereby a liquid increases the rate of passage of dissolved insecticide across the wax/water interface is purely physical and should obtain with any liquid or mixture of liquids possessing the requisite properties. Webb & Green pointed out that some of the solvents they had used, which showed little or no carrier efficiency because they were deficient in one or other of the properties essential for high carrier efficiency, could be mixed so that the mixture lacked none of these properties. A number of such mixtures were tested, notably carbitol + methyl benzoate and aniline + dimethyl-aniline, and carrier efficiency above that of either constituent was obtained in almost all cases. Where mixtures such as aniline + benzonitrile were used and no increased carrier efficiency resulted, the failure was found to be due, at least in part, to a much lower rate of penetration of the mixture through beeswax than had been anticipated from the physical properties of the constituents. Similarly, if the physical explanation is true, the mechanism should be capable of extension to insecticides other than diphenylamine. Accordingly, tests were carried out with three other contact insecticides, namely, dixanthogen, ω-nitrostyrene dibromide and rotenone, incorporating a selection of solvents known to show high, medium or low carrier efficiency with diphenylamine. It was found in every case that those solvents possessing high carrier efficiency with diphenylamine behaved in a similar manner with the other three insecticides. There is, therefore, good evidence to suggest that the mechanism proposed is substantially correct. We may infer then that insecticides of the diphenylamine type normally pass directly across the cuticle by diffusion either as far as the ends of the pore canals or to the epidermal cells beneath the endocuticle, and that this process can be greatly speeded up by purely physical means, namely, the addition of a suitable solvent. The most important aspects of this concept, so far as formulation of new insecticides is concerned, lies in the realization, first, that the ability of the insecticide to dissolve in water is as important if not more so than its solubility in oils and fats, and secondly, that insecticides such as rotenone and dixanthogen are capable of exerting their toxic effects when present in extremely small quantities. With regard to this last point it should be noted that rotenone and dixanthogen are only soluble in water to the extent of *c.* 0·1 and *c.* 0·25 mg./100 c.c. respectively. This means that the tendency for these insecticides to remain in an oil phase would be so great that only extremely minute quantities could pass into the aqueous phase.

It is worth mentioning that, although the Ferguson concept of phase distribution evidently applies to solutions of insecticides and liquids with carrier efficiency, it is impossible to estimate the *effective* chemical potential of the insecticide in that liquid at the time of equilibrium. Clearly the initial chemical potentials or percentage saturations of diphenylamine in the liquids tested by Webb & Green bore no relationship to the toxic effects of the mixtures (see Fig. 3, A, B), since both the liquid and the insecticide were passing into the insect at different rates. Burtt has already added a rider to the Ferguson theory covering the depletion of insecticide from bulk solutions in which the insecticide is sparingly soluble. A further rider should be added covering reduction of volume of solvent, where this is initially small, by passage into the insect body. Thus, in a case such as a small quantity of insecticidal solution spread upon a powder, the chemical potential in the circumambient phase at the time of equilibrium, if this is ever reached, cannot be measured and cannot be used as a toxic index.

A number of experiments showing increase in permeability of insect cuticle to alcohols, water and insecticides when insects are immersed in mixtures of a fat solvent such as kerosene + ethyl alcohol or when they are immersed in light or heavy oils, has been described by Hurst (1940, 1941 and 1945) and Wigglesworth (1942). For instance, when blowfly larvae are immersed in equal parts of kerosene and ethyl alcohol, penetration of alcohol into the insect and passage of water out through the cuticle are very rapid, whereas larvae immersed in either ethyl alcohol or kerosene alone are relatively unaffected. Hurst (1945) attributes this increase in permeability of cuticle to the participation of kerosene in the spatial arrangement of the epicuticle causing 'swelling and increase in phase volume of the lipophilic radially arranged aggregates'. This he holds is accompanied by displacement of lipid from the general fabric of the epicuticle; thus the lipophilic elements become more permeable to fat solvents such as ethyl alcohol, and the remainder of the cuticle, owing to removal of lipid, more permeable to water. Although some such changes may take place the mode of penetration will be essentially the same as that outlined by Webb & Green, namely, a three-dimensional diffusion through oil and water phases, and will be governed by the Ferguson-Burtt concept of phase-distribution relationships. The difference between Hurst's (1945) theory and that of Webb & Green lies in the suggestion by Hurst that fat solvents such as ethyl alcohol penetrate mainly through the lipophilic elements. Webb & Green hold that ethyl alcohol passes through the aqueous phase after the lipophilic elements have been rendered permeable to it by kerosene.

The possibility of mixing solvents to give a liquid possessing requisite properties for carrier efficiency was utilised in an attempt to improve existing media for fly sprays. As the usual base for fly sprays is kerosene, a liquid capable of penetrating cuticular wax readily but almost insoluble in water, it was thought probable that, by adding a suitable liquid providing the remaining physical properties, the rate of penetration of the toxic agent used in the spray would be increased and an improved rate of knockdown of flies would result. Among other liquids tested, benzyl alcohol was found to possess the correct physical properties and, moreover, was only 5% soluble in fly-spray kerosene, an important point, for the percentage saturation of benzyl alcohol in kerosene must be high if it is to leave the kerosene readily and enter the aqueous phase of the cuticle. Thus a fly spray consisting of kerosene saturated with benzyl alcohol and using D.D.T. as toxic agent was found to increase the rate of knockdown of *Musca domestica* very considerably in comparison with sprays lacking either benzyl alcohol or D.D.T. In spite of this success, however, the rate of knockdown of flies with a kerosene-benzyl alcohol-D.D.T. spray was not nearly so rapid as that obtained with pyrethrins in kerosene. Furthermore, the addition of benzyl alcohol to a pyrethrin-kerosene spray did not appreciably improve its rate of knockdown. It appears, therefore, that pyrethrins are outside the scope of action of liquids with carrier ability, and, as carrier efficiency seems to depend on a physical mechanism, it can only be concluded that pyrethrum is an insecticide in a separate class penetrating cuticle in a manner different from that of diphenylamine.

VII. PENETRATION BY INTERFACIAL DIFFUSION

The failure of pyrethrins to show increased rate of penetration through cuticle in the presence of solvents with high carrier efficiency was held to suggest that transport through the cuticle in that instance was not governed by phase-distribution relationships of the type envisaged for diphenylamine penetration. Apparently anomalous results of a similar kind were obtained by Hurst (1943*a*, *b*). Hurst examined the effect of homologous series of normal primary alcohols and fatty acids on blowfly larvae (*Phormia terraenovae*) both by external application in a bulk of liquid considerably greater in volume than that of the larva and by injection into the body. He found that, when the drugs were injected, as the chain length of carbon atoms in the alcohol increased from C_1 to C_5, so the toxicity of the drug increased, the molecular activity being roughly proportional to chain length (see Fig. 4). Now with injections of fatty acids, as the series proceeds from C_1 to C_5, there is little difference in toxicity between the members, showing that the isoactive molar concentrations remain relatively constant. With blowfly larvae immersed in the drug, however, the alcohol series shows an increase in toxicity, as measured by survival time, from chain length C_1 to

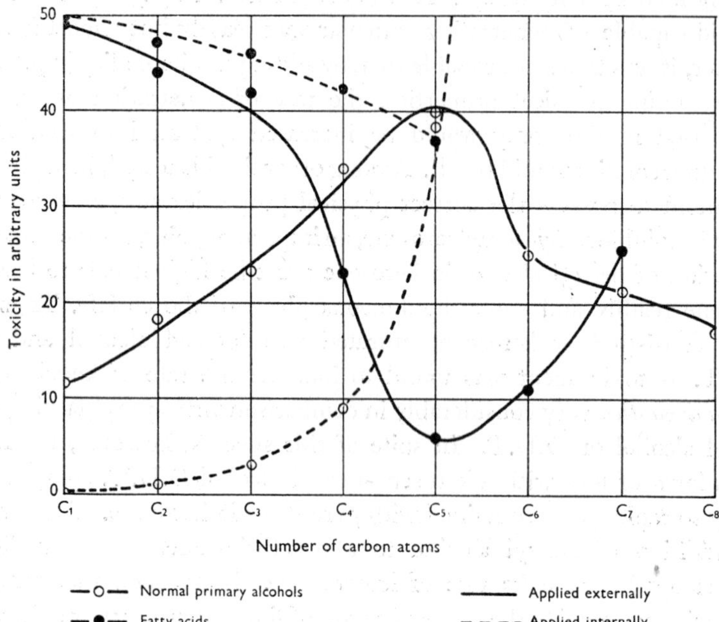

Fig. 4. Graph showing the effect of homologous series of normal primary alcohols and fatty acids on blowfly larvae (*Phormia terraenovae*) after external application and after injection (after Hurst).

C_5 and thereafter from C_5 to C_8, a corresponding decrease. On the other hand, the fatty acids showed an enormous decrease in toxicity from C_1 to C_5 and thereafter from C_5 to C_7, a corresponding increase. From a comparison of these two sets of results from internal and external drug application, it is evident that the disparity between them is due to the relative permeability of the cuticle not only to members within each series but also to alcohols and fatty acids as a whole. Hurst noted that, in homologous series, physical properties such as differential oil/water solubility, capillary activity, viscosity and vapour pressure change uniformly and in the same direction as the series are ascended. He omitted to mention, however, that, in penetration of a two-phase system of wax/water, such as appears to obtain in insect cuticle, only those members of the series capable not only of penetrating the wax but also of leaving the wax and entering the underlying aqueous medium would be expected to pass through the cuticle. Thus, whatever the intrinsic toxicity of members of the series, only those penetrating the cuticle would reach the site of toxic action and be capable of exerting their effect. Now with differential oil/water solubility changing uniformly as the series is ascended, only those members with a *partition coefficient* between wax and water approaching unity would be expected to penetrate the cuticle if phase-distribution relationships held. This condition obtains in the alcohol series at about the C_5 alcohol, which showed greatest toxicity with external application but not with internal application. It seems, therefore, that the alcohol series obeys the Ferguson-Burtt concept of phase distribution, and the phenomena observed by Hurst after both internal and external application of the members of that series are readily explicable by this means. The fact that similar results were not obtained with the fatty acid series is important, for it suggests that here penetration depends on factors outside those governing phase distribution.

In considering this new method of penetration it is advisable first to examine the type of cuticle with which Hurst was working. Hurst provided evidence to show that the surface of the cuticle in blowfly larvae consists of a mosaic of heterogeneous protein and lipoid associations, while the bulk of the cuticle comprises a chitin-protein framework permeated by fatty secretion from the epidermal cells. Thus, although the lipoid elements appear to be predominantly in the outer epicuticle, the entire cuticle consists of a complex system of heterogeneous phases of lipophilic and hydrophilic elements. This picture of cuticle structure may not be that found in all insects. Indeed, the disparity between the accounts of cuticle structure given by Wigglesworth (1947) for *Rhodnius* and Dennell (1946) for *Sarcophaga* larvae leads to the assumption that insect cuticle may differ considerably from one order of insects to another and possibly even between different instars of a single insect, particularly in the Endopterygota.

Now it is well known that certain substances insoluble in water, but with a molecule, part of which has an affinity for water such as a hydroxyl group, form a monomolecular layer when spread on a water surface. The molecules in such a layer become highly orientated with the water-soluble portion projecting at right angles to the water surface. The molecules in a cholesterol film on water behave in this manner. In such films cross-linkages between the molecules of varying degrees of strength may also be formed. Rideal (1945) and his co-workers have shown that, if a very thin film is made on water and a third substance is injected into the water beneath that film, one of three effects may be observed. First, the injected substance may have no effect on the original film; secondly, it may replace the original film and itself form a layer on the surface; or, thirdly, it may enter the original film and form a mixed film. Changes such as those envisaged in the second and third possibilities are detected by changes in the strength of the surface film. In the formation of mixed films the added substance must possess a molecule with (*a*) a water-soluble group capable of linkage with the water-soluble group of the molecule forming the original film, and (*b*) a water-insoluble group highly compatible with the water-insoluble groups in the original film. This concept may be extended from monomolecular layers on water to the interface between such immiscible phases as paraffin or wax and water. Here, the third substance becomes orientated with the water-soluble portion of the molecule lying in the aqueous phase and the oil-soluble portion in the oil phase. Furthermore, diffusion of such orientated molecules along the interface can take place and may, in certain circumstances, be extremely rapid.

Considering the results obtained by Hurst (1943*a*, *b*) with the fatty acid series on blowfly larvae, in the light both of his view of the structure of the cuticle in that larva and also interfacial diffusion, it is possible to formulate a theory of cuticle penetration without involving phase-distribution relationships in the sense already applied. According to Cockbain & Schulman (1939), fatty acids are capable of penetrating and participating in synthetic monomolecular amine and protein films because of the strong polar interaction of fatty acids with proteins. Hurst's account of cuticle structure provides for a network of protein/lipoid interfaces at which fatty acid molecules could be orientated extending throughout the cuticle. He visualizes a competition for the fatty acid molecules between the protein and lipoid phases of the cuticle. With fatty acids of short-chain length the attraction to the protein phase of the polar group of the molecule is the stronger, whereas with fatty acids of long-chain length attraction of the non-polar group to the lipoid phase predominates. In either case rapid diffusion can take place along the interface (Gibbs' layer), on the one hand chiefly in the protein phase and, on the other, chiefly in the lipoid phase. At a chain length of C_5, however, a balance obtains between the opposing attractions of protein and lipoid phases, and a stable lipoprotein-fatty acid association is formed in which the fatty acid molecules become anchored at the interface and are not free to diffuse to any extent. Thus, unlike the alcohol series, in the fatty acid series C_1–C_7, toxicity of drugs applied externally is greatest at each end of the series, falling to a minimum at C_5. This is because ability to diffuse along lipoid/protein interfaces, and hence to penetrate the cuticle, is at a minimum at C_5 and tends to a maximum at C_1 and C_7. Here then is a method of penetration through cuticle distinct from the three-dimensional diffusion across the bulk of the cuticle obtaining with diphenylamine and other similar insecticides and presumably also with the alcohol series. Hurst (1943 *a*, *b*) also shows that pyrethrins behave in much the same manner as fatty acids and are capable of rapid diffusion along two-dimensional boundary layers of fat and water, in fact, so rapid is penetration in this instance, that the speed of action of pyrethrins cannot be rivalled by insecticides penetrating through the bulk of the cuticle even in the presence of solvents of high carrier efficiency.

It has been suggested that the structure of the cuticle in blowfly larvae may be atypical in some respects. As pyrethrins, however, show rapid toxic effects with a wide variety of insects other than blowfly larvae, it is assumed that the structure of the cuticle in those insects affected by pyrethrins is such as to permit two-dimensional diffusion. There must, according to this theory, be a network of oil/water interfaces extending inwards from the wax layer of the epicuticle at least as far as the tips of the pore canals and probably through the entire bulk of the cuticle in such insects. Penetration of the cuticle, therefore, presumably can take place in at least one or both of two ways in the majority of insects, that is, either by two-dimensional or by three-dimensional diffusion.

Hurst (1943 *a*, *b*) gives a number of factors held to influence two-dimensional diffusion along oil/water interfaces in the cuticle, but, in attempting to fit into the alcohol series his theory, the behaviour of which is here believed to be explained by partition coefficients between wax and water, he tends to reach conclusions some of which may be suspect. Nevertheless, it seems true to say that two-dimensional diffusion of a drug with the appropriate type of molecule will be decreased if the interfacial region is blocked by preferential absorption at the oil/water interfaces of some substances of 'moderately high capillary activity (e.g. cholesterol, cetyl alcohol, decyl alcohol)'. Thus the addition of, say, cetyl alcohol to a drug entering by two-dimensional diffusion reduces the rate of penetration. Hurst further suggests that an increase in two-dimensional diffusion rates occurs in the presence of certain non-polar fat solvents which serves to reduce the viscosity of the lipoid phase. He instances hexane, octane and dodecane as solvents behaving in this manner. Theoretically, too, it must be presumed that any substance reducing the viscosity of the aqueous (protein) phase would increase penetration of insecticides passing chiefly through that phase.

VIII. THE IMPORTANCE OF CUTICLE PERMEABILITY TO INSECTICIDE ACTION

The essence of chemical control of insects lies in the extermination of large numbers of individuals without undue waste either of time and labour or of insecticidal material. A very great variety of substances will kill insects provided that they are applied in sufficiently large quantities under particular conditions, but their use is uneconomical or in other ways impracticable. To be of value as an insecticide a substance must first possess a high degree of toxicity, so that, when broadcast over large areas in a suitably dilute form, an insect in that area will pick up a dose large enough to prove fatal. Secondly, it must be toxic preferentially to arthropods and certainly not to man or domestic animals. Now most toxic substances in contact with living tissue will exert their toxic effect whether the tissue be that of an insect or a mammal. Thus, where a toxic substance shows specificity, it usually does so by virtue of its ability or failure to penetrate the body. As far as contact insecticides are concerned ability to penetrate insect cuticle together with failure to penetrate the body surface of other animals are clearly desirable features. The permeability of the cuticle, therefore, is the largest single factor other than intrinsic toxicity contributing to the efficiency of a contact insecticide. This is well illustrated in Hurst's (1943*a*, *b*) experiment with series of normal primary alcohols and fatty acids. When injected into the body of a blowfly larva the alcohol and the fatty acid, each with a chain length of C_5, prove equally toxic. When they are applied externally to the surface of the cuticle, on the other hand, the fatty acid penetrates with extreme slowness and therefore shows little apparent toxicity, whereas the alcohol penetrates rapidly and proves much more toxic. Thus any estimate of the efficiency of an insecticide should take into account both its intrinsic toxicity and also its mode of entry into the body. As mode of entry into the insect will depend entirely on structural and physiological features, such as type of mouthparts, form of spiracles, method of breathing (whether by diffusion or by mechanical ventilation of the tracheal system) and permeability of cuticle, which in turn may vary from one region of the body to another, it is clear that, within a range of insects, there is ample scope for differences in susceptibility even to one insecticide. The difficulty of correlating such data has led those seeking new insecticides to employ purely empirical methods of trial and error. Nevertheless, the problem is not so complex as it seems, for sufficient is known now to permit formulation of insecticides along broad lines according to the structure of the insect to be controlled.

The type of insecticide employed in any one instance will not always be the same. In some cases, such as fly sprays, it is desirable that the insecticide shall act with the greatest rapidity possible, but in others it matters little how long the insect takes to die. Although it is preferable that insecticides shall penetrate cuticle rapidly, for, in so doing, it is usually possible to obtain a given effect with a smaller amount of material, in choosing an insecticide for a given purpose, rapidity of action is by no means the only consideration. Such matters as cost, availability and residual action may be of greater importance. The real issue at stake is whether an insecticide penetrates cuticle or does not and, within limits, the speed at which penetration is accomplished is usually a secondary matter.

In the foregoing account two separate methods of penetration have been postulated. The first involves three-dimensional diffusion through the bulk of the cuticle and obeys the Ferguson-Burtt concept of phase distribution, while the second, according to Hurst's theory, is a two-dimensional diffusion along oil/water interfaces. All insecticides probably penetrate by the first method to a greater or lesser extent, but only to those whose molecular structure permits orientation at an oil/water interface will the more rapid penetration by the second method be possible. The existence of these two methods of penetration at least seems probable.

REFERENCES

BEAMENT, J. W. L. (1945). *J. Exp. Biol.* **21**, 115.
BURTT, E. T. (1945). *Ann. Appl. Biol.* **32**, 247.
COCKBAIN, E. G. & SCHULMAN, J. H. (1939). *Trans. Faraday Soc.* **35**, 716.
DENNELL, R. (1946). *Proc. Roy. Soc.* B, **133**, 348.
FERGUSON, J. (1939). *Proc. Roy. Soc.* B, **127**, 387.
HURST, H. (1940). *Nature, Lond.*, **145**, 462.
HURST, H. (1941). *Nature, Lond.*, **147**, 388.
HURST, H. (1943*a*). *Nature, Lond.*, **152**, 292.
HURST, H. (1943*b*). *Trans. Faraday Soc.* **39**, 390.
HURST, H. (1945). *Brit. Med. Bull.* **3**, 132.
MEYER, K. H. & HEMMI, H. (1935). *Biochem. Z.* **277**, 39.
OVERTON, C. E. (1900). *Jb. wiss. Bot.* **34**, 669.
OVERTON, C. E. (1901). *Studien über Narkose*, Jena.
RAMSAY, I. A. (1935). *J. Exp. Biol.* **12**, 373.
RIDEAL, E. K. (1945). *Endeavour*, **4**, 83.
TRAUBE, J. (1904). *Pflüg. Arch. ges. Physiol.* **105**, 541.
TRAUBE, J. (1908). *Pflüg. Arch. ges. Physiol.* **123**, 419.
TRAUBE, J. (1909). *Verh. dtsch. phys. Ges.* **10**, 800.
TRAUBE, J. (1924). *Biochem. Z.* **153**, 358.
WEBB, J. E. (1945*a*). *Bull. Ent. Res.* **36**, 15.
WEBB, J. E. (1945*b*). *Proc. Zool. Soc. Lond.* **115**, 218.
WEBB, J. E. (1947). *Parasitology*, **38**, 70.
WEBB, J. E. & GREEN, R. A. (1945). *J. Exp. Biol.* **22**, 8.
WIGGLESWORTH, V. B. (1942). *Bull. Ent. Res.* **33**, 205.
WIGGLESWORTH, V. B. (1944). *Nature, Lond.*, **153**, 493.
WIGGLESWORTH, V. B. (1945). *J. Exp. Biol.* **21**, 97.
WIGGLESWORTH, V. B. (1947). *Proc. Roy. Soc.* B, **134**, 163.

SELECTIVITY OF DRUG ACTION IN PROTOZOAL BLOOD INFECTIONS

By E. M. LOURIE

Warrington Yorke Department of Chemotherapy, Liverpool School of Tropical Medicine

I. INTRODUCTION

This contribution is intended to be no more than a broad background on which to base the discussion of specific problems involved.

Selectivity of drug action is the antithesis of the concept of the 'general protoplasmic poison', a term which modern text-books of pharmacology have carefully left behind in their earlier editions. The main emphasis of modern theory in chemotherapy is on the specific, or selective, nature of the effect produced by an active compound. This is, indeed, a re-emphasis, supported by the knowledge gained in the intervening generation (and especially during the past decade), of a viewpoint clearly conceived, and expounded with characteristic imagery, by Paul Ehrlich, as a result of his studies on the chemotherapy of infections caused by blood Protozoa, namely trypanosomes, these studies having led the way to his similar researches on spirochaetoses. Ehrlich taught that the ideal chemotherapeutic agent must have the minimum degree of 'organotropism', or affinity for the body cells of the host, together with maximum 'parasitotropism', or affinity for the parasite cell. The requirements for this selectivity of drug action he frequently described in terms similar to the following, delivered at a Congress in London in 1913:

'I have explained above that the parasites possess a whole series of chemoreceptors which are specifically different from one another. Now if we can succeed in discovering among them a grouping which has no analogue in the organs of the body, then we should have the possibility of constructing an ideal remedy if we select a haptophoric group which is specially adjusted to the functions of the parasites. A remedy provided with such a haptophoric group would be entirely innocuous in itself, as it is not fixed by the organs; it would, however, strike the parasites with full intensity, and in this sense it would correspond to the immune productions, to the anti-substances discovered by Behring, and which, after the manner of the bewitched balls, fly in search of the enemy.'* In another part of this address, Ehrlich wrote: 'A complete exhaustive knowledge of all the different

* This extract has recently been quoted in an interesting review by Marshall (1947).

chemoreceptors of a certain definite parasite is what I should like to characterize as the "therapeutic physiology of the parasite cell", and this is a *sine qua non* of any successful chemotherapeutic treatment. I should like to emphasize the fact that many observations indicate that certain chemoreceptors are due (*sic*) to several different kinds of parasites, not only to a single one. The knowledge of these is of special practical importance, because remedies which are adjusted to these have a healing influence on a very large series of the most various pathogenic agents. The larger the number of different chemoreceptors, therefore, which can be demonstrated, the greater is the possibility of a successful chemotherapy.'

As a general forecast of modern theory all this is remarkably correct. Almost all that is necessary is to read 'enzymes or their substrates', in the place of 'chemoreceptors'. The restatement of these aims, by Fildes (1940) and his associates, in the precise terms of current developments in biochemistry and bacterial chemistry, gave a very considerable impetus to chemotherapeutic research. Much effort, in recent years, has accordingly been devoted to isolating and identifying the enzyme systems and essential metabolites of bacterial species. So far as the discovery of chemotherapeutic agents be the objective, the first fruits of this work, following the demonstration that pantothenic acid is essential for the growth of haemolytic streptococci (McIlwain, 1939), was the demonstration that the structurally related pantoyltaurine can, by competitive inhibition, cure mice infected with these bacteria (McIlwain, 1942; McIlwain & Hawking, 1943). Here, for the first time, sound rationale of drug action was expounded in advance of, and led the way to, the discovery of a new chemotherapeutic agent. The rational basis on which this work was conceived was fully supported by showing that the chemotherapeutic effect of pantoyltaurine in mice could be annulled by concomitant administration of pantothenic acid, and that no therapeutic action could be produced in rats, whose pantothenate blood level is normally considerably lower than that of mice. This work illustrates at least one of the difficulties inherent in such an approach, namely, that many of the substances essential for the parasite are essential also for the host, so that an attack on this chink in the parasite's armour may likewise be an attack on a similar chink in the host's armour (vide Ehrlich's aim to discover 'among the chemoreceptors a grouping which has no analogue in the organs of the body'). Undoubtedly, many substances shown to be essential for micro-organisms will later be shown to be essential also for higher animals. The case of *p*-aminobenzoic acid is an early example of this sequence of discovery, its indispensability having been demonstrated for bacteria in the first instance (Woods, 1940; Rubbo & Gillespie, 1940) and then for laboratory animals (Ansbacher, 1941).

Nevertheless, it need hardly be doubted that there must be a very considerable range of substances which are essential—not as 'growth factors', but as 'essential metabolites', in the definitions of these terms used by Fildes (1940)—to various parasitic (and other) micro-organisms without being essential to the host (or to other higher organisms). This is surely axiomatic, in face of the vast panorama of morphological and physiological variability among living things.

II. CHEMOTHERAPY BY BLOCKING SPECIFIC METABOLITES

In spite of the very rapid advances now being made in bacterial chemistry, it is not very likely that highly effective chemotherapeutic agents for bacterial infections will be discovered, in the present state of our knowledge, as a direct result of aiming at a single specific growth factor or essential metabolite. It is too soon entirely to discount the traditional methods of trial and error among compounds selected empirically, or at best on tenuous reasoning, and whose activity, if found, is to be explained after the event instead of before. Still less likely is it that useful chemotherapeutic agents for protozoal infections will, at the present time, be found as a direct result of this approach, since our knowledge of the nutritional requirements of parasitic Protozoa is far less developed than similar knowledge in respect of bacteria. Even such scanty information as is available on the nutritional needs of Protozoa refers almost exclusively to free-living forms. It is, no doubt, the more complex cellular organization of parasitic Protozoa, as compared with parasitic bacteria, which accounts for their greater dependence on the environment provided by the host and for the consequent difficulty of cultivating them *in vitro*.

In any case, as far as some species of parasitic Protozoa are concerned, such forms as are cultivable *in vitro* are, in the well-known and commonly used culture techniques, apt to differ very conspicuously in morphology and in physiological requirements from the forms which occur in the vertebrate host. This is particularly so when the normal developmental cycle involves an intermediate host, as is the case with non-intestinal protozoal infections. Thus, in *Leishmania* cultures the parasites adopt the leptomonas, and in *Trypanosoma cruzi* the crithidia form, these being the forms which the parasites assume in their respective insect vectors. Nutritional studies on these stages of parasitic Protozoa can contribute little or nothing to an understanding of nutritional requirements within the vertebrate host. In the two examples mentioned, *Leishmania* and *Trypanosoma cruzi*, the developmental stages of the parasite responsible for the actual disease processes in the vertebrate host are essentially intracellular.

Techniques which would maintain or perpetuate these intracellular forms of the vertebrate host (e.g. Hawking, 1946) should prove most suitable for the purposes in view, but it would be particularly difficult to identify the nutrients essential to the parasites in their complex environment within the tissue cells.

In many of those Trypanosomidae which, as far as one knows, live an exclusively, or at least predominantly, extracellular life in the blood and other body fluids, it should not be very difficult to devise *in vitro* techniques for maintaining the organisms indefinitely in the forms in which they occur *in vivo*, and under conditions which would enable an analysis to be made of their essential nutritional requirements *in vivo*. The important pathogens, *Trypanosoma gambiense* and *T. rhodesiense* of human sleeping sickness, and *T. congolense* of cattle trypanosomiasis, come within the category of organisms for which a wide field is open for further research along these lines.

Culture of the malaria parasite *in vitro*, unlike that of many of the Trypanosomidae, cannot by any simple technique reproduce the series of developmental forms encountered in the insect vector. The developmental cycle is too complicated and probably too intimately associated with the mosquito's tissues. As for *in vitro* culture of the forms characteristic of infection in the vertebrate host, which is, of course, the kind of culture in which we are here most interested, this has been extraordinarily difficult to achieve, the most notable success so far with the asexual blood forms being that of Ball, Anfinsen, Geiman, McKee & Ormsbee (1945), Geiman, Anfinsen, McKee, Ormsbee & Ball (1946), Anfinsen, Geiman, McKee, Ormsbee & Ball (1946) with *Plasmodium knowlesi* of the monkey. The fact that these forms live and grow within the red blood corpuscles must (as in the case of the intracellular forms of other parasites) necessarily hamper attempts at a fine analysis of their basic nutritional needs. Nevertheless, some progress in this direction has already been made, even with relatively crude cultures, in which the parasites could be maintained at body temperature for no more than a few weeks, and without any appreciable increase in numbers. Trager (1943) found with such cultures of *P. lophurae* that calcium pantothenate favours viability of the parasites. This suggested the trial of pantothenate analogues as antimalarials, and a number of these derivatives were, in fact, then shown to possess antimalarial activity, the most effective being 'phenyl pantothenone' (Mead, Rapport, Senear, Maynard & Koepfli, 1946; Brackett, Waletzky & Baker, 1946; Blanchard & Schmidt, 1946; Marshall, 1946, 1947). It is interesting that both for bacterial and for protozoal infections the first chemotherapeutic agents to be discovered by testing analogues of growth factors, or essential metabolites, should be derivatives of pantothenic acid. Just as the mechanism of antibacterial activity by pantoyltaurine was supported by finding that the therapeutic effect occurs in mice, whose blood pantothenate level is low, and not in rats, whose level is high, so the mechanism of antimalarial action of 'phenyl pantothenone' was supported by demonstrating activity in chicks, whose pantothenate blood level is low, and not in ducks, whose level is high. Common ground between mechanisms of antibacterial and antimalarial action has also been established by the finding that *p*-aminobenzoic acid antagonizes the antimalarial activity of sulphanilamide (Marshall, Litchfield & White, 1942; Maier & Riley, 1942), and that it is a growth factor for malaria parasites (Ball *et al.* 1945; Anfinsen *et al.* 1946).

However, the likelihood of discovering useful new chemotherapeutic agents for malaria by testing analogues of metabolites essential for the blood forms would seem to be particularly remote. The cardinal requirement in antimalarial chemotherapy is to find substances which will attack the parasite in its exoerythrocytic stages, the occurrence of which in mammalian malaria has long been suspected on unexceptionable circumstantial evidence, and has now been brilliantly proved by Shortt and Garnham (1948*a*, *b*). Nutritional studies on the blood forms *in vitro* are most unlikely to give results referable in any way to the exoerythrocytic forms, with their entirely different morphology and metabolism, entrenched as they are in a cell environment radically different from that of the blood forms, and already proved to be obstinately refractory to many compounds which are highly active against the asexual blood forms. As one example, among many, of this type of selective drug action, differentiating between the exoerythrocytic and the asexual blood forms, Gingrich, Schoch & Taylor (1946) have shown, in *P. cathemerium* infections in the canary, that the maximum tolerated dose of 'Metachloridine' (2-metanilamido-5-chloropyrimidine) is 2000 times the minimum dose effective against the blood forms, and yet it does not cure, that is, it does not eradicate the exoerythrocytic parasites.

Cultures of the exoerythrocytic forms (Hawking, 1945) are perhaps likely to provide more information than cultures of blood forms, in nutritional studies designed to lead towards curative antimalarials, but here again, even more than in cultures of blood forms, precise analysis of the parasites' nutritional requirements in their complex intracellular environment would present very great difficulties.

III. DIVERSITY OF MECHANISMS OF CHEMOTHERAPEUTIC ACTION

There are, of course, many mechanisms of chemotherapeutic action alternative to a simple intracellular competitive interference with a metabolite essential to the parasite but not to the host. For instance, various physico-chemical influences may prevent the parasite from assimilating some simple substance, whilst not interfering—or not interfering appreciably—with the host in the same way. Thus, Gale & Taylor (1946, 1947) have shown that penicillin has a considerable limiting effect on the assimilation of glutamic acid by actively growing *Staphylococcus aureus*, and this may well be the underlying basis of its chemotherapeutic power.

The diversity of possible modes of chemotherapeutic action is, in fact, very considerable. An extreme and unusual example is where the therapeutic agent (it can perhaps hardly be called a drug) is itself an enzyme, such as the one which hydrolyses the capsular polysaccharide of specific types of pneumococci, thereby facilitating phagocytic destruction of the organisms (Avery & Dubos, 1931). This is, indeed, a high degree of specificity of therapeutic action, being even more highly selective between types of pneumococci than is the activity of type-specific antisera (Sickles & Shaw, 1934).

IV. PARTICIPATION OF IMMUNITY MECHANISMS IN THE CHEMOTHERAPEUTIC EFFECT

As in the above example, so in the more common types of chemotherapeutic activity, participation of the host's immunity mechanisms often provides a substantial—sometimes the preponderant—quota in the sum total of influences which counteract the invading organisms. In the case of compounds whose activity is primarily bacteriostatic rather than bactericidal, it obviously devolves upon the host to complete the curative process, and much has been written about this in connexion with the chemotherapy of bacterial infections. As for protozoal infections, in the chemotherapy of malaria, where a superficial judgement might place all the credit for eliminating an infection on the effects which the drug produces upon the parasite, the part played by the host is, indeed, very considerable. For example, in the early stages of acute infections of *Plasmodium cathemerium* in the canary the asexual blood forms increase in number about fifteen-fold every 24 hr., but the body defences normally destroy about two-thirds of the daily crop of new parasites almost immediately (Taliaferro, 1925). The effect of quinine treatment, in appropriate dosage, was found to be not a simple destruction of the parasites, but merely that asexual development of the blood forms was so retarded, that they increased only about five-fold over periods of 24–48 hr., an increase with which the normal body defences are able quite easily to contend (Lourie, 1934*a*). At a later stage, these combined effects of drug and of natural immunity upon the parasite are further supplemented by the development of a high degree of acquired immunity, which prevents any further accumulation of parasites in the blood, without, however, eliminating the infection entirely. Problems of specificity or selectivity of action against species or strains of parasite obviously apply in the operation of this acquired immunity; and the whole problem of relapse and of reinfection in malaria after non-curative treatment depends not only on the intricacies of the parasite's life cycle, to which so much attention has been paid in recent years, but on the degree of immunity acquired by the host, the strain- or species-specificity of this immunity, its duration, and the antigenic stability of strains of malaria parasite.

In the pathogenic trypanosome infections the supervention of acquired immunity also plays a fundamentally important role. This immunity is generally specific to a high degree in respect of the particular immunological variant of the trypanosome responsible for infection at the time of treatment. It commonly persists, often for long periods after treatment, but, even while still operative, it may be unable to prevent relapses or reinfections, for the reason that the trypanosomes may be of such a labile antigenic constitution that they readily assume a changed immunological character, which enables them to circumvent the host's immune state (Fulton & Lourie, 1946).

Selectivity of action upon the parasite consequent on administration of a chemotherapeutic agent is, in this context, obviously a manifestation of immunological or serological activity.

V. SELECTIVITY OF DRUG ACTION AMONG DIFFERENT PARASITE SPECIES OF THE SAME GENUS

Returning now to consideration of the more direct effects of drug upon parasite, it is clear that selectivity of drug action is to be conceived as differentiating not only between host and parasite and between different stages of the same parasite, as discussed earlier in this paper, but between different parasite species and between different strains of the same species. Antimalarial chemotherapy provides innumerable instances of differences in response to the same drug on the part of different species of parasite. Obvious and familiar examples are the differences between different forms of human malaria in respect of amenability to treatment with standard antimalarial compounds; and the wartime studies on avian malaria in Britain and the United States, in connexion with the search for improved antimalarials, have provided many further and more striking examples (Marshall, 1946; Davey, 1946*a*, *b*). Among trypanosome strains also, it remains to be explained, for example, why arsenicals in general are effective against *Trypanosoma gambiense* but not against *T. rhodesiense* and *T. congolense*, and why phenanthridinium compounds are, conversely, so much more active against *T. congolense* than against *T. gambiense* and *T. rhodesiense*.

The bases for these various instances of selective drug action must remain unexplained until much more fundamental knowledge has been acquired about the parasites' physiology and metabolism.

VI. SELECTIVITY OF DRUG ACTION AMONG DIFFERENT STRAINS OF THE SAME PARASITE SPECIES

Strains of different geographical origins. It will also be a long time before any rational understanding can be gained of the differences in drug susceptibilities, when tested under identical conditions, between different strains of the same species originating from different geographical localities. In malaria such differences had been more than suspected since the earliest times of exact clinical observation in different parts of the world, and they were eventually established beyond doubt by controlled experiment in human malaria (*Plasmodium falciparum*) by James, Nicol & Shute (1932), and in avian malaria (*P. relictum*) by Lourie (1934*b*).

Drug-resistant strains. The development of acquired drug resistance in parasitic Protozoa, as a result of subcurative treatment, has received a great deal of attention since Ehrlich's pioneer work on the subject in trypanosome infections (Ehrlich, 1907). Acquired drug resistance obviously provides special instances of selective drug action—selective as between the parent, or normal strain, on the one hand, and the resistant strain on the other. Ehrlich's original hypothesis that the development of resistance is due to a loss of affinity between the parasite's chemoreceptor and the corresponding chemotherapeutic agent received solid experimental backing in the case of arsenic resistance in pathogenic trypanosomes, by Yorke, Murgatroyd & Hawking (1931). They showed, and it has since been confirmed by many others, that the resistant trypanosomes have, in fact, acquired the property of preventing the arsenical from entering, or becoming fixed by, the cell substance. Acquired arsenic resistance does not, however, imply resistance to *all* trypanocidal arsenicals, and, as was pointed out by Yorke & Murgatroyd (1930), the term 'arsenic-resistance' is therefore misleading. So-called arsenic-resistant trypanosomes, produced by subcurative treatment with the commonly used arsenicals, such as mapharside (or halarsol), neoarsphenamine, or tryparsamide (type *a*), remain sensitive to at least the following other types of arsenical:

(*b*) unsubstituted phenylarsenoxide, *p*-xylylarsenoxide, and other arsenoxides devoid of markedly polar or hydrophilic groups (Hawking, 1937; King, 1943; Eagle & Magnuson, 1944);

(*c*) carboxyl-substituted phenylarsenoxides and their corresponding arsenobenzenes and phenylarsonates (Ehrlich, 1909; Yorke & Murgatroyd, 1930; King, 1943; Eagle, 1945); and (*d*) melaminyl-substituted phenylarsenoxide and phenylarsonate (Van Hoof, 1947; Williamson & Lourie, 1948).

Mapharside	Xylyl arsenoxide	Eagle's compound	Melarsen oxide
As=O; NH_2; OH	As=O; CH_3; CH_3	As=O; $(CH_2)COOH$	As=O; NH; N; NH_2; N; N; NH_2
TYPE (*a*)	TYPE (*b*)	TYPE (*c*)	TYPE (*d*)

Representatives of four types of trypanocidal arsenical

King (1943) has made a strong case for the view that the commonly used arsenicals (type *a*) and those of types (*b*) and (*c*) differ from one another in their mode of fixation on the trypanosome cell, but that the mechanism of the final lethal effect on the trypanosome is the same in all three cases, depending on inactivation of essential SH groups. This theory has been supported in respect of compounds of type (*c*), and extended to include those of type (*d*), by Williamson & Lourie (1946, 1948) in experiments which showed that *p*-aminobenzoic acid may interfere selectively with the trypanocidal action of arsenoxides of type (*c*) and melamine with those of type (*d*), whilst glutathione interferes equally with both these and with the mapharside type of arsenoxide.

P. gallinaceum in chicks has recently been shown to be capable of developing a high degree of resistance to paludrine (Bishop & Birkett, 1947; Williamson, Bertram & Lourie, 1947; Williamson & Lourie, 1947). Parasites which have acquired such resistance also exhibit resistance to close biguanide analogues of paludrine which, like paludrine, do not contain the pyrimidine ring. They remain sensitive, however, to those pyrimidine antimalarials, such as 3349, which share with paludrine certain physico-chemical features which Curd, Davey & Rose (1945) considered to be essential for antimalarial activity in this general group of compounds.

Paludrine — Cl–C₆H₄–NH.C(=NH).NH.C(=NH)–NH.CH$(CH_3)_2$

3349, a pyrimidine precursor of paludrine — Cl–C₆H₄–NH.C(=NH).NH–(pyrimidine: N, N, CH_3, $NH(CH_2)_2N(C_2H_5)_2$)

We have suggested that here also the ultimate mechanism of lethal action is common to paludrine and its biguanide analogues on the one hand, and to the pyrimidine precursors of paludrine on the other, but that these two types of compounds differ in their route of entry into, or mode of fixation on, the parasite cell (Williamson & Lourie, 1947); and we further suggest that it is this difference which is associated, as in the above example among trypanosomes, with the fact that malaria parasites made resistant to the one type of compound retain their sensitivity to the other.

The underlying bases for selectivity of drug action in these cases is therefore to be sought in physical or physico-chemical terms, involving such factors as surface activity, polarity, the various phenomena associated with cell interfaces, and the electrical charges borne by the parasite or by different parts of its protoplasm, rather than in more strictly chemical terms. Such an approach to the problem of acquired drug resistance in trypanosomes has recently been made by Schueler (1947), who has produced evidence that the acquirement of resistance involves a significant shift in the isoelectric points of the trypanosome's proteins, particularly in certain localized portions of the trypanosome cell.

It need hardly be said that the above examples must be representative of only one kind among innumerable possible mechanisms of acquired drug resistance, which may be as diverse in type as are the many possible mechanisms of drug action.

VII. SUMMARY

1. Selectivity of chemotherapeutic drug action is to be conceived as discriminating between (*a*) host and parasite, (*b*) different developmental stages of the same parasite, (*c*) different parasite species of the same genus, and (*d*) different strains of the same parasite species. These aspects of selective drug action are broadly discussed, with particular reference to protozoal blood infections.

2. The acquirement of drug resistance provides special cases of drug action selective as between different strains of the same parasite species. The most familiar examples in protozoal blood infections are probably to be explained in physical or physico-chemical terms, rather than along more strictly chemical lines.

3. In discussions on modes of drug action the contribution towards the total chemotherapeutic effect provided by the host's immunity mechanisms, both natural and acquired, tends very often to be neglected. In many cases this is very considerable, and obviously involves problems of selective action, of an immunological character, in addition to those attributable to more direct action of the drug on the parasite.

REFERENCES

ANFINSEN, C. B., GEIMAN, Q. M., McKEE, R. W., ORMSBEE, R. A. & BALL, E. G. (1946). Studies on malarial parasites; factors affecting growth of *Plasmodium knowlesi in vitro*. *J. Exp. Med.* **84**, 607.

ANSBACHER, S. (1941). *p*-Aminobenzoic acid, a vitamin. *Science*, **93**, 164.

AVERY, O. T. & DUBOS, R. J. (1931). The protective action of a specific enzyme against type III pneumococcus infection in mice. *J. Exp. Med.* **54**, 73.

BALL, E. G., ANFINSEN, C. B., GEIMAN, Q. M., MCKEE, R. W. & ORMSBEE, R. A. (1945). *In vitro* growth and multiplication of the malaria parasite *Plasmodium knowlesi*. *Science*, **101**, 542.

BISHOP, A. & BIRKETT, B. (1947). Acquired resistance to paludrine in *Plasmodium gallinaceum*. *Nature, Lond.*, **159**, 884.

BLANCHARD, K. C. & SCHMIDT, L. H. (1946). Chapter III. Chemical series of potential interest. *A Survey of Antimalarial Drugs*, 1941–1945, p. 73. Ann Arbor, Michigan: J. W. Edwards.

BRACKETT, S., WALETZKY, E. & BAKER, M. (1946). The relation between pantothenic acid and *Plasmodium gallinaceum* infections in the chicken and the antimalarial activity of analogues of pantothenic acid. *J. Parasitol.* **32**, 453.

CURD, F. H. S., DAVEY, D. G. & ROSE, F. L. (1945). Studies on synthetic antimalarial drugs. X. Some biguanide derivatives as new types of antimalarial substances with both therapeutic and causal prophylactic activity. *Ann. Trop. Med. Parasit.* **39**, 208.

DAVEY, D. G. (1946*a*). The use of avian malaria for the discovery of drugs effective in the treatment and prevention of human malaria. I. Drugs for clinical treatment and clinical prophylaxis. *Ann. Trop. Med. Parasit.* **40**, 52.

DAVEY, D. G. (1946*b*). The use of avian malaria for the discovery of drugs effective in the treatment and prevention of human malaria. II. Drugs for causal prophylaxis and radical cure or the chemotherapy of exo-erythrocytic forms. *Ann. Trop. Med. Parasit.* **40**, 453.

EAGLE, H. (1945). A new trypanocidal agent: (*p*-arsenosophenyl)-butyric acid. *Science*, **101**, 69.

EAGLE, H. & MAGNUSON, H. J. (1944). The spontaneous development of arsenic-resistance in *Trypanosoma equiperdum*, and its mechanism. *J. Pharmacol.* **82**, 137.

EHRLICH, P. (1907). Chemotherapeutische Trypanosomen-Studien. III. *Berl. klin. Wschr.* **44**, 313.

EHRLICH, P. (1909). Über die neuesten Ergebnisse auf dem Gebiete der Trypanosomenforschung. *Arch. Schiffs- u. Tropenhyg.* **13**, Bhft., 321.

EHRLICH, P. (1913). Chemotherapeutics: scientific principles, methods and results. *Lancet*, **2**, 445.

FILDES, P. (1940). A rational approach to research in chemotherapy. *Lancet*, **1**, 955.

FULTON, J. D. & LOURIE, E. M. (1946). The immunity of mice cured of trypanosome infections. *Ann. Trop. Med. Parasit.* **40**, 1.

GALE, E. F. & TAYLOR, E. S. (1946). Action of penicillin in preventing the assimilation of glutamic acid by *Staphylococcus aureus*. *Nature, Lond.*, **158**, 676.

GALE, E. F. & TAYLOR, E. S. (1947). The assimilation of amino-acids by bacteria. 5. The action of penicillin in preventing the assimilation of glutamic acid by *Staphylococcus aureus*. *J. Gen. Microbiol.* **1**, 314.

GEIMAN, Q. M., ANFINSEN, C. B., MCKEE, R. W., ORMSBEE, R. A. & BALL, E. G. (1946). Studies on malarial parasites; methods and techniques for cultivation. *J. Exp. Med.* **84**, 583.

GINGRICH, W., SCHOCH, E. W. & TAYLOR, C. A. (1946). Therapeutic, prophylactic and curative tests in avian malaria (*Plasmodium cathemerium*) with metachloridine. *J. Parasitol.* **32** (Supplement), no. 13, p. 9.

HAWKING, F. (1937). Studies on chemotherapeutic action. I. The absorption of arsenical compounds and tartar emetic by normal and resistant trypanosomes and its relation to drug-resistance. *J. Pharmacol.* **59**, 123.

HAWKING, F. (1945). Growth of protozoa in tissue culture. I. *P. gallinaceum*, exoerythrocytic forms. *Trans. R. Soc. Trop. Med. Hyg.* **39**, 245.

HAWKING, F. (1946). Growth of Protozoa in tissue culture. III. *Trypanosoma cruzi*. *Trans. R. Soc. Trop. Med. Hyg.* **40**, 345.

JAMES, S. P., NICOL, W. D. & SHUTE, P. G. (1932). A study of induced malignant tertian malaria. *Proc. Roy. Soc. Med.* **25**, 1153.

KING, H. (1943). Chemical structure of arsenicals and drug resistance of trypanosomes. *Trans. Faraday Soc.* **39**, 383.

LOURIE, E. M. (1934*a*). Studies on chemotherapy in bird malaria. II. Observations bearing on the mode of action of quinine. *Ann. Trop. Med. Parasit.* **28**, 255.

LOURIE, E. M. (1934*b*). Studies on chemotherapy in bird malaria. III. Difference in response to quinine treatment between strains of *Plasmodium relictum* of widely separated geographical origins. *Ann. Trop. Med. Parasit.* **28**, 513.

McILWAIN, H. (1939). Pantothenic acid and the growth of *Staphylococcus haemolyticus*. *Brit. J. Exp. Path.* **20**, 330.

McILWAIN, H. (1942). Bacterial inhibition by metabolite analogues. 3. Pantoyltaurine. The antibacterial index of inhibitors. *Brit. J. Exp. Path.* **23**, 95.

McILWAIN, H. & HAWKING, F. (1943). Chemotherapy by blocking bacterial nutrients. Antistreptococcal activity of pantoyltaurine. *Lancet*, **1**, 449.

MAIER, J. & RILEY, E. (1942). Inhibition of antimalarial action of sulfonamides by *p*-aminobenzoic acid. *Proc. Soc. Exp. Biol., N.Y.*, **50**, 152–4.

MARSHALL, E. K., JR. (1946). Chemotherapy of malaria, 1941–1945. *Fed. Proc.* **5**, 298.

MARSHALL, E. K., JR. (1947). Scientific principles, methods, and results of chemotherapy, 1946. *Medicine*, **26**, 155.

MARSHALL, E. K., JR., LITCHFIELD, J. T. & WHITE, H. J. (1942). Sulfonamide therapy of malaria in ducks. *J. Pharmacol.* **75**, 89.

MEAD, J. F., RAPPORT, M. M., SENEAR, A. E., MAYNARD, J. T. & KOEPFLI, J. B. (1946). The synthesis of potential antimalarials. Derivatives of pantoyltaurine. *J. Biol. Chem.* **163**, 465.

RUBBO, S. D. & GILLESPIE, J. M. (1940). Para-amino-benzoic acid as a growth factor. *Nature, Lond.*, **146**, 838.

SCHUELER, F. W. (1947). The mechanism of drug-resistance in trypanosomes. II. A method for the differential staining of normal and drug resistant trypanosomes and its possible relation to the mechanics of drug-resistance. *J. Infect. Dis.* **81**, 139.

SHORTT, H. E. & GARNHAM, P. C. C. (1948*a*). The pre-erythrocytic development of *Plasmodium cynomolgi* and *Plasmodium vivax*. *Trans. R. Soc. Trop. Med. Hyg.* **41**, 785.

SHORTT, H. E. & GARNHAM, P. C. C. (1948*b*). Demonstration of a persisting exoerythrocytic cycle in *Plasmodium cynomolgi* and its bearing on the production of relapses. *Brit. Med. J.* **1**, 1225.

SICKLES, G. M. & SHAW, M. (1935). A microorganism which decomposes the specific carbohydrate of pneumococcus type VIII. *Proc. Soc. Exp. Biol., N.Y.*, **32**, 857.

TALIAFERRO, L. G. (1924). Infection and resistance in bird malaria, with special reference to periodicity and rate of reproduction of the parasite. *Amer. J. Hyg.* **5**, 742.

TRAGER, W. (1943). Further studies on the survival and development *in vitro* of a malarial parasite. *J. Exp. Med.* **77**, 411.

VAN HOOF, L. M. J. (1947). Observations on trypanosomiasis in the Belgian Congo. *Trans. Roy. Soc. Trop. Med. Hyg.* **40**, 728.

WILLIAMSON, J., BERTRAM, D. S. & LOURIE, E. M. (1947). Acquired resistance to paludrine in *Plasmodium gallinaceum*. *Nature, Lond.*, **159**, 885.

WILLIAMSON, J. & LOURIE, E. M. (1946). Interference with the trypanocidal action of γ-(*p*-arsenosophenyl)-butyric acid by *p*-aminobenzoic acid. *Ann. Trop. Med. Parasit.* **40**, 255.

WILLIAMSON, J. & LOURIE, E. M. (1947). Acquired paludrine-resistance in *Plasmodium gallinaceum*. I. Development of resistance to paludrine and failure to develop resistance to certain other antimalarials. *Ann. Trop. Med. Parasit.* **41**, 278.

WILLIAMSON, J. & LOURIE, E. M. (1948). 'Melarsen' and 'Melarsen oxide': Activity against tryparsamide-resistant trypanosomes, and interference by 'Surfen C'. *Nature, Lond.* (in the Press).

WOODS, D. D. (1940). The relation of *p*-aminobenzoic acid to the mechanism of the action of sulphanilamide. *Brit. J. Exp. Path.* **21**, 74.

YORKE, W. & MURGATROYD, F. (1930). Studies in chemotherapy. III. The action *in vitro* of certain arsenical and antimonial compounds on *T. rhodesiense* and on atoxyl- and acriflavine-resistant strains of this parasite. *Ann. Trop. Med. Parasit.* **24**, 449.

YORKE, W., MURGATROYD, F. & HAWKING, F. (1931). Studies in chemotherapy. V. Preliminary contribution on the nature of drug resistance. *Ann. Trop. Med. Parasit.* **25**, 351.

ASPECTS OF THE SELECTIVE TOXICITY OF SULPHONAMIDES AND OTHER ANTI-METABOLITE INHIBITORS

By D. D. WOODS and R. H. NIMMO-SMITH

From the Department of Biochemistry, University of Oxford

I. INTRODUCTION

During the past eight years considerable progress has been made in elucidating, in biochemical terms, the mode of action of the sulphonamide drugs on bacteria. But the reason for their activity against the invading micro-organism, and relative lack of toxicity to the cells of the host, is by no means clear; yet it is upon this selective toxicity that the therapeutic success of the sulphonamides depends. It is the main purpose of the present communication to examine what light is thrown on this problem by recent work bearing on the mode of action of the sulphonamides. Some consideration will also be given to the question of the selective action (*in vitro*) of a given sulphonamide against different species of bacteria. Other aspects of the selective action of the sulphonamides cannot, for reasons of space, be considered here. There is, for example, the question of the relative activity of different members of this family of drugs against the same bacterium. Satisfactory physico-chemical explanations appear to be available though there is not yet complete agreement; this aspect has been reviewed fairly recently by Roblin (1946). Differential toxic effects of the various sulphonamides on the animal do not seem to be correlated with their relative activities against bacteria; indeed amongst the drugs which have the least ill effect on the animal are to be found those with the greatest *in vitro* anti-bacterial potency (e.g. sulphathiazole, sulphadiazine). Again there is the greater therapeutic activity of some sulphonamides than others against a given bacterial infection, differences which do not always run parallel with *in vitro* findings. Here questions of the selective absorption of the drugs by different tissues may be of importance.

The findings to be described concerning the mechanism of the anti-bacterial effect of the sulphonamides stimulated a search for new chemo-therapeutic agents active in an analogous manner; i.e. by inhibiting the metabolism of substances essential for bacterial growth by virtue of their chemical similarity to such substances. A large number of compounds with anti-bacterial activity have been found (see Roblin, 1946; Woolley, 1947), but very few have appreciable therapeutic value. Certain aspects of their relative toxicity to host and micro-organism will be considered briefly since these may indicate a more profitable use of this approach.

II. *p*-AMINOBENZOIC ACID AND SULPHONAMIDE ACTION

The discovery that *p*-aminobenzoic acid overcame, in a competitive manner, the *in vitro* inhibition of bacterial growth by sulphonamides led directly to the hypothesis (Woods, 1940) that *p*-aminobenzoic acid is an 'essential metabolite' for bacteria, and that sulphanilamide, by virtue of a structural similarity, competes for the enzyme concerned in its utilization. Since then the sulphonamide-*p*-aminobenzoic acid relationship has been found to occur with all micro-organisms tested which are susceptible to sulphonamides except *Bact. tularense* (Tamura, 1944). Furthermore, the same effect is obtained with a wide variety of the sulphonamide drugs, exceptions in this case being the marfanil group and 'V' drugs (Shreus, 1942; Lawrence, 1945; Evans, Fuller & Walker, 1944) and a series of sulphanilylanilides (Goetchius & Lawrence, 1945). *p*-Aminobenzoic acid has been isolated from natural sources (Blanchard, 1941). But perhaps the most striking confirmation of the hypothesis has been the fulfilment of the prediction that *p*-aminobenzoic acid is an essential metabolite; it has been found to be an essential growth factor for a wide variety of micro-organisms (listed by Knight, 1945; Woods, 1947). It has also been shown to be synthesized by many organisms which do not require it added to the medium. It has been argued (Sevag, 1946) that this substance is not a growth factor in the usual sense, since it may, at higher concentrations, inhibit growth of some organisms. Such behaviour is not, however, unique amongst growth factors; aneurin and riboflavin for example, show this effect with *Rhizobium trifolium* (West & Wilson, 1939). The status of *p*-aminobenzoic acid as a true growth factor was most convincingly demonstrated by the isolation of induced mutants of *Esch. coli* (Lampen, Roepke & Jones, 1946) and *Neurospora crassa* (Tatum & Beadle, 1942) which require it for growth, for the same treatment produced also mutants requiring one or other of most of the well-established B-group vitamins.

III. PRODUCTS OF THE UTILIZATION OF *p*-AMINOBENZOIC ACID IN RELATION TO SULPHONAMIDE INHIBITION

If the above hypothesis is accepted the next step is obviously to discover for what chemical processes of the cell *p*-aminobenzoic acid is required, since these are presumably the reactions inhibited by sulphonamides. Considerable evidence is now to hand that this metabolite is involved, either directly or indirectly, in the synthesis of folic acid, purines, thymine and certain amino-acids (notably methionine and lysine).

Folic acid is a member of the vitamin B complex and is an essential growth factor for a number of micro-organisms, mainly, but not exclusively, enterococci and lactobacilli; it is also involved in the nutrition of animals and birds (for detailed literature see Jukes & Stokstad, 1948). Other bacteria which have been tested appear to synthesize the factor for themselves. Folic acid is of wide occurrence in animal, vegetable and bacterial tissues, but there is evidence (summarized and extended by Hall, 1947) that folic

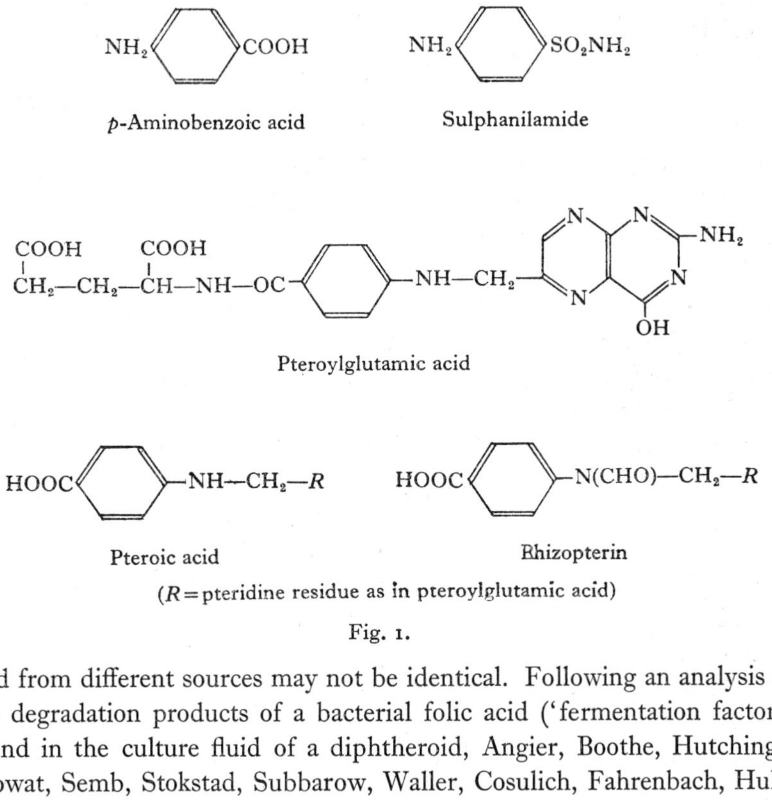

NH_2—C₆H₄—COOH

p-Aminobenzoic acid

NH_2—C₆H₄—SO_2NH_2

Sulphanilamide

COOH COOH
CH_2—CH_2—CH—NH—OC—C₆H₄—NH—CH_2—(pteridine: N, N, N, N; NH_2; OH)

Pteroylglutamic acid

HOOC—C₆H₄—NH—CH_2—*R*

Pteroic acid

HOOC—C₆H₄—N(CHO)—CH_2—*R*

Rhizopterin

(*R* = pteridine residue as in pteroylglutamic acid)

Fig. 1.

acid from different sources may not be identical. Following an analysis of the degradation products of a bacterial folic acid ('fermentation factor') found in the culture fluid of a diphtheroid, Angier, Boothe, Hutchings, Mowat, Semb, Stokstad, Subbarow, Waller, Cosulich, Fahrenbach, Hultquist, Kuh, Northey, Seeger, Sickels & Smith (1946) succeeded in synthesizing a compound (pteroylglutamic acid (Fig. 1))* which had the full biological activity of the factor present in liver and active for *Lb. casei* and other folic acid-requiring organisms. There is strong evidence (Pfiffner, Calkins, Bloom & O'Dell, 1946) that vitamin B_c from yeast and liver is also pteroylglutamic acid. The compound also exists naturally in the form of

* In this article the term 'pteroylglutamic acid' will be used only for the synthetic product; 'folic acid' will be used as a general term in reference to natural forms of the free factor.

conjugates with extra glutamic acid residues; thus 'fermentation' *Lb. casei* factor and vitamin B_c conjugate are respectively pteroyltriglutamic acid and pteroylheptaglutamic acid. Jukes & Stokstad (1948) have summarized the evidence that the differing responses of various organisms to the conjugated forms depend upon their ability to produce free pteroylglutamic acid by peptic hydrolysis.

The occurrence of a *p*-aminobenzoic acid residue in the pteroylglutamic acid molecule is strong presumptive evidence that one purpose for which the former is utilized by the bacterial cell is for the synthesis of folic acid. Earlier work had indeed suggested this possibility. The production of material with folic acid activity had been observed by Mills, Briggs, Luckey & Elvehjem (1944) in cultures of *Mycobact. tuberculosis* on a medium containing *p*-aminobenzoic acid, whilst the latter stimulated folic acid production by mixed cultures from fowl intestine (Briggs, Luckey, Mills, Elvehjem & Hart, 1943). More recently Sarett (1947) has studied the production of *Lb. casei* factor by growing cultures of *Lb. arabinosus* (an organism requiring *p*-aminobenzoic acid for growth) and has found some stimulation (not proportional) with increasing concentration of *p*-aminobenzoic acid. Finally, Nimmo-Smith, Lascelles & Woods (1948) have been able to obtain synthesis of *Lb. casei* factor by suspensions of *Streptobacterium plantarum* in a system containing only buffer, glucose, glutamate and *p*-aminobenzoic acid, and in which no growth occurred. A quantitative relationship was found between the concentration of *p*-aminobenzoic acid added and *Lb. casei* factor synthesized; furthermore, significant synthesis began at just that concentration of *p*-aminobenzoic acid which is limiting for optimal growth of this organism. The synthesis was inhibited by sulphonamides but was restored by increasing the *p*-aminobenzoic acid concentration, i.e. the inhibition was competitive. Quantitative relationships between the two competitors were closely similar to those obtained by Nimmo-Smith & Woods (1948) for the effect on growth. With *Esch. coli* (an organism that does not require preformed *p*-aminobenzoic acid), Miller (1944) has found that folic acid synthesis by growing cultures on a complex medium is inhibited by sulphonamides.

Now if the only function of *p*-aminobenzoic acid in cell metabolism is its utilization for the synthesis of folic acid, and if this function is, as seems likely from the above, inhibited by the sulphonamides, then two things would be expected. First, folic acid should replace *p*-aminobenzoic acid for the growth of organisms requiring the latter factor. Secondly, folic acid should overcome sulphonamide inhibition of growth and in a non-competitive manner; that is, in the presence of folic acid the organism should be relatively insensitive to the drug, for the product of the reaction which it inhibits has already been provided.

Nimmo-Smith & Woods (1948) have examined the properties in these respects of a number of organisms requiring *p*-aminobenzoic acid; the source of folic acid used was synthetic pteroylglutamic acid. With *Clostridium acetobutylicum* (two strains) and *Streptobacterium plantarum* (three strains) growth in the presence of pteroylglutamic acid was insensitive to sulphonamides, the concentration necessary for growth over a wide range of sulphonamide concentration being no greater than in the absence of the drug. This concentration was, however, 10–100 times greater on a molar basis than that of *p*-aminobenzoic acid required for growth in the absence of drug. Essentially similar results both in regard to sulphonamide inhibition and growth factor activity of pteroylglutamic acid were obtained independently by Lampen & Jones (1947) with *Sbm. plantarum*, and also with *Lb. arabinosus* (see also Sarett, 1947). The lower growth-factor activity of pteroylglutamic acid compared with *p*-aminobenzoic acid might be due (as suggested by Lampen & Jones, 1947) either to a less efficient absorption of the former or to the possibility that it is not identical with the folic acid normally produced by the organism.

Morgan (1948) has also found recently that psittacosis virus is unaffected by sulphonamides when pteroylglutamic acid is present.

With other organisms requiring *p*-aminobenzoic acid (e.g. *Acetobacter suboxydans* 621, induced mutants of *Neurospora crassa* and *Esch. coli*) the concentration of pteroylglutamic acid required to replace *p*-aminobenzoic acid is proportionately still greater; furthermore, reversal of sulphonamide inhibition, if it occurs at all, is strictly competitive (Lampen, Roepke & Jones, 1946; Nimmo-Smith & Woods, 1948). It is doubtful whether the effects can be wholly explained by the known small free arylamine content of the substance, for this is probably *p*-amino-benzoylglutamic acid which is even less active for these organisms. These results will be further discussed below.

The case of organisms which cannot synthesize folic acid and require it preformed in the medium is of particular interest. Lampen & Jones (1946) have shown that such organisms (e.g. *Lb. casei*, *Strept. faecalis* R) are essentially insensitive to sulphonamides. This would be expected on the present hypothesis since it is necessary to add the product of the inhibited reaction in order to obtain growth; the metabolic lesion induced by sulphonamides is in fact already present in folic acid-requiring organisms. Lampen & Jones also studied a number of enterococci for which pteroylglutamic acid was not essential, though it stimulated growth, i.e. the power to synthesize was limited. These organisms were very susceptible to sulphonamides in the absence of pteroylglutamic acid; in its presence, however, they were essentially insensitive, the level of the factor required being the same as that needed by closely related organisms for which folic acid is essential.

It may be concluded therefore that with certain organisms inhibition of the synthesis of folic acid from *p*-aminobenzoic acid represents the main, if not the only, point of attack of the sulphonamides, and that this accounts adequately for the effect on growth. Amongst the organisms so far tested these are those enterococci with limited powers of synthesizing folic acid, and certain of the organisms requiring *p*-aminobenzoic acid. Apart from the other members of the latter group mentioned earlier, Lampen & Jones (1946) have reported that they are unable to find non-competitive antagonism by pteroylglutamic acid with *Esch. coli*, *Staph. aureus* or *Diplococcus pneumoniae*. The position with regard to such organisms is not yet clear. Within the bounds of the present hypothesis there are several possible explanations:

(1) The organisms cannot assimilate preformed pteroylglutamic acid; no data is available on this point.

(2) Pteroylglutamic acid may differ slightly in chemical structure from the folic acid required (and normally synthesized); some organisms may be unable to convert the former to the latter. In this connexion, in addition to the evidence presented by Hall (1947), it is interesting to note that natural *Strept. faecalis* R factor has been identified as a formylpteroic acid (rhizopterin, Fig. 1) and has greater growth factor activity than pteroic acid (Rickes, Trenner, Conn & Keresztesy, 1947). Again the analogous formylpteroylglutamic acid is more active than pteroylglutamic acid in overcoming growth inhibition of *Strept. faecalis* R by 7-methylpteroylglutamic acid (Gordon, Ravel, Eakin & Shive, 1948). Finally, the folic acid synthesized by *Strept. faecalis* R from rhizopterin is less stable than pteroylglutamic acid (Stokes & Larsen, 1945).

Another possibility under this general heading is that the final substance with functional activity is an even more complex molecule than pteroylglutamic acid, and that the latter is not a normal intermediate in its synthesis by some organisms.

(3) *p*-Aminobenzoic acid may have functions other than its requirement for the synthesis of folic acid; these functions also being inhibited by sulphonamides. There is considerable circumstantial evidence that *p*-aminobenzoic acid is concerned in some way in the synthesis of (*a*) purines and thymine, (*b*) methionine, (*c*) possibly other amino-acids especially lysine. This evidence is derived mainly from the fact that these substances either alone or in conjunction, can, under certain conditions, replace *p*-aminobenzoic acid, wholly or in part, as growth factor or anti-sulphonamide agent with various micro-organisms. It is not possible here to review this work

in full; detailed references to the literature are given by Henry (1944) and Woods (1947). For these effects the substances need to be added in much higher concentration than *p*-aminobenzoic acid or folic acid, which would indicate that they are end products of synthetic systems in which the latter factors have a catalytic function. Since purines plus thymine can replace folic acid as growth factor and as anti-sulphonamide agent, it is probable that the function of *p*-aminobenzoic acid in the synthesis of these substances occurs via the intermediate formation of folic acid (Stokes, 1944; Snell, 1946; Lampen & Jones, 1946, 1947). On the other hand there is no evidence so far that the function in the synthesis of methionine or lysine occurs via folic acid. This may therefore represent a separate type of reaction involving *p*-aminobenzoic acid which is inhibited by sulphonamide, and may explain the ineffectiveness of pteroylglutamic acid alone as sulphonamide antagonist with some organisms. Lampen *et al.* (1946) have found that the *p*-aminobenzoic acid requirement of a mutant strain of *Esch. coli* was completely replaced by a mixture of purines, thymine and amino-acids (of which methionine and lysine were the most important). Pteroylglutamic acid alone was ineffective. Growth in the presence of the purines-thymine-amino-acid mixture was highly resistant to sulphonamides and this was also true of the parent strain (not requiring *p*-aminobenzoic acid) from which the mutant was derived. More recently, Winkler & de Haan (1948) have shown that for *Esch. coli* and *Salmonella typhimurium* the successive addition of methionine, xanthine and serine progressively lowered the amount of *p*-aminobenzoic acid needed to overcome a given concentration of sulphonamide. The further addition of pteroylglutamic acid (or of thymine) to a simple medium containing these three compounds permitted growth in the presence of high concentrations of sulphonamides and in the absence of added *p*-aminobenzoic acid. The further addition of valine allowed growth at optimal rate. It would seem that if *p*-aminobenzoic acid has other functions than for the synthesis of folic acid (and this is not yet certain) then these are also inhibited by sulphonamides.

On balance therefore it is probable that inhibition of bacterial growth by sulphonamides can be satisfactorily accounted for by their effect upon the utilization of *p*-aminobenzoic acid alone. It has been argued by Henry (1944) and Sevag (1946) that the primary effect of sulphonamides is on respiratory mechanisms. Since the purine, adenine, is a component of the structure of several respiratory coenzymes, and since these suffer some destruction while functioning which must be made good (Morel, 1943), it is more likely that any effect on respiration is secondary and due to the effect, via *p*-aminobenzoic acid, on adenine synthesis. Similarly effects of sulphonamide on peptide and protein synthesis (Havinga, Julius, Veldstra

& Winkler, 1946; Tatum & Giese, 1946; Gale, 1947) may well be secondary to the involvement of *p*-aminobenzoic acid in the synthesis of certain amino-acids.

IV. SELECTIVE TOXICITY OF SULPHONAMIDES

The question of the relative toxicity of the sulphonamides as between (*a*) different species of bacteria, and (*b*) host and invading bacterium, may now be considered against the above background.

(*a*) *To different bacteria*

It is at once apparent that it is a very difficult matter to establish experimentally a true spectrum of the relative activity *in vitro* of a given sulphonamide against a wide variety of organisms, and no real information is available on this point. Different organisms have mostly been tested in the medium most suitable for their growth, but such media may contain widely different amounts of *p*-aminobenzoic acid, folic acid and other sulphonamide antagonists. Even if the same complex test medium were used for all organisms the results might be invalidated by the differing ability of various organisms to release antagonists from inactive bound forms. *p*-Aminobenzoic acid is known, for example, to exist in natural extracts partly as inactive complexes (see, for example, Blanchard, 1941). To establish a true comparison all organisms would have to be grown on media of defined chemical composition in which the concentration of sulphonamide antagonists was strictly controlled, and this has not yet been done with a wide range of organisms. Wyss, Grubaugh & Schmelkes (1942) compared the susceptibility of *Esch. coli* and *Staph. aureus* on the same synthetic medium to a number of sulphonamides and found the critical concentration closely similar for each organism with a given sulphonamide. They also found an overall variation of only a factor of seven in the critical sulphonamide concentration required for seven organisms (from different genera) all on defined media to which a constant level of *p*-aminobenzoic acid had been added. Since none of these organisms required *p*-aminobenzoic acid as an added growth factor, this small difference may have been due to quantitative differences in the ability of the different organisms to add to the *p*-aminobenzoic acid present by their own synthesis. Such differences are known to exist with other organisms (Landy, Larkum & Oswald, 1943).

Sulphonamides are active irrespective of whether the organism requires added *p*-aminobenzoic acid for growth or not. In the latter case it might be expected that the relative susceptibility would depend on the amount of *p*-aminobenzoic acid or other antagonists synthesized by the organism. Here again experimental observations are scanty. Landy & Gerstung (1945) examined the properties of 175 strains of *Neisseria gonorrhoeae* isolated from patients showing variable response to sulphonamide therapy. With very few exceptions, a positive correlation was found between *in vivo* and *in vitro* sensitivity to sulphonamide and amount of *p*-aminobenzoic acid (or material with the same biological activity) synthesized on a defined medium free of this factor. More information is available in respect to organisms 'trained' to sulphonamide resistance. In several cases this is correlated with an increased capacity to synthesize *p*-aminobenzoic acid but in other cases it is not; synthesis of folic acid, purines, pyrimidines, etc. has not yet been tested (see, for example, Landy, Larkum, Oswald & Streightoff, 1943; Housewright & Koser, 1944; Mirick, 1943; Lemberg, 1946). There may be more than one mechanism of adaptation to sulphonamides (e.g. Cole & Hinshelwood, 1948; James & Hinshelwood, 1948). It is clear, however, that quantitative differences in capacity to synthesize *p*-aminobenzoic acid are one factor in determining sensitivity to sulphonamides, and that this may vary even within a species.

It is also clear from the last section that capacity to synthesize folic acid or rhizopterin from *p*-aminobenzoic acid or simpler substances is another factor. Organisms which cannot do this and require the preformed factor for growth are substantially unaffected by sulphonamides. Lampen & Jones (1946) suggest that the observed clinical insensitivity of the enterococci (some of which require these substances) may be explained along these lines, if there is sufficient folic acid in the tissue fluids.

(*b*) *Host and invading bacterium*

It has been demonstrated in the case of micro-organisms that an important feature of the action of sulphonamides is their effect on the metabolism of the *p*-aminobenzoic acid/folic acid group of essential growth factors. In common with other members of the vitamin-B complex, these substances are also involved in the nutrition of higher creatures (e.g. animals, birds) and are probably required for fundamental reactions common to all types of cells (Knight, 1945). It remains to be explained therefore why sulphonamides do not have a drastic effect on the host, rather than the relatively mild toxicity compared with that for bacteria which makes possible their therapeutic use.

The role of *p*-aminobenzoic acid in vertebrate nutrition is by no means clear; it has been claimed to alleviate a multiplicity of symptoms of widely different types. The literature has been reviewed by Ansbacher (1944), who concludes that in most cases where the lesion was frankly of nutritional origin, the effect of *p*-aminobenzoic acid was probably due to its conversion to folic acid by micro-organisms of the gut. This seems very probable as such synthesis is known to occur (vide supra).

The relation of the folic acid group of factors to vertebrate nutrition has recently been comprehensively and admirably reviewed by Jukes & Stokstad (1948) in the light of the present chemical knowledge. Experimental nutritional deficiency diseases of monkeys and poultry respond to treatment with pteroylglutamic acid, as does also anaemia of nutritional origin in man (tropical macrocytic anaemia). In all these diseases a characteristic symptom is the development of a megaloblastic anaemia. Furthermore, treatment with pteroylglutamic acid leads to remission of the haematological (though not the neurological) symptoms of Addisonian pernicious anaemia and the anaemia of sprue; these are again megaloblastic anaemias, though not of nutritional origin except perhaps indirectly and in part in the case of sprue. All this points to folic acid as an important factor in the metabolism of bone marrow, one of the few tissues in the adult animal in which active cell proliferation occurs. This is in accord with the findings with micro-organisms that this factor is involved directly with synthetic mechanisms of the cell.

The available evidence indicates that animal cells are unable to synthesize their own folic acid from *p*-aminobenzoic acid or simpler compounds, and require a source of preformed factor. This may derive directly from the food or from bacterial synthesis in the gut. The latter may play a substantial part in some animals. In the rat, for instance, blood dyscrasia which can be relieved by folic acid, is induced by feeding sulphonamides (which inhibit bacterial synthesis), but not by feeding a diet purified from folic acid (Wright & Welch, 1943). *p*-Aminobenzoic acid cannot replace pteroylglutamic acid in curing the nutritional deficiency disease of monkeys (Wilson, Saslaw & Doan, 1946). Again, rhizopterin gives no haemotological response in rats treated as above with sulphonamides, and neither this compound nor pteroic acid are active for folic acid-deficient chicks (Keresztesy, 1947; Angier *et al.* 1946). If, as seems likely, the animal cell requires intact folic acid, then its relative insensitivity to sulphonamides is readily explained. The position would be the same as for those bacteria which require preformed folic acid; i.e. the product of the sulphonamide-inhibited reaction has to be provided in any case, so no inhibition can occur. This possibility has been stated by Woolley (1947). If the animal is partly dependent on bacterial synthesis of folic acid in the gut, then some toxic effect due to partial deficiency of folic acid might occur indirectly. Leucopaenia and agranulocytosis are among the toxic effects sometimes found in sulphonamide treatment (van Dyke, 1943).

Evidence has been given earlier that folic acid is concerned with synthetic processes leading to substances (nucleic acid derivatives, amino-acids) which must be essential for reduplication of cell material. Even if there is some deficiency of folic acid in the animal as a result of sulphonamide therapy, a short interruption of cell proliferation is likely to do less harm to the host tissues than to the parasite. The former may also concentrate folic acid (in free or bound form) in the tissues where it may be particularly required; more data is required concerning this with regard to bone marrow, testis, etc. But a short-term interruption may hold the parasite in check for a sufficient period for phagocytosis to occur.

Finally the *in vivo* susceptibility of the micro-organism must depend, amongst other things, upon the concentration of sulphonamide antagonists in the host tissues. Of these, *p*-aminobenzoic acid and folic acid will be the most important since they are active at very low concentration. McIlwain (1942) has shown that the blood-concentration of sulphonamide aimed at for successful therapy is of the order that would be anticipated from the concentration of free *p*-aminobenzoic acid present as estimated microbiologically.

Since folic acid is a vitamin for animals it is not surprising that it is widely distributed in foodstuffs and in animal tissues. It is therefore necessary to assess what effect it may have on sulphonamide bacteriostasis *in vivo*. Pteroylglutamic acid overcomes inhibition non-competitively only with some organisms (vide supra); and those with which this is not the case include the bacteria which are usually attacked by sulphonamide therapy. Nevertheless, with these organisms, too, it is probable that the presence or absence of folic acid is one important factor in determining the sulphonamide sensitivity of the organism. Jukes & Stokstad (1948) have briefly reviewed the distribution of folic acid as estimated by microbiological response of *Lb. casei* or *Strept. faecalis* R, and conclude that it exists naturally to a considerable extent in the form of conjugates (see p. 179). Such conjugates appear to be an available source of folic acid to the animal which has enzymes in certain tissues which release the free factor (Bird, Binkley, Bloom, Emmett & Pfiffner, 1945; Bird, Bressler, Brown, Campbell & Emmett, 1945; Laskowski, Mims & Day, 1945). Although very few organisms have so far been tested this does not seem to be the case always with bacteria. Thus vitamin B_c conjugate (pteroylheptaglutamate) is inactive for both *Lb. casei* and *Strept. faecalis* R, whilst 'fermentation factor' (pteroyltriglutamate) has activity for the former organism but much less for the latter (Pfiffner, Calkins, O'Dell, Bloom, Brown, Campbell & Bird, 1945; Hutchings, Stokstad, Bohonos & Slobodkin, 1944; Lampen & Jones, 1946). Pteroyltriglutamate is less active than pteroylglutamic acid for *Lb. arabinosus* (Lampen & Jones, 1947). More data are required with other organisms, but such considerations taken in conjunction with the possibility that folic acid may be concentrated in certain tissues of the host, may mean that there is very little folic acid in general circulation and in a form available to bacteria.

The possibility that the form of free folic acid found in animal tissues (pteroylglutamic acid) may not be identical with 'bacterial folic acid' or convertible to it by some micro-organisms has been discussed in the previous section.

Finally, one must consider whether sulphonamides inhibit bacterial reactions other than the utilization of *p*-aminobenzoic acid, and which do not occur in animal metabolism. It has been pointed out already that effects on bacterial respiration are probably secondary; even if they were not it is difficult to see how this could explain selective toxicity since the respiratory mechanisms of bacteria are known to be closely similar to those of animal cells, and to involve the same type of coenzymes.

V. OTHER ANTI-METABOLITE GROWTH INHIBITORS

The hypothesis that sulphonamides inhibit the utilization of *p*-aminobenzoic acid by bacteria, and that the latter is essential for their growth, stimulated an intensive search for analogues of other known growth factors which might have antibacterial action (and possibly therapeutic use) for the same reasons. The literature is far too extensive to review in detail here and there are several comprehensive and excellent reviews (McIlwain, 1944; Welch, 1945; Roblin, 1946; Woolley, 1947). As a result of the application of what Fildes (1940) has termed the rational approach to research in chemotherapy, a considerable array of such analogues have been found which inhibit the growth of bacteria *in vitro*. Very few, however, have proved to have significant therapeutic value, and all these so far have been pantothenic acid analogues. McIlwain & Hawking (1943) have found pantoyltaurine to protect rats (but not mice) against streptococcal infection; the dibromoanilide derivative of this compound was effective with mice (White, Lee, Jackson, Himes & Alverson, 1946). Phenyl pantothenone (Woolley & Collyer, 1945) has been found by Marshall (1946) to be of value in the treatment of malarial infestations (blood-induced) of birds and man. The rational approach can of course only suggest possible antibacterial substances and cannot predict whether they will be suitable from the point of view of absorption, excretion and so on by the host: the potential value of pantoyltaurine, for example, is reduced by its rapid excretion by the host (McIlwain & Hawking, 1943). It may, however, be possible in the light of more recent work to make some estimate of whether the substance is likely to be active *in vivo*.

In general growth-inhibition by analogues is overcome by the addition of excess of the factor in question. The concentration of factor (in a form available to bacteria) in the host tissues is therefore of importance. In the case of pantoyltaurine quoted above, the lack of activity in mice compared with rats was shown to be correlated with the concentration of free pantothenate in the blood of these animals.

Many of the analogues tested have been modelled on growth factors which are equally important for the normal metabolism of the host cells (e.g. the vitamin B group). It is not surprising therefore that such analogues have often proved to be equally as toxic to the host as to the bacterium. Thus, for example, pyrithiamine (an analogue of vitamin B_1), besides inhibiting the growth of a variety of micro-organisms, was found by Woolley & White (1943) to induce typical symptoms of vitamin B_1 deficiency in mice; there have been similar observations with certain riboflavin analogues. It might be more profitable therefore to model inhibitors on factors which are essential only to micro-organisms. But it is doubtful whether such factors exist, for all evidence so far points to a basic similarity of fundamental reactions in all types of cell (Knight, 1945). One should therefore seek factors involved in processes quantitatively more important to the parasite than to the host. As Roblin (1946) has pointed out, these may well be those associated more directly with the anabolic mechanisms of the cell, for the rate of cell proliferation of the invading bacteria (and synthesis of new cell material) is much greater than that of the host cells, except possibly in certain tissues, e.g. bone marrow, testis. A short-term interruption to such processes is likely therefore to do relatively little harm to the host, whilst holding the parasite in check sufficiently long for phagocytosis to occur. Such considerations may account in part (vide supra) for the success of the sulphonamides. Unfortunately, knowledge as to which bacterial growth factors are more directly concerned in anabolic processes is very limited, and such a modification of the rational approach may have to await further research on the function of those growth factors whose role in metabolism is at present obscure.

Analysis of the recent work with sulphonamides suggests that it might also be profitable to give consideration to the relative availability to host and parasite respectively of different members of a series of growth factors of increasing complexity but of the same ultimate function. Many of the growth factors or vitamins represent only part of the molecule which finally functions in cell metabolism (e.g. nicotinic acid and the respiratory coenzymes), and the need for the factor reflects the failure of the organism to synthesize that part of the final molecule; if provided with this part it is able to complete the synthesis, and growth occurs. Different organisms often fail at different stages of the chain of reactions leading to the final molecule, and consequently series of organisms are frequently found with increasingly complex requirements (Knight, 1945). If in such a series, therefore, the requirement of the host is for a higher form than that needed by the parasite, it may be possible to inhibit the utilization of the less complex form by the appropriate analogue. There should be no toxicity for the host cell because it cannot in any case carry out the inhibited reaction. It has been shown above that such considerations may explain the selective toxicity of the sulphonamides. From this viewpoint it would be expected that analogues which inhibit the utilization of folic acid would prove toxic both to animal and bacterium. This would seem to be the case, for certain synthetic pterins inhibit the growth of folic acid-requiring bacteria, and also induce the typical signs of folic acid deficiency in chicks; another analogue of pteroylglutamic acid giving a similar effect with bacteria induced a deficiency syndrome in rats (Daniel, Norris, Scott & Heuser, 1947; Daniel, Scott, Norris & Heuser, 1948; Franklin, Stokstad, Belt & Jukes, 1947). Detailed knowledge of the specificity of requirement of host and bacterium for a chain of essential metabolites may therefore be helpful in designing therapeutic agents. Such an approach could of course only suggest antibacterial agents likely to be non-toxic to the host; it could not ensure that these would be active *in vivo*. This would depend on the concentration of antagonists present and these would include presumably the higher form of the factor required by the host itself.

The degree of success attained with pantothenate analogues merits further analysis. It is probable that the functionally active form of pantothenate in both micro-organisms and animals is a higher compound (coenzyme A). Micro-organisms requiring pantothenate convert it to this compound and such conversion is inhibited by pantothenate analogues (McIlwain, 1946; Novelli & Lipmann, 1947). Bacteria which do not require an exogenous source of pantothenate are not susceptible, and since simpler explanations (e.g. synthesis of excess pantothenate or destruction of the inhibitor) do not appear to be valid, it is possible that synthesis of coenzyme A by these organisms does not proceed via free pantothenate as an intermediate. This may apply also to animal cells and thus explain the lack of toxicity of the analogues here also; alternatively the animal cell may require intact coenzyme A, and the effects of pantothenate in animal nutrition follow from bacterial conversion to this substance in the gut. In so far as tested coenzyme A cannot replace pantothenate for the growth of bacteria requiring this factor; the reason for this is not known, possibly it cannot be assimilated.

The nutritional approach to the chemotherapy of bacterial infections has not so far yielded a high dividend in new and potent drugs. But the attempt has led to a much increased knowledge of the complex metabolic relationships between micro-organism, host, essential metabolite and inhibitor, and sufficient success has been obtained to merit continuing the search along these lines. Other aspects of these problems have been considered by other contributors to this Symposium (e.g. Dr H. N. Rydon). Perhaps the greatest difficulty is that of predicting relative toxicity to parasite and host. A close analysis of the failures, as well as of the partial successes, coupled with a deeper understanding of the metabolic function of the various growth factors and vitamins, may well point the way to a more profitable use of this approach.

REFERENCES

Angier, R. B., Boothe, J. H., Hutchings, B. L., Mowat, J. H., Semb, J., Stokstad, E. L. R., Subbarow, Y., Waller, C. W., Cosulich, D. B., Fahrenbach, M. J., Hultquist, M. E., Kuh, E., Northey, E. H., Seeger, D. R., Sickels, J. P. & Smith, J. M., Jr. (1946). *Science*, **103**, 667.

Ansbacher, S. (1944). *Vitamins and Hormones*, **2**, 215.

Bird, O. D., Binkley, S. B., Bloom, E. S., Emmett, A. D. & Pfiffner, J. J. (1945). *J. Biol. Chem.* **157**, 413.

Bird, O. D., Bressler, B., Brown, R. A., Campbell, C. J. & Emmett, A. D. (1945). *J. Biol. Chem.* **159**, 631.

Blanchard, K. C. (1941). *J. Biol. Chem.* **140**, 919.

Briggs, G. M., Luckey, T. D., Mills, R. C., Elvehjem, C. A. & Hart, E. B. (1943). *Proc. Soc. Exp. Biol., N.Y.*, **52**, 7.

Cole, E. H. & Hinshelwood, C. N. (1948). *Trans. Faraday Soc.* **14**, 266.

Daniel, L. J., Norris, L. C., Scott, M. L. & Heuser, G. F. (1947). *J. Biol. Chem.* **169**, 689.

Daniel, L. J., Scott, M. L., Norris, L. C. & Heuser, G. F. (1948). *J. Biol. Chem.* **173**, 123.

Evans, D. G., Fuller, A. T. & Walker, J. (1944). *Lancet*, **2**, 523.

Fildes, P. (1940). *Lancet*, **1**, 955.

Franklin, A. L., Stokstad, E. L. R., Belt, M. & Jukes, T. H. (1947). *J. Biol. Chem.* **169**, 427.

Gale, E. F. (1947). *J. Gen. Microbiol.* **1**, 327.

Goetchius, G. R. & Lawrence, C. A. (1945). *J. Bact.* **49**, 575.

Gordon, M., Ravel, J. M., Eakin, R. E. & Shive, W. (1948). *J. Amer. Chem. Soc.* **70**, 878.

Hall, D. A. (1947). *Biochem. J.* **41**, 287, 294.

Havinga, E., Julius, H. W., Veldstra, H. & Winkler, K. S. (1946). *Modern Development of Chemotherapy*. Amsterdam: Elsevier Publishing Co., Inc.

Henry, R. J. (1944). *The Mode of Action of Sulphonamides*, **2**, no. 1. Publications of the Josiah Macy Foundation, Review Series.

Housewright, R. D. & Koser, S. A. (1944). *J. Infect. Dis.* **75**, 113.

Hutchings, B. L., Stokstad, E. L. R., Bohonos, N. & Slobodkin, N. H. (1944). *Science*, **99**, 371.

James, A. M. & Hinshelwood, C. N. (1948). *Trans. Faraday Soc.* **14**, 274.

Jukes, T. H. & Stokstad, E. L. R. (1948). *Physiol. Rev.* **28**, 51.

Keresztesy, J. C. (1947). Cited by Stokstad, E. L. R. & Hutchings, B. L. *Biol. Symp.* **12**, 239.

Knight, B. C. J. G. (1945). *Vitamins and Hormones*, **3**, 105.

Lampen, J. O. & Jones, M. J. (1946). *J. Biol. Chem.* **166**, 435.

Lampen, J. O. & Jones, M. J. (1947). *J. Biol. Chem.* **170**, 133.

LAMPEN, J. O., ROEPKE, R. R. & JONES, M. J. (1946). *J. Biol. Chem.* **164**, 789.
LANDY, M. & GERSTUNG, R. B. (1945). *J. Immunol.* **51**, 269.
LANDY, M., LARKUM, N. W. & OSWALD, E. J. (1943). *J. Bact.* **45**, 25.
LANDY, M., LARKUM, N. W., OSWALD, E. J. & STREIGHTOFF, F. (1943). *Science*, **97**, 265.
LASKOWSKI, M., MIMS, V. & DAY, P. L. (1945). *J. Biol. Chem.* **157**, 731.
LAWRENCE, C. A. (1945). *J. Bact.* **49**, 149.
LEMBERG, R. (1946). *Nature, Lond.*, **157**, 103.
MCILWAIN, H. (1942). *Brit. J. Exp. Path.* **23**, 265.
MCILWAIN, H. (1944). *Biol. Rev.* **19**, 135.
MCILWAIN, H. (1946). *Biochem. J.* **40**, 269.
MCILWAIN, H. & HAWKING, F. (1943). *Lancet*, **1**, 459.
MARSHALL, E. K. (1946). *Fed. Proc.* **5**, 298.
MILLER, A. K. (1944). *Proc. Soc. Exp. Biol., N.Y.*, **57**, 151.
MILLS, R. C., BRIGGS, G. M., LUCKEY, T. D. & ELVEHJEM, C. A. (1944). *Proc. Soc. Exp. Biol., N.Y.*, **56**, 240.
MIRICK, G. S. (1943). *J. Bact.* **45**, 66.
MOREL, M. (1943). *L'acide nicotinique, Facteur de croissance pour 'Proteus vulgaris'.* Monographies de l'Institut Pasteur. Paris: Masson et Cie.
MORGAN, H. R. (1948). *J. exp. Med.* **88**, 285.
NIMMO-SMITH, R. H., LASCELLES, J. & WOODS, D. D. (1948). *Brit. J. Exp. Path.* **29**, 264.
NIMMO-SMITH, R. H. & WOODS, D. D. (1948). *J. Gen. Microbiol.* **2**, x.
NOVELLI, G. D. & LIPMANN, F. (1947). *Arch. Biochem.* **14**, 23.
PFIFFNER, J. J., CALKINS, D. G., BLOOM, E. S. & O'DELL, B. L. (1946). *J. Amer. Chem. Soc.* **68**, 1392.
PFIFFNER, J. J., CALKINS, D. G., O'DELL, B. L., BLOOM, E. S., BROWN, R. A., CAMPBELL, C. J. & BIRD, O. D. (1945). *Science*, **102**, 228.
RICKES, E. L., TRENNER, N. R., CONN, J. B. & KERESZTESY, J. C. (1947). *J. Amer. Chem. Soc.* **69**, 2751.
ROBLIN, R. O. (1946). *Chem. Rev.* **38**, 255.
SARETT, H. P. (1947). *J. Biol. Chem.* **171**, 265.
SEVAG, M. G. (1946). *Advances in Enzymol.* **6**, 33.
SHREUS, H. T. (1942). *Klin. Wschr.* **30**, 671.
SNELL, E. E. (1946). *Ann. Rev. Biochem.* **15**, 375.
STOKES, J. L. (1944). *J. Bact.* **48**, 201.
STOKES, J. L. & LARSEN, A. (1945). *J. Bact.* **50**, 219.
TAMURA, J. T. (1944). *J. Bact.* **47**, 529.
TATUM, E. L. & BEADLE, G. W. (1942). *Proc. Nat. Acad. Sci., Wash.*, **28**, 234.
TATUM, E. L. & GIESE, A. C. (1946). *Arch. Biochem.* **9**, 15.
VAN DYKE, H. B. (1943). *Ann. N.Y. Acad. Sci.* **44**, 477.
WELCH, A. D. (1945). *Physiol. Rev.* **25**, 687.
WEST, P. M. & WILSON, P. W. (1939). *J. Bact.* **37**, 161.
WHITE, H. J., LEE, M., JACKSON, E. R., HIMES, A. & ALVERSON, C. M. (1946). Cited by ROBLIN, R. O., *Chem. Rev.* **38**, 255.
WILSON, H. E., SASLAW, S. & DOAN, C. A. (1946). *J. Lab. Clin. Med.* **31**, 631.
WINKLER, K. C. & DE HAAN, P. G. (1948). *Arch. Biochem.* **18**, 97.
WOODS, D. D. (1940). *Brit. J. Exp. Path.* **21**, 74.
WOODS, D. D. (1947). *Ann. Rev. Biochem.* **16**, 115.
WOOLLEY, D. W. (1947). *Physiol. Rev.* **27**, 308.
WOOLLEY, D. W. & COLLYER, M. L. (1945). *J. Biol. Chem.* **159**, 263.
WOOLLEY, D. W. & WHITE, A. G. C. (1943). *J. Biol. Chem.* **149**, 285.
WRIGHT, L. D. & WELCH, A. D. (1943). *Science*, **98**, 179.
WYSS, O., GRUBAUGH, K. K. & SCHMELKES, F. C. (1942). *Proc. Soc. Exp. Biol., N.Y.*, **49**, 618.

THE EFFECTS OF LONG-CHAIN FATTY ACIDS ON THE GROWTH OF *HAEMOPHILUS PERTUSSIS* AND OTHER ORGANISMS

By M. R. POLLOCK (Leverhulme Research Fellow)
The Medical Research Council Unit for Bacterial Chemistry, Lister Institute, London

I. INTRODUCTION

The remarkable biological properties of long-chain fatty acids attracted our attention in an unexpected manner. An attempt was being made to determine the growth requirements of *Haemophilus pertussis*, an organism which is well known to be difficult to cultivate in artificial media, and which, on first isolation from cases of whooping cough, can only be grown (though slowly and unsatisfactorily) on the complex Bordet-Gengou agar containing a nutrient peptone base plus potato extract and at least 30% blood (preferably human). Strains of *H. pertussis* can sometimes be trained on to simpler media, but always at the expense of a reduction in virulence and protective antigenicity for laboratory animals. Cultures that are able to grow on plain nutrient agar without blood are of very low virulence (Lawson, 1939).

The strain of *H. pertussis* used was one of slightly reduced virulence (Pollock, 1947) and able to grow on ordinary 10% blood agar, though not on plain nutrient agar without blood (Pl. 1 *a*, *b*). Fractionation of the blood showed that the active factor was albumen. Further experiments showed that blood could also be replaced by 0·2% activated charcoal, previously roasted at 900° to destroy organic matter and twice treated with conc. HCl to remove Fe (Pl. 1 *c*). Moreover, this growth-promoting effect of charcoal could be demonstrated through cellophane placed on a charcoal-agar plate (Pl. 1 *e*). It appeared likely, therefore, that the function of the charcoal and blood was to remove some toxic material from the culture. The analogy between this phenomenon and the results reported by Dubos and his colleagues in their work with the tubercle bacillus was clearly apparent. Davis & Dubos (1946) found that the growth-promoting effect of albumen for the tubercle bacillus in their 'Tween' medium was due to its power of combining with the minute traces of free oleic acid present in the Tween 80 (a water-soluble derivative of sorbitan mono-oleate), and found to be very highly toxic, preventing growth at a conc. of 1 μg./ml. or less.

Tests with *H. pertussis* showed it to be likewise highly sensitive to the unsaturated fatty acids oleic, linoleic and linolenic, concentrations of

4μg./ml. inhibiting growth of heavy inocula; but since, even in the absence of added fatty acid, albumen was necessary for growth, the true sensitivity was considered likely to be much higher. The substance to be tested was incorporated in an agar plate which was then heavily inoculated with a suspension of *H. pertussis*, and albumen added to porcelain cylinders sealed on the surface as used in penicillin assay. The albumen diffused out into the agar, and there developed, around the cups, circular plaques of growth the diameters of which were found to depend upon the amount of albumen added and the antibacterial activity of the compound tested (Pl. 2*a–c*). In this way it was possible to determine the sensitivity of *H. pertussis* to different fatty acids, and to assay roughly the activity of unknown extracts with the minimum of trouble. More elaborate experiments in liquid media, titrating down both albumen and fatty acid, confirmed results obtained by the simpler technique. Palmitic and stearic acids were found to be without action at a conc. of 8μg./ml., and at 40μg./ml. were slightly stimulatory.

Confirmation that at least part of the function of the charcoal in promoting growth was due to its power of absorbing toxic material from the culture was provided by testing a methanol extract of the charcoal used for the growth of *H. pertussis* in tryptic meat broth. This extract proved to be highly toxic to *H. pertussis* itself, its activity being equivalent to about 20μg. oleic acid per ml. of medium (Pl. 2*c*).

In order to complete the picture, repeated attempts were made to train this strain of *H. pertussis* to grow on agar without albumen, in order to test its fatty acid sensitivity, but they were quite unsuccessful. However, two different strains of 'Phase IV', avirulent *H. pertussis* were obtained from another source. Both were able to grow in broth or upon nutrient agar without albumen, blood or charcoal although growth was improved in their presence. One strain grew as well in the presence of 160μg./ml. oleic acid as without it, and growth was only slightly delayed at the highest conc. (320μg./ml.) tested; while the other was not completely inhibited by concentrations below 400μg./ml. It is thus fairly clear that the ability of strains of *H. pertussis* to grow on plain agar is due, as would be expected, to their ability to resist the action of unsaturated fatty acids.

II. ANTIBACTERIAL ACTION OF UNSATURATED FATTY ACIDS

At this point it may be relevant to emphasize that the potent antibacterial action of long-chain fatty acids and their soaps has long been realized (see Bayliss, 1936). The most sensitive organisms are pneumococci (Walker, 1924), haemolytic streptococci (Burtenshaw, 1942), meningococci and gonococci (Walker, 1926), *Corynebacterium diphtheriae* (Walker, 1925; Burtenshaw, 1942), *Erysipelothrix rhusiopathiae* (Hutner, 1942), lactobacilli (Kodicek & Worden, 1945; Williams, Broquist & Snell, 1947) and *Mycobacterium tuberculosis* (Drea, 1942; Dubos & Davis, 1946). In all these cases it is the unsaturated fatty acids which are, in general, by far the most active. Very recently, Foster & Wynne (1948) have shown that germination of the spores of *Clostridium botulinum* is inhibited by oleic acid at concentrations down to 0·1 μg./ml. although the vegetative forms are relatively resistant. Certain viruses are also reported to be susceptible to fatty acids. According to Stock & Francis (1940, 1943) influenza virus is rapidly inactivated by M/1000 oleic, linoleic and linolenic acids; and Casals & Olitsky (1947) have claimed that Japanese B encephalitis is inactivated by serum lipids.

III. SUBSTANCES ABLE TO PROTECT AGAINST EFFECTS OF FATTY ACIDS

The antagonistic action of serum in protecting bacteria against the action of fatty acids is also well known (Lamar, 1911; Walker, 1924; Eggerth, 1927). What, however, has not perhaps been fully appreciated is the extremely low concentrations at which these fatty acids may act; and therefore the possibility of growth of sensitive organisms being inhibited by traces that may be present in many bacteriological media, or be introduced as contamination during their preparation. The protecting and growth-promoting effect of serum even in the absence of added lipoid (McLeod & Wyon, 1921) is well known; and it is, in fact, just those organisms which are particularly sensitive to unsaturated fatty acids which appear to derive most benefit from blood or serum in the culture medium. Serum, however, is not the only protecting colloid used in this way. Starch was first reported as being useful for promoting growth of gonococci and tubercle bacilli over 30 years ago (Vedder, 1915), and has more recently been used and found necessary for the growth of *Haemophilus pertussis* itself in semi-synthetic liquid media (Hornibrook, 1939) and on nutrient agar (Pollock, 1947). Gould, Kane & Mueller (1943) also found starch necessary for the growth of gonococci and meningococci on agar, an effect which Ley & Mueller (1946) attributed to the adsorption of toxic fatty acids present in the agar itself.

The use of charcoal, instead of blood, for promoting bacterial growth was first described (for meningococci) by Wyon (1923), who attributed its action to the absorption of toxic material. This was later confirmed for both meningococci and gonococci by Glass & Kennett (1939). More recently Dubos & Davis (1946) discovered a similar effect with serum albumen for *Mycobacterium tuberculosis*. Like albumen, both charcoal (Linner &

Gortner, 1934; Cassidy, 1940; Klaesson, 1946) and starch (Schoch & Williams, 1944) are known to be highly efficient adsorbers of long-chain fatty acids, and, again, the organisms that benefit are those which are especially sensitive to the common long-chain unsaturated fatty acids. It is even probable that the growth-promoting effect of the saturated fatty acids (palmitic and stearic) for *Haemophilus pertussis* (Pollock, 1947) and for lactobacilli (Kodicek & Worden, 1945) is due to micelle formation at higher concentrations and to the adsorption of the unsaturated acids on to the micelles. Amongst other substances which also probably act in this manner, the most striking is perhaps cholesterol, reported as being able to replace serum for the growth of *Moraxella lacunata* and gonococci in peptone broth (Lwoff, 1947). Cholesterol, too, is able to antagonize the antibacterial action of unsaturated fatty acids (Eggerth, 1927; Kodicek & Worden, 1945).

Although there is no reliable information concerning the amounts of free fatty acid present in the ordinary basal bacteriological media (broth and peptone, etc.), it seemed, in the first instance, to be not unreasonable to assume that the traces inevitably present might be sufficient to account for the failure of such media to support growth of certain fatty acid-sensitive micro-organisms. Indeed, Lwoff (1947) actually found that ether extraction of broth or even simple dilution with water reduced its toxic effect for *M. lacunata*. But if this were the whole explanation, it might be expected that thorough treatment of broth with charcoal, followed by its removal, would render it capable of supporting growth. However, as first shown by Glass & Kennett (1939) for gonococci and meningococci, and later by Pollock (1947) for *Haemophilus pertussis*, this is not so. In the latter case it was also shown that (in the presence of charcoal) growth in charcoal-treated broth was just as good as in the untreated broth, thus disposing of the possible objection that charcoal might have removed some essential growth factors as well as the fatty acids.

IV. MECHANISM OF PROTECTIVE EFFECT OF CHARCOAL, ETC., AGAINST FATTY ACIDS

There were several possible explanations for the fact that the charcoal had apparently to be present *during* growth in order to exert its effect, and most of our subsequent work with *H. pertussis* (so far unpublished) has been directed towards its solution.

The hypotheses which appeared possible are as follows:

(1) *The adsorptive powers of the charcoal might be altered by conditions associated with growth* (*change of* pH, *metabolites, etc.*), *so that fatty acids were only completely taken up while the organisms were growing*.

(2) *The media, glassware, etc., might be contaminated by traces of fatty acids introduced after treatment of the broth with charcoal.*

(3) *The charcoal or albumen might be acting as a sort of buffer for the continued supply, in minute concentrations, of an unsaturated fatty acid which (like oleic acid for tubercle bacilli) was stimulatory in small amounts and inhibitory at higher concentrations.*

(4) *Toxic fatty acids might actually be produced by the organisms during growth and, when liberated into the medium, produce auto-intoxication by action at the external surface of the cell membrane.*

Clearly, the solution of these possibilities would have been greatly facilitated had it been possible to grow *H. pertussis* in a truly synthetic medium. However, it was found that reliable, heavy and constant growth from small inocula could not be obtained even with eighteen amino-acids and sixteen growth factors (see Appendix) with or without charcoal, unless yeast extract was added as well. Therefore, while attempts were being made to determine the nature of the responsible substance or substances in yeast, it was decided to approach the problem from a rather different angle.

Hypothesis (1) appeared to be theoretically unlikely. There was certainly no change in pH during growth of the organisms. As far as is known, there is little evidence for the possibility of changed adsorptive powers of charcoal under such conditions, and this hypothesis has not been further pursued.

Hypothesis (2). The possibility of extraneous fatty acid contamination of the medium needed careful consideration in view of the ubiquitous nature of substances such as oleic and linoleic acids, and their very great antibacterial activity. In the first place, fatty acids are known to form complexes with proteins and carbohydrates, and to be essential components of all organic tissues. Whether or no they have any antibacterial action when so adsorbed is another matter which will be dealt with later. Bacteriological agar, for instance, has been found by Ley & Mueller (1946) to contain, even after prolonged dialysis, an extractable substance (probably fatty acid) demonstrably toxic to gonococci at a concentration of 0·5 μg./ml. which was neutralized by starch. Fatty acid contamination may also arise from inefficiently cleaned glassware, though this can easily be overcome by treating all flasks, tubes and pipettes with hot chromic acid. In liquid media a much more serious source of contamination comes from cotton-wool plugs (Pollock, 1948). Wright (1934) found that a greasy film was deposited on the inside of the glass of a test-tube plugged with cotton-wool and sterilized in an oven at 170° for $1\frac{1}{2}$ hr., and that this greasy material was inhibitory to pneumococci. Later Drea (1942) described the same phenomenon in relation to the growth of tubercle bacilli, and recommended the use of glass, cellophane or aluminium caps instead of cotton-wool plugs.

Sufficient attention has not yet been directed to these important observations which may be the explanation of a number of difficulties associated with the production of satisfactory culture media for micro-organisms. Before glass-wool plugs (fat extracted and subsequently acid-cleaned) were used in experiments on *H. pertussis* the dry-heat sterilization of cotton-wool plugged flasks led to the formation of an invisible film on the inside of the glass which rendered the surface completely unwettable. A rough bio-assay of the oleic acid-equivalent concentration in this greasy film, carried out by means of an oleic acid-requiring bacterium (to be described later), showed that: (*a*) in each cotton-wool plug there was from 0·5 to 1·0 mg. (at least 100 times the amount needed to inhibit organisms such as *Mycobacterium tuberculosis* or *Haemophilus pertussis* in 5 ml. of medium) and (*b*) after hot-air sterilization (180° for $1\frac{1}{2}$ hr.) with cotton-wool plugs the flasks contained on an average 10 μg. deposited free on the inside of the glass (at least twice the amount necessary completely to inhibit the growth of the most sensitive bacteria in 5 ml.). On substituting glass-wool for cotton-wool it was found that, with the same batch of broth, it was possible to reduce the albumen content necessary for the growth of *H. pertussis* from 0·015 to 0·005%. The binding power of albumen for oleic acid tested on the growth of *Mycobacterium tuberculosis* has been calculated by Davis & Dubos (1947) to be 3–6 molecules of oleic acid to 1 of albumen. On these figures, a reduction of 0·010% albumen would be roughly equivalent to the addition of from 6 to 12 μg. oleic acid per flask, a figure which is of the same order as that calculated by bio-assay to be deposited on the glass from the cotton-wool plugs.

Using glass-plugged flasks it was possible to obtain moderate growth, using a relatively large inoculum (10^7 organisms) in a casein hydrolysate medium (see Appendix) in the absence of charcoal. With small inocula, however, both yeast and charcoal were still essential.

It is quite clear that extraneous fatty acid can very easily contaminate culture media, often in unexpected ways, and that extreme care must be taken to prevent such contamination when dealing with organisms known to be highly susceptible. It could not, however, be accepted as a complete explanation of the growth peculiarities of *Haemophilus pertussis*.

Hypothesis (3). This hypothesis needs some further explanation. Dubos (1946) found that oleic acid stimulated the growth of *Mycobacterium tuberculosis* in the presence of albumen, while without albumen it was highly inhibitory. Later, Davis & Dubos (1947) suggested that the growth-promoting power of albumen was due to its action as a buffer for oleic acid, never allowing the concentration to rise to a toxic level while the complex with oleate formed a reserve for its continued supply as it was used up during growth. For this to take place it is necessary to assume that the combination between albumen and oleic acid is reversible. No direct evidence for this has yet been produced, although it would seem to be highly probable on theoretical grounds, and Boyer, Baillou & Luck (1947) have shown that the saturated fatty acids up to caprylic do in fact combine reversibly with albumen. On this hypothesis there would always be a minute quantity of free fatty acid present uncombined (the equilibrium being overwhelmingly in favour of the combined complex), but, as this was utilized by the cells during growth, more would be released by dissociation of the complex to form the equilibrium mixture. The necessity for the continued presence of the adsorbing agent during growth is thus obvious.

Davis & Dubos (1947) also found that purified Tween 80 acted, like albumen, as a protective agent against oleic acid; and it seems not impossible that charcoal might also act in this way, though the adsorption complex might be expected to be more stable.

The tubercle bacillus is not the only organism on which this 'double action' of oleic acid is manifested. The growth-promoting properties of unsaturated fatty acids for certain micro-organisms are now fairly well established and will be dealt with more fully later on. The point here to be emphasized is that in many of these cases the oleic or linoleic acids have an inhibitory effect as well (Table 1). In some instances the nature of the action appears to depend simply upon the concentration; in others, the fatty acid is inhibitory by itself but stimulatory in the presence of some 'protecting' agent such as albumen (for *Mycobacterium tuberculosis*) or saponin (for *Erysipelothrix rhusiopathiae*) or Tween 40 (for lactobacilli). In the case of certain lactobacilli (Williams *et al.* 1947) and of *Mycobacterium*

Table 1. *Growth-promoting and inhibiting properties of unsaturated fatty acids for bacteria*

Fatty acid	Organism	Stimulating conc. (μg./ml.) (*a*)	Inhibiting conc. (μg./ml.) (*b*)	References
Oleic	*M. tuberculosis* (human and avian)	Up to 100 (with albumen	1·0	Dubos & Davis (1946), Dubos (1947)
Oleic	*Micrococcus* 'C'	1·0–10	100	Dubos (1947)
Oleic	*E. rhusiopathiae*	50 (with saponin)	20	Hutner (1942)
Oleic	*C. diphtheriae*	10	50 (lethal)	(*a*) Cohen *et al.* (1941), (*b*) Walker (1925)
Oleic	*L. plantarum* *L. leishmanii*	1·0	10	Hutchings & Boggiano (1947)
Oleic	*L. bulgaricus* *Str. lactis*	0·2–1·0 (alone) or at least up to 10 (+Tween 40)	2·0	Williams *et al.* (1947)
Linoleic	*L. casei*	0·2–1·0	10	Strong & Carpenter (1942)
Linoleic	*L. casei*	0·5–2·0	17	Bauernfeind *et al.* (1942)

tuberculosis (Davis & Dubos, 1947), the water-soluble oleate Tween 80 can act itself as a non-toxic source of the necessary fatty acid growth factor.

The buffer theory is thus certainly applicable to organisms such as *M. tuberculosis* and lactobacilli, but all attempts to demonstrate any growth-promoting activity of unsaturated fatty acids, or even the unidentified toxic charcoal extract previously referred to, for *Haemophilus pertussis* have so far failed; and it does not at the moment appear even to contribute to a solution of this particular problem.

Hypothesis (4). The possibility of toxic fatty acids being actually produced and excreted by the cells themselves was one which occupied our attention from the beginning. In the first place, it seemed reasonable to assume that the organisms must themselves contain fatty acids (free or combined), and that if they did not need them as growth factors they must synthesize them from other constituents in the medium. This, of course, might be achieved *de novo*, or by hydrolysis of fatty acid esters already present in the broth.

An attempt was therefore made, in collaboration with G. A. Howard, to fractionate the extract from 'used' charcoal in order to determine the nature of, and possibly isolate, the substance active against *H. pertussis*. Complete purification and exact identification of the compound or compounds has not yet been achieved, but it was soon found that the activity resided wholly in the small fraction (about 2·5% of the total methanol extract) soluble in petroleum-ether and containing the free fatty acids (iodine number, 51). Earlier in this investigation it had been thought that the activity of this charcoal extract probably represented the free unsaturated fatty acids originally present in the broth. It should, however, be explained that the method of removing the charcoal from the medium (after autoclaving) was to filter through paper or 'hyflo supercel', so that a proportion at least, if not most, of the cells would be retained along with the charcoal and would be extracted with it by the methanol. There was no way of avoiding this difficulty, and the so-called 'charcoal extract' in fact represents the combined extracts from charcoal and organisms. Although, however, it was not possible to extract charcoal and cells separately in any one culture, it was possible to extract charcoal which had been in contact with uninoculated broth for an equivalent time under similar conditions to that in the growing culture, and to extract cells alone washed off the surface of a charcoal-agar medium. The anti-*H. pertussis* activities, expressed in terms of oleic acid per ml. of medium, and in some cases the actual amounts of total free fatty acids obtained (petrol-soluble fraction) from these extracts (Table 2) show (*a*) that from 4 to 8 times more active material was obtained from charcoal extracts of growing cultures than from the sterile medium, and (*b*) that the cells themselves contained material (either adsorbed on the surface or actually within the cell wall) which possessed very high activity in inhibiting their own growth.

This, in itself, was fairly good evidence that *H. pertussis* produces in broth a large proportion of the toxic substance which inhibits its own growth. But in the absence of any reliable data on the fatty acid content of tryptic meat broth, it was impossible to arrive at a firm conclusion, and actual proof that this was in fact the case was eventually supplied, with an entirely different method, largely through the good fortune of a rather remarkable coincidence.

Table 2. *Comparison of quantity and anti-*Haemophilus pertussis *activity of material extracted by methanol from charcoal in inoculated and uninoculated tryptic meat broth, and from cells alone grown on charcoal-agar*

Whether inoculated	mg./l. of medium		
	Methanol extract	Petrol-soluble fraction	Approx. anti-*H. pertussis* activity (expressed as oleic acid)
+	4060	~	20
−	748	~	2·2
+	615	19·0	~
−	270	2·5	~
+	750	11·2	15
−	210	3·8	4·2
Cells alone extracted (grown on charcoal agar)	44	~	14

Incubation: For 5 days at 35°.
Inoculum: *H. pertussis* (same strain as that used for testing activity).

V. ISOLATION AND PROPERTIES OF AN OLEIC-ACID-REQUIRING BACTERIUM (*CORYNEBACTERIUM* 'Q')

In all experiments with *Haemophilus pertussis* it was the practice to incubate a control-inoculated flask of plain broth as a check on the stability of the culture. There should, of course, be no growth in this tube. However, on one occasion slow uniform growth somewhat characteristic of *H. pertussis* occurred in this flask, the contents of which were therefore plated out on to 10% blood agar on the suspicion of contamination. But a heavy and apparently pure growth of *H. pertussis* resulted on this subculture; and, what was more striking, repeated subcultures from the original broth tube into further broth resulted in rapid heavy growth of what was apparently pure *H. pertussis*. However, on continued incubation of the blood plates used for plating out these cultures there appeared some minute pin-point colonies (barely visible to the naked eye, even after 7 days) of an organism which proved to be a Gram-positive *Corynebacterium*, easily distinguishable from *Haemophilus pertussis* in a stained film.

This micro-organism (not yet fully identified, but provisionally labelled *Corynebacterium* 'Q') proved to be an obligatory aerobic diphtheroid, fermenting glucose, but not lactose, sucrose, mannitol or dulcitol. Its source was never established, although there seems no reason for believing that it appeared in any other way than as a chance contaminant from the air.

Corynebacterium 'Q' proved to have several interesting properties. In almost every respect its growth requirements were the exact opposite to those of *Haemophilus pertussis*. Growth in broth was scanty, and on nutrient agar small, thin, grey colonies appeared after 48 hr. incubation. On 10% blood agar, pin-point, barely visible colonies appeared after 5–7 days, while on 0·2% charcoal agar no visible growth occurred at all, even with the heaviest inoculum. On agar containing 1:25,000 oleic acid, however, growth was luxuriant.

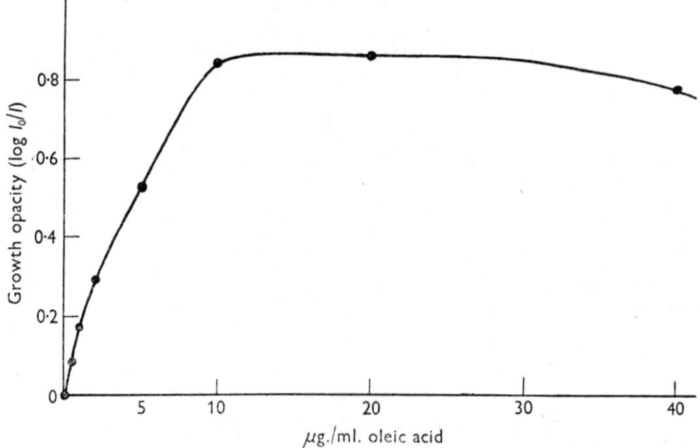

Text-fig. 1. Relation between growth of *Corynebacterium* 'Q' and oleic acid content of the medium. Inoculum: approx. 10,000 cells. Medium: vitamin-free casein hydrolysate + 16 growth factors. Incubation: 35° for 3 days.

Experiments in a synthetic medium containing eighteen amino-acids and sixteen growth factors (see Appendix) proved that 'Q' was, in fact, an oleic acid-requiring organism: or more correctly, an unsaturated fatty acid-requiring organism, for linoleic and linolenic acids had the same activity as oleic acid (visible growth occurring in all cases at a concentration of 0·5 μg./ml. or less), while palmitic and stearic acids were both quite inactive. Moreover, the extent of growth (measured by opacity) was proportional to the amount of oleic acid added (see Text-fig. 1), and it was found possible to use the organism for biological assay of minute quantities of oleic acid.

Finally, 'Q' had the striking property of stimulating the growth of *H. pertussis*, not only in plain broth and on agar, but in synthetic media and from minute inocula. Moreover, this symbiosis was mutual. Pl. 1 (*b* and *d*) shows that the growth of *H. pertussis* on an agar plate previously heavily inoculated with 'Q' was as good as on a blood plate. Colonies of 'Q' arising from single cells inoculated on to agar previously heavily inoculated with *H. pertussis* promoted satellite growth of *H. pertussis* colonies around them (Pl. 3*a*). In broth, 'Q' and *H. pertussis* inoculated into separate flasks led to feeble growth of the former and none of the latter, while if both were inoculated together into the same flask, growth was very heavy (Table 3), and on plating out could be shown to be a mixture of both organisms, *H. pertussis* predominating. This effect was even more striking in the casein hydrolysate media where neither organism grew visibly when inoculated alone. The reciprocity of this stimulation was finally confirmed by demonstrating satellitism of 'Q' colonies around those of *H. pertussis* on 0·2% charcoal agar (Pl. 3*b*).

Table 3. *Symbiosis of an oleophobe* (Haemophilus pertussis) *and an oleophile* (Corynebacterium 'Q')

Medium	Inoculated organism	Growth	
		4 days	7 days
Tryptic meat broth	'Q'	tr. +	tr. +
	H. pertussis	—	—
	'Q' + *H. pertussis*	—	+ + + +
Synthetic medium	'Q'	—	—
	H. pertussis	—	—
	'Q' + *H. pertussis*	—	+ + +*

* Plated out and found to be a mixture of *H. pertussis* and 'Q' in the proportions of approximately 50:1.

Inocula: 'Q', *c.* 15,000 cells; *H. pertussis*, *c.* 5000 cells.

VI. PRODUCTION AND LIBERATION OF 'OLEIC-ACID' BY *HAEMOPHILUS PERTUSSIS*

The simplest, and indeed most probable explanation of these phenomena is that oleic acid, or some similar compound, is in fact formed by *H. pertussis*, and needed and removed by 'Q'. A simple transference of oleic acid from *H. pertussis* to 'Q' would explain all the observed results. That this, or something extremely similar, is what in fact occurs is proved by the experiments summarized in Text-fig. 2. A large inoculum (10^7 cells) of *H. pertussis* was added to the basal medium of vitamin-free casein hydrolysate plus sixteen growth factors (*a*) with and (*b*) without 0·04% charcoal. Viable counts show how, after 2 days, growth without charcoal ceased and the cells died rapidly, while in the presence of charcoal growth continued and viability was maintained for at least 4 days. Over this period samples from the two cultures were removed and the supernatant separated after spinning off the cells and charcoal, autoclaved and assayed with 'Q', in the same medium, for oleic acid content. The onset of the appearance of 'oleic acid' in the medium outside the cells in the absence of charcoal to a final concentration of 4·8 μg./ml. coincided with cessation of growth and death of the cells. In the presence of charcoal a relatively small release of fatty acid occurred. The 7-day supernatant from the culture without charcoal was also tested, in the same medium, against *H. pertussis*, using the same-sized inoculum, and found to inhibit growth completely at a concentration of 1:5 and delay growth at 1:20. These dilutions correspond to a concentration of oleic acid of 1·2 and 0·3 μg./ml. respectively, as assayed by 'Q'.

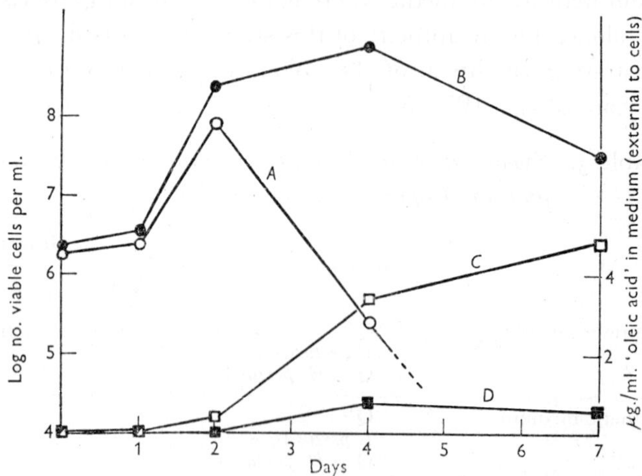

Text-fig. 2. Comparison of viable counts and of 'oleic-acid' liberated into the medium during growth of *Haemophilus pertussis* in casein hydrolysate plus growth factors with and without charcoal: A, viable count without charcoal; B, viable count with charcoal; C, 'oleic-acid' in medium without charcoal; D, 'oleic-acid' in medium with charcoal. *Note.* Viable count of cells without charcoal after 7 days was < 10/ml.

Under exactly similar conditions, pure oleic acid itself completely inhibited *H. pertussis* at 1·0 μg./ml. and delayed growth at 0·1 μg./ml. The active material liberated from the cells in cultures of *H. pertussis* has, therefore, the same relative activity on *H. pertussis* and 'Q' as oleic acid, and it appears not unlikely that they are identical or very closely related compounds.

It can thus be accepted not only that *H. pertussis* does synthesize an oleic acid-like substance during growth, but also that this substance is actually liberated into the medium and, in the absence of charcoal or some suitable colloid capable of adsorbing it, is enabled to exert its effect at the external surface of the cell. Whether the substance is actually oleic acid, or some other fatty acid known or unknown, or a mixture, and whether it is actively secreted by the cells or merely liberated by lysis are problems that have not yet been solved, and will not be further considered here. However, it must be pointed out that this demonstration of synthesis and liberation of fatty acids by micro-organisms is not entirely new. It has already been observed, in the case of *Pseudomonas pyocyanea*, by Hettche (1934) who attributed the antagonism shown by this organism against gram-positive bacteria to a fatty acid secreted into the medium. An analogous case is described by Hölzl (1940) who gave the same explanation of the inhibitory effect of an unidentified coccus upon the growth of *C. diphtheriae*. The case of *H. pertussis* is exceptional only in so far as the fatty substance formed and liberated is highly toxic to the producer organism itself.

VII. UNSATURATED FATTY ACIDS AS GROWTH FACTORS FOR MICRO-ORGANISMS

The need for certain long-chain unsaturated fatty acids by some species of bacteria has only recently been firmly established, although Fleming (1909) reported the growth-stimulating properties of oleic acid for *Corynebacterium acnes* nearly 40 years ago. Oleic acid has since been shown to be necessary for the growth of *C. diphtheriae* (Cohen, Snyder & Mueller, 1941), *Clostridium tetani* (Feeney, Mueller & Miller, 1943), *Cl. sporogenes* in the absence of biotin (Shull, Thoma & Peterson, 1948), the unidentified *Micrococcus* 'C' (Dubos, 1947) and certain lactobacilli, in some cases only in the absence of biotin (Williams & Fieger, 1946, 1947; Williams *et al.* 1947; Hutchings & Boggiano, 1947; Whitehill, Oleson & Subbarow, 1947). It is also known to stimulate the growth of *Mycobacterium tuberculosis* (Dubos, 1946) in the presence of albumen or Tween and that of *Erysipelothrix rhusiopathiae* in the presence of saponin (Hutner, 1942). Linoleic acid has been reported as stimulating certain lactobacilli (Strong & Carpenter, 1942; Bauernfeind, Sotier & Boruff, 1942), including many also able to grow with oleic acid; and linolenic and arachidonic acids are able to replace oleic acid for *Micrococcus* 'C' (Dubos, 1947).

In these cases, higher concentrations of the unsaturated fatty acids have, where tested, been found to have an inhibitory effect on the same organisms which they stimulate at low concentrations (Table 1). Perhaps the most striking case of this 'double action' is that of oleic acid on *Lactobacillus bulgaricus* (Williams *et al.* 1947), where the stimulatory effect has a sharp optimum at a concentration of 1·0μg./ml. above which the oleic acid becomes inhibitory, and no growth occurs at all (at pH 6·5) above 2·0μg./ml. However, in the presence of Tween 40 (the water-soluble palmitate ester of the polyoxyethylene derivative of sorbitan), itself inactive, the toxic effect of oleic acid is abolished and the usual type of relationship between mass of growth and concentration of growth factor supervenes, with optimal growth at 4μg./ml. and no inhibitory effect at least up to 10μg./ml. The analogous water-soluble oleate ester (Tween 80) behaves like the mixture of Tween 40 and oleic acid itself; that is, it stimulates growth without any inhibitory action at higher concentrations and in the absence of added oleic acid. Although unpurified Tween 80 contains a small proportion of free oleic acid and even the pure compound undergoes spontaneous hydrolysis on storage (Davis, 1947), the effect mentioned above is greater than could be accounted for by contamination with unesterified oleic acid, and it appears probable that Tween 80 supplies the oleic acid needs of *L. bulgaricus* without possessing the disadvantages of the free acid.

Oleic acid has also been reported as necessary for the growth of certain fungi (e.g. *Pityrosporon ovale*) by Benham (1941); and linoleic acid is suggested as the probable fatty compound present in serum which is necessary for the growth of a protozoon, *Trichomonas vaginalis* (Sprince & Kupferberg, 1947). Like bacteria, some fungi are also inhibited by fatty acids (e.g. *Trichophyton interdigitale*, Wyss, Ludwig & Joiner, 1945; and *Microsporon audouinii*, Rothman, Smiljanic & Weitkamp, 1946). In general, however, although little work has yet been done on the action of fatty acids on non-bacterial micro-organisms, they do not appear to show such a high order of sensitivity as bacteria.

VIII. SUMMARY AND DISCUSSION

It is a difficult task to attempt to summarize and co-ordinate the scattered and in some cases unrelated data presented in this somewhat historical account of the subject. Inevitably, a discussion such as this leads to putting forward more questions than are answered. It is a temptation, when first confronted with the range of biological activities of the long-chain unsaturated fatty acids, to try and explain too much; and the ensuing observations must be treated largely as speculations upon the many possible ways in which these substances might play a part in the economy of micro-organisms.

One of the most striking properties of these unsaturated fatty acids is what has previously been referred to as their 'double action'. For *Lactobacillus bulgaricus*, for instance, oleic acid is a growth factor at a concentration of 1·0 μg./ml. and a growth inhibitor at 2·0 μg./ml. (Williams *et al.* 1947). Or, again, oleic acid exerts some inhibitory effect on *Mycobacterium tuberculosis* at a concentration of 0·1 μg./ml. (Drea, 1942; Dubos & Davis, 1946), yet is stimulatory in the presence of albumen or Tween 80 up to a concentration of 100 μg./ml. The margin between the stimulating and inhibiting concentration is small or non-existent for all bacteria shown to need these substances.

What, in particular, makes the antibacterial action of oleic acid remarkable is that it is presumably a normal constituent of every living cell, in most cases probably in considerable concentration. The same may apply to linoleic and linolenic acids. It is true, of course, that most of the fatty acids in the cells are combined as the microbiologically inactive esters—neutral fats, cholesterol esters, etc.—and it is probable that a large part of the remaining 'free' fatty acids are adsorbed, to a greater or lesser degree, on to large molecules or colloids—presumably mainly proteins and carbohydrates. To take one example, namely, blood: all protein fractions are known to contain lipoid (Blix, Tiselius & Svensson, 1941), and it is probable that the small quantities of unesterified fatty acid are bound in some way or other with protein. Although blood and serum are known to have antibacterial activity, which may partly be due to traces of free fatty acid, their main effect in bacteriological media is to antagonize the action of unsaturated fatty acids.

It is probable that most 'purified' proteins and polysaccharides contain adsorbed fatty acid. For instance, it is possible by cold ether extraction of plasma to remove a portion of the fat, and the albumen therefrom can be crystallized several times and thoroughly dialysed, yet still contain up to 2·9% of free fatty acid which is only removed by hot ethanol extraction after denaturation of the protein (Kendall, 1941). The twice crystallized horse-serum albumen used for the growth of *Haemophilus pertussis* (Pollock, 1947) was found to yield 1·4% fat by this method. This fatty material proved to have an inhibitory effect on *H. pertussis* which corresponded to an oleic acid content of approximately 10%. The addition of albumen to a final concentration of 0·5% thus inevitably entails the simultaneous addition of 4·0μg./ml. bound oleic acid (or its biological equivalent). The higher concentrations of albumen are not, however, inhibitory to *H. pertussis* growing in broth, 0·5% actually promoting slightly more rapid and more profuse growth than the concentration of 0·1% usually employed. Thus the firmly bound fatty acid did not apparently destroy the effectiveness of the albumen in protecting against external fatty acid during growth. Purified, 'analar' starch, used successfully to promote the growth of meningococci, gonococci and *H. pertussis*, was found by us to contain 0·05% lipoid on extraction with hot methanol. This fatty material also proved to be inhibitory to *H. pertussis* at a concentration of 20μg./ml. without further purification. Most cereal starches contain a considerable concentration of fat. Maize starch is said to contain 0·65% fat (Schoch & Williams, 1944), of which 75% is reputed to be oleic or linoleic acid (Taylor & Lehrman, 1926). Rice starch contains 0·83% fatty matter (Taylor & Lehrman, 1926). These impurities have given rise to erratic results in the assay of biotin (Williams & Fieger, 1945) and riboflavin (Bauernfeind *et al.* 1942; Strong & Carpenter, 1942) with certain lacto-bacilli for which oleic acid may partially or completely replace biotin or riboflavin as necessary growth factors. Potato starch, however, is reported as being free from fat (Lehrman & Kabat, 1933), and it is a matter for speculation whether this fact bears any relation to its use in the Bordet-Gengou medium for primary isolation of *H. pertussis*. However, unless the binding power of the protein or carbohydrate for fatty acid is already saturated by fat originally present, it is more likely that small amounts of previously bound lipoid will not materially affect their ability to protect against additional fatty acid.

Another polysaccharide, ordinary bacteriological agar, has been found to contain fatty acids. Ley & Mueller (1946) extracted 30 g. of dialysed agar with methanol, and, after partial purification, obtained 3 mg. of a fatty substance, similar to oleic acid, which inhibited the growth of gonococci in liquid media at a concentration of 0·5 μg./ml. This gives a final concentration of 2·0 μg./ml. in the ordinary 2% agar plate, quite sufficient in itself to inhibit a number of fatty acid-sensitive bacteria. These workers also showed that agar, thus extracted, could be used in solid media for gonococci without the addition of starch that is usually necessary for sensitive strains. But the question arises as to whether this fatty acid which is bound to agar and not removed by prolonged dialysis is in fact under such conditions biologically active against micro-organisms. The fatty acids bound to albumen and starch do not, apparently, adversely affect the organisms which are stimulated by these colloids. An alternative possibility is that the combining power of agar for fatty acid is anyway small, and that the removal of the original fat bestows upon it a feeble power to act as a starch substitute and adsorb the small quantities of fatty acid originally present in the medium or possibly produced, as in the case of *H. pertussis*, by the cells themselves. The fact that gonococci would grow in the liquid medium without starch does not really dispose of this hypothesis because if, in fact, gonococci liberate fatty acid into the medium during growth, the localized concentration within a micro-colony on agar is likely to be much greater than that attained in liquid cultures.

The use of fat-extracted agar was found to make little difference to the growth of *H. pertussis*, the addition of charcoal still being essential. In fact, the fatty extract from this agar (prepared by the method of Ley & Mueller) was found to be inhibitory to *H. pertussis*, but at a concentration which would be equivalent to less than 0·2 μg./ml. oleic acid in 2% agar.

The important factor, therefore, appears to be not so much the fatty acid content of a material, but the extent and firmness of the complexes formed between the fatty acids and the adsorbing substances. As far as the fatty acid-sensitive bacteria are concerned, the fatty acid is harmless (or maybe even beneficial) as long as it is 'in its place'. But if the conditions are altered, if the protecting colloid is destroyed or removed, or if the micro-organisms are cultivated in an environment where the ubiquitous fatty acids are not detoxified by adsorption on to larger molecules or particles, then a very powerful antibacterial agent is unleashed.

If these points are borne in mind it is possible to offer some constructive criticism of some of the methods of artificial culture of bacteria. As previously pointed out, the role of serum in promoting growth appears to depend very largely upon its power of 'detoxifying' free fatty acids. The question as to where this fatty acid comes from in specific cases is largely unanswered. In complex, 'natural' media there is clearly some in the medium from the start. It may also appear by contamination of glassware with extraneous fat (e.g. from cotton-wool); and it may, as has been shown in the case of *H. pertussis*, be formed by the growing cells themselves and escape into the medium by lysis or secretion and so exert its effect externally on the cell membrane.

A further possibility is the action of lipase, present either in the medium or in the cells, which would liberate free fatty acids from non-toxic esters. Davis & Dubos (1948) have shown that uncrystallized bovine serum albumen originally used in their Tween medium for *Mycobacterium tuberculosis* was contaminated by serum lipase which liberated free oleic acid from the Tween 80, thus proving very much less satisfactory than purified serum albumen. They also found that a sterile filtrate of the Tween albumen medium, after growth of *Mycobacterium phlei*, was toxic for *M. tuberculosis* in the presence of Tween 80. Treatment of this filtrate for 30 min. at 56° destroyed its growth-inhibiting property, and they proved the effect to be due to the presence of a lipase derived from *M. phlei*. Humfield (1947) has reported the case of an unidentified bacterium apparently producing a powerful antibiotic in a medium containing wheat bran. The 'antibiotic', however, proved to have chemical and biological properties very similar to linoleic acid, and the very reasonable suggestion was made (unfortunately not yet supported by direct experimental evidence) that the organisms had merely hydrolysed the fats present in wheat bran releasing water-soluble soaps which were easily extractable and of very high activity against certain susceptible bacteria (*Staphylococcus aureus* and *Streptococcus faecalis*).

Enough has been said concerning the parts played by fatty acids in bacteriological media to show their importance. It can now be appreciated how far from ideal certain culture media may be for sensitive micro-organisms. These points may have some significance in the production of satisfactory vaccines, for instance. In the case of *Haemophilus pertussis*, where the development of a satisfactory vaccine may well depend upon the method of culture of the organisms, an analysis of the Bordet-Gengou medium, universally used hitherto for vaccine production, may serve to emphasize the problem. The basal medium of peptone or tryptic meat broth will already contain some free fatty acid, and the beneficial effects of the blood and potato starch may be due largely to their ability to adsorb this free fat. But, while the more blood is added the more albumen is available, at the same time the concentration of possibly toxic serum lipoid is raised. In addition, the serum lipase may increase the toxicity of the medium by liberating further free fatty acids. It should, incidentally, be mentioned that Bordet & Gengou (1906) advised heating serum (used in their liquid medium for *H. pertussis*) for $\frac{3}{4}$ hr. at 56°. They gave no reason for this recommendation, but the effect would certainly have been to destroy the serum lipase and possibly considerably reduce the toxicity of the medium. The beneficial effect of human blood (as opposed to the more usually employed horse blood) in maintaining virulence and antigenicity (Toomey & Takacs, 1938) may be due to the much higher content of lipase in horse serum (Davis & Dubos, 1948). Even on the slender existing knowledge, therefore, the Bordet-Gengou medium would appear to be far from ideal.

The slow growth of *H. pertussis* on the Bordet-Gengou agar might in itself be due partly to the inhibitory action of free fatty acids; but in any case it is possible that the growing cells would be progressively modified and attenuated in the presence of toxic fatty acids, with diminution of virulence and antigenicity. Stress is usually laid upon the necessity for young cultures in vaccine production, and Gray (1946) has reported that 24 hr. cultures on Bordet-Gengou agar have much greater virulence and form more efficient vaccines (Gray, 1947) than those grown for 48 hr. or longer.

The consideration of these problems of artificial culture thus leads directly to the question of the possible effects of fatty acids on virulence and pathogenicity of micro-organisms in the animal body.

It has already been shown that strains of *H. pertussis* which have been trained to grow on plain nutrient agar without blood do so because they have developed a resistance to the action of unsaturated fatty acids. These strains are also known to have greatly reduced virulence and protective antigenicity, and this change of phase is usually regarded as analogous to the S→R variation studied so extensively in the typhoid-Salmonella group. Although there is so far little direct evidence, it is not impossible that the association of reduced virulence and insusceptibility to fatty acids is more than coincidental. An analogous case of S→R variation being associated with loss of sensitivity to oleic acid has been reported by Morel (1945) for *Proteus*; although this is, as she pointed out, certainly not true of all S→R variations in micro-organisms.

Consideration should also be given to the possibly related phenomena of virulence enhancement of bacteria by mucin. Nungester, Wolf & Jourdonais (1932) found that the virulence of pneumococci, haemolytic streptococci and *Staphylococcus aureus* injected intraperitoneally into mice was considerably increased if the organisms were suspended in gastric mucin; and that agar, starch and gelatin also provided a 'temporary protective coating' for the bacteria. In subsequent work there appeared to be some difference of opinion both as to which species of bacteria respond to mucin in this way, and also what other substances, besides mucin, are effective. But there seems no doubt that *Haemophilus pertussis* is one of those which are affected, either by starch (Powell & Jamieson, 1936) or mucin (Silverthorne, 1938; Mishulow, Klein, Liss & Leifer, 1939). The fatty acid combining power of the active fraction of mucin is unknown, but there can be little doubt that at least one of the effects of the starch would be to protect the bacteria against the action of unsaturated fatty acids.

The possible effects of fatty acids upon the antigenicity of micro-organisms is an even more difficult problem to discuss. Middlebrook & Dubos (1947) have succeeded in producing specific anti-Tween 80 anti-serum in rabbits by immunization with tubercle bacilli grown in, or simply treated with, a solution of Tween 80. By this they have demonstrated that Tween is actually adsorbed on to the surface of the cell. It would have been particularly interesting to know whether this adsorption of Tween caused any alteration in the antigenic properties of the normal proteins in the cell, by blocking their active surfaces, etc. It seems likely that such surface adsorption may take place with naturally occurring fatty acids, such as oleic acid; and if this is so, the effective surface antigenic pattern may be sufficiently distorted to cause a reduced production of protective antibodies in the animal immunized with organisms treated in such a way. Davis (1948) has, in fact, claimed to have demonstrated the adsorption of oelic acid by living tubercle bacilli, but unfortunately his experiments do not really eliminate the possibility of some sort of neutralization effect by large molecules liberated from the cells into the medium. In any case, one may, perhaps, visualize, along lines suggested by Hotchkiss (1946), that the first stage of action is likely to be a combination of fatty acid with the external cell surface, and might involve partial or complete inhibition of growth and modification of antigenicity before the effect on cell metabolism ended in irreversible inactivation of all vital functions. Thus the coating of bacteria with fatty acids during artificial cultivation might conceivably be responsible for poor antibody response on vaccination.

To what extent the animal body makes use of its natural fatty acids in combating infection within the tissues is completely unknown. But there is no doubt that these defences come into action in skin infections. Burtenshaw (1942) has prepared extracts from human skin and hair and shown them to be highly inhibitory to haemolytic streptococci, *Corynebacterium diphtheriae*, *Streptococcus viridans* and some strains of pneumococci and staphylococci. He identified the active fractions as fatty acids. Ricketts, Squire & Topley (personal communication) have confirmed Burtenshaw's results and further shown, using *Str. pyogenes* and *Staphylococcus aureus* (coagulase +ve) as test organisms, that the antibacterial activity of skin lipoid lies in the unsaturated fatty acid fraction. Rothman *et al.* (1946) ascribe the spontaneous cure of ringworm (*Microsporon audouinii*) at puberty to the change in type of fatty acids in the sebaceous secretion which occurs at that time. They have shown that ether extracts of adult hair are five times more effective *in vitro* against the fungus than are similar extracts of child hair.

It is thus probable that the sebaceous secretions are largely responsible for the self-disinfection of the skin. But whether the same processes play any part in combating infections of the mucous membranes and glands or other parts devoid of sebaceous glands is as yet unknown.

On the other hand, there are some micro-organisms, as might be expected, which are stimulated by the natural skin fat. Pl. 3*c* and *d* shows this 'double-action' effect of fatty acids in a human hair lying on the surface of a 0·2% charcoal agar plate heavily inoculated (*a*) with 'Q' and (*b*) with *Haemophilus pertussis*.

The question thus arises whether some of these fatty acid-requiring organisms might act in the body, as 'Q' has been shown to act on *H. pertussis in vitro*, protecting the invading bacteria against the natural fatty acid defences, and thus contributing to the spread of the disease. This effect, of course, would be more marked if the symbiosis were truly mutual.

The therapeutic possibilities of unsaturated fatty acids have apparently at times been vaguely considered, but from what is already known it would appear rather unlikely that they could have much action *in vivo*. Chaulmoogra oil has been used for leprosy with reputed success, and of course soaps are well recognized as highly effective skin disinfectants, at least for certain bacteria. It is on the external surface of the body, and perhaps in the upper respiratory tract, that one might expect fatty acids to be still left unattached, and capable of acting against micro-organisms.

It is probable that the full significance and implications of the biological activities of long-chain fatty acids are only just beginning to be appreciated. They have not so far been studied as extensively as the corresponding component parts of proteins and carbohydrates, and the only systematic work on bacterial lipids has been that of Anderson and his co-workers on tubercle bacilli. But, even at this stage it is recognized that fungi and bacteria are not the only living entities profoundly affected by specific fatty acids. Fraenkel & Blewett (1946, 1947) have shown that linoleic and linolenic acids are essential growth factors for the proper development of the moth *Ephestia kuehniella*. Oleic acid was found here to be ineffective, and has been reported to be highly toxic for mosquito larvae (Golberg, 1946), 0·1% in the diet suppressing pupation and 0·5% causing rapid death. The probably essential nature of some of the higher unsaturated fatty acids in mammalian nutrition (see Bloor, 1943) is also recognized.

It is not the purpose of this article to speculate upon mode of action. The most striking property of unsaturated fatty acids appears to be what has been referred to as 'double action'—the combined stimulating and inhibiting action, even in many cases upon the same micro-organism. This might, of course, be simply the reflexion of two completely unrelated characteristics of the same molecule. Nevertheless, it is remarkable that, with the possible exception of certain lipophilic organisms that appear to thrive on large quantities of fatty acid (e.g. *Corynebacterium acnes* (Fleming, 1909), *P. ovale* (Benham, 1941)) all micro-organisms which have so far been shown to need fatty acids for growth, have also, where tested, been found to be inhibited by only a slightly higher concentration of the same compound. This suggests that in these cases the property of the fatty acid which confers the ability to stimulate growth may be closely related to, or identical with, the property responsible for inhibition.

I am indebted to H. Proom and A. F. B. Standfast for strains of avirulent ('Phase IV') *Haemophilus pertussis*; to Dr W. T. J. Morgan for gifts of pure fatty acids; to Dr R. A. Kekwick for a gift of crystalline horse-serum albumen; and to the Curator and staff of the National Type Collection of Cultures for doing some tests on an unidentified diphtheroid bacterium.

I also wish to express my appreciation to Sir Paul Fildes, F.R.S., for his constant advice and encouragement.

APPENDIX

The casein hydrolysate medium referred to in the text was made up as follows:

'Vitamin-free' casein hydrolysate (Ashe)	0·5 %
KH_2PO_4	0·45 %
NaOH	to pH 7·6

+M/6000-Mg^{++}, M/200,000-Fe^{++}, M/10,000-cystine, M/10,000-tryptophan, and the following growth factors:

Biotin	(5·0 mμg./ml.)	Guanine sulphate	(5·5 μg./ml.)
Nicotinamide	(0·5 μg./ml.)	Inositol	(4·5 μg./ml.)
Ca pantothenate	(0·4 μg./ml.)	Thymine	(3·25 μg./ml.)
Pyridoxin	(1·25 μg./ml.)	Uracil	(2·75 μg./ml.)
Riboflavin	(0·6 μg./ml.)	Adenine sulphate	(5·0 μg./ml.)
Thiamin HCl	(0·25 μg./ml.)	Cytosine	(2·75 μg./ml.)
p-Aminobenzoic acid	(0·25 μg./ml.)	Pimelic acid	(0·4 μg./ml.)
Haemin	(1·0 μg./ml.)	Folic acid	(0·5 μg./ml.)

In some experiments with *Corynebacterium* 'Q' the casein hydrolysate, tryptophan and cystine were replaced by M/4000 of each of eighteen amino-acids.

REFERENCES

BAUERNFEIND, J. C., SOTIER, A. L. & BORUFF, C. S. (1942). *Industr. Engng Chem.* (Anal. ed.), **14**, 666.
BAYLISS, M. (1936). *J. Bact.* **31**, 489.
BENHAM, R. W. (1941). *Proc. Soc. Exp. Biol., N.Y.*, **46**, 176.
BLIX, G., TISELIUS, A. & SVENSSON, H. (1941). *J. Biol. Chem.* **137**, 485.
BLOOR, W. R. (1943). *Biochemistry of the Fatty Acids.* New York: Reinhold Publishing Corporation.
BORDET, J. & GENGOU, O. (1906). *Ann. Inst. Pasteur*, **20**, 731.
BOYER, P. D., BALLOU, G. A. & LUCK, J. M. (1947). *J. Biol. Chem.* **167**, 407.
BURTENSHAW, J. M. L. (1942). *J. Hyg., Camb.*, **42**, 184.
CASALS, J. & OLITSKY, P. K. (1947). *Science*, **106**, 267.
CASSIDY, H. G. (1940). *J. Amer. Chem. Soc.* **62**, 3073.
COHEN, S., SNYDER, J. C. & MUELLER, J. H. (1941). *J. Bact.* **41**, 581.
DAVIS, B. D. (1947). *Arch. Biochem.* **15**, 359.
DAVIS, B. D. & DUBOS, R. J. (1946). *Arch. Biochem.* **11**, 201.
DAVIS, B. D. & DUBOS, R. J. (1947). *J. Exp. Med.* **86**, 215.
DAVIS, B. D. & DUBOS, R. J. (1948). *J. Bact.* **55**, 11.
DREA, W. F. (1942). *J. Bact.* **44**, 149.
DUBOS, R. J. (1946). *Proc. Soc. Exp. Biol., N.Y.*, **63**, 56.
DUBOS, R. J. (1947). *J. Exp. Med.* **85**, 9.
DUBOS, R. J. & DAVIS, B. D. (1946). *J. Exp. Med.* **83**, 409.
EGGERTH, A. H. (1927). *J. Exp. Med.* **46**, 671.
FEENEY, R. E., MUELLER, J. H. & MILLER, P. A. (1943). *J. Bact.* **46**, 559.
FLEMING, A. (1909). *Lancet*, **1**, 472.
FOSTER, J. W. & WYNNE, E. S. (1948). *J. Bact.* **55**, 495.
FRAENKEL, G. & BLEWETT, M. (1946). *J. Exp. Biol.* **22**, 172.
FRAENKEL, G. & BLEWETT, M. (1947). *Biochem. J.* **41**, 475.
GLASS, V. & KENNETT, S. J. (1939). *J. Path. Bact.* **49**, 125.
GOLBERG, L. (1946). D.Sc. Thesis, University of Witwatersrand.
GOULD, R. G., KANE, L. W. & MUELLER, J. H. (1943). *J. Bact.* **47**, 287.
GRAY, D. F. (1946). *Austr. J. Exp. Biol. Med. Sci.* **24**, 301.

Gray, D. F. (1947). *J. Path. Bact.* **59**, 235.
Hettche, H. O. (1934). *Z. ImmunForsch.* **83**, 499.
Hölzl, H. (1940/42). *Z. Hyg. InfektKr.* **123**, 500.
Hornibrook, J. W. (1939). *Publ. Hlth Rep., Wash.*, **54**, 1847.
Hotchkiss, R. D. (1946). *Ann. N.Y. Acad. Sci.* **46**, 479.
Humfield, H. (1947). *J. Bact.* **54**, 513.
Hutchings, B. L. & Boggiano, E. (1947). *J. Biol. Chem.* **169**, 229.
Hutner, S. H. (1942). *J. Bact.* **43**, 629.
Kendall, F. E. (1941). *J. Biol. Chem.* **138**, 97.
Klaesson, S. (1946). *Ark. Kemi. Min. Geol.* **23**, 75.
Kodicek, E. & Worden, A. N. (1945). *Biochem. J.* **39**, 78.
Lamar, R. V. (1911). *J. Exp. Med.* **14**, 256.
Lawson, G. McC. (1939). *Amer. J. Hyg.* **29**, 119.
Lehrman, L. & Kabat, E. A. (1933). *J. Amer. Chem. Soc.* **55**, 850.
Ley, H. L., Jun. & Mueller, J. H. (1946). *J. Bact.* **52**, 453.
Linner, E. R. & Gortner, R. A. (1934). *J. Phys. Chem.* **39**, 35.
Lwoff, A. (1947). *Ann. Inst. Pasteur*, **73**, 735.
McLeod, J. W. & Wyon, G. A. (1921). *J. Path. Bact.* **24** (Pt. II), 207.
Middlebrook, G. & Dubos, R. J. (1947). *J. Immunol.* **56**, 301.
Mishulow, L., Klein, I. F., Liss, M. M. & Leifer, L. (1939). *J. Immunol.* **37**, 17.
Morel, M. (1945). *Ann. Inst. Pasteur*, **71**, 225.
Nungester, W. J., Wolf, A. A. & Jourdonais, L. F. (1932). *Proc. Soc. Exp. Biol., N.Y.*, **30**, 120.
Pollock, M. R. (1947). *Brit. J. Exp. Path.* **28**, 295.
Pollock, M. R. (1948). *Nature, Lond.* **161**, 853.
Powell, H. M. & Jamieson, W. A. (1936). *Proc. Soc. Exp. Biol., N.Y.*, **34**, 435.
Rothman, S., Smiljanic, A. M. & Weitkamp, A. W. (1946). *Science*, **104**, 201.
Schoch, T. J. & Williams, C. B. (1944). *J. Amer. Chem. Soc.* **66**, 1232.
Shull, G. M., Thoma, R. W. & Peterson, W. H. (1948). *Fed. Proc.* **7**, 188.
Silverthorne, N. (1938). *Canad. J. Publ. Hlth*, **29**, 233.
Sprince, H. & Kupferberg, A. B. (1947). *J. Bact.* **53**, 441.
Stock, C. C. & Francis, T. (1940). *J. Exp. Med.* **71**, 661.
Stock, C. C. & Francis, T. (1943). *J. Immunol.* **47**, 303.
Strong, E. M. & Carpenter, L. E. (1942). *Industr. Engng Chem.* (Anal. ed.), **14**, 909.
Taylor, T. C. & Lehrman, L. (1926). *J. Amer. Chem. Soc.* **48**, 1739.
Toomey, J. A. & Takacs, W. S. (1938). *J. Infect. Dis.* **62**, 297.
Vedder, E. B. (1915). *J. Infect. Dis.* **16**, 385.
Walker, J. E. (1924). *J. Infect. Dis.* **35**, 357.
Walker, J. E. (1925). *J. Infect. Dis.* **37**, 181.
Walker, J. E. (1926). *J. Infect. Dis.* **38**, 127.
Whitehill, A. R., Oleson, J. H. & Subbarow, Y. (1947). *Arch. Biochem.* **15**, 31.
Williams, W. L., Broquist, H. P. & Snell, E. E. (1947). *J. Biol. Chem.* **170**, 619.
Williams, V. R. & Fieger, E. A. (1945). *Industr. Engng. Chem.* (Anal. ed.), **17**, 127.
Williams, V. R. & Fieger, E. A. (1946). *J. Biol. Chem.* **166**, 335.
Williams, V. R. & Fieger, E. A. (1947). *J. Biol. Chem.* **170**, 399.
Wright, H. D. (1934). *J. Path. Bact.* **38**, 499.
Wyon, G. A. (1923). *J. Path. Bact.* **26**, 441.
Wyss, O., Ludwig, B. J. & Joiner, R. R. (1945). *Arch. Biochem.* **7**, 415.

EXPLANATION OF PLATES

Plate 1

Growth of drop-inocula of *Haemophilus pertussis* (containing 10^8, 10^6, 10^4 and 10^2 viable cells respectively starting at the bottom left-hand quadrant of the plate and passing anticlockwise) on 2 % agar plates containing the following:

(*a*) Tryptic meat medium.
(*b*) Tryptic meat medium + 10 % human blood.
(*c*) Tryptic meat medium + 0·2 % charcoal.
(*d*) Tryptic meat medium alone, previously heavily inoculated with a suspension of *Corynebacterium* 'Q'.

(*e*) Growth of *Haemophilus pertussis* from an inoculum spread by wire on the upper surface of cellophane lying on a 0·2 % charcoal-nutrient agar plate.

Plate 2

'Albumen-cup' method of titration of anti-*Haemophilus pertussis* activity of an unknown fatty acid preparation, showing growth plaques, on heavily inoculated tryptic meat agar containing the test substance, around cylinders containing 1, 2 and 3 drops respectively of albumen solution (for details see text):

(*a*) Control (no addition).
(*b*) Control + 4·0 μg./ml. oleic acid.
(*c*) Control + 2·0 mg./ml. 'used' charcoal extract.

Plate 3

(*a, b*) Symbiosis of *H. pertussis* and *Corynebacterium* 'Q' showing: (*a*) satellitism of *Haemophilus pertussis* colonies around colonies of 'Q', on plain nutrient agar (× 4½); (*b*) satellitism of 'Q' colonies around colonies of *H. pertussis* on 0·2 % charcoal agar (× 16).

(*c, d*) 'Double action' of sebaceous material on human hair, showing: (*c*) stimulation of *Corynebacterium* 'Q' (× 22); (*d*) inhibition of *Haemophilus* (× 16); *pertussis* in both cases on 0·2 % charcoal agar.

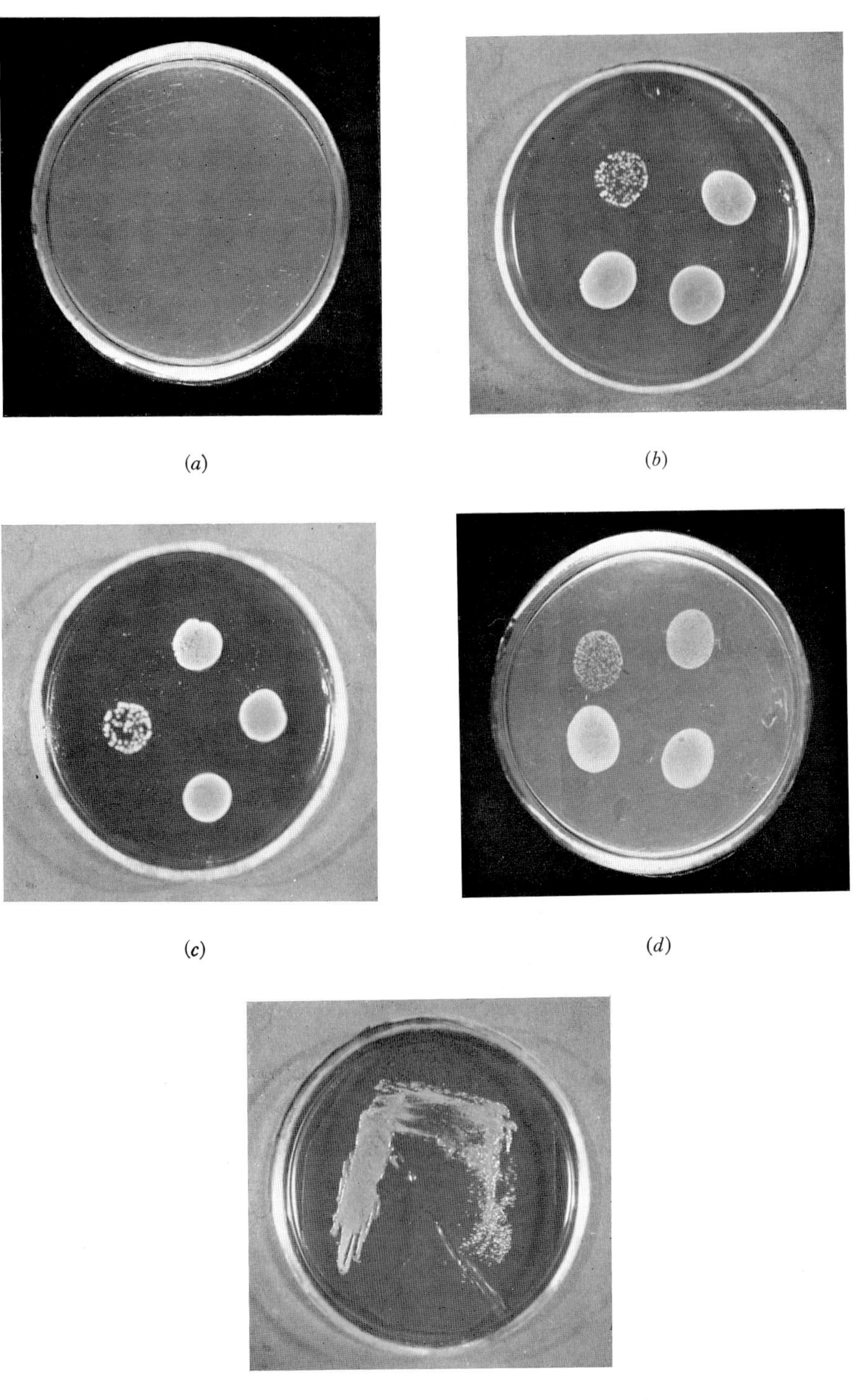

(a) (b) (c) (d) (e)

For explanation see p. 216

PLATE 2

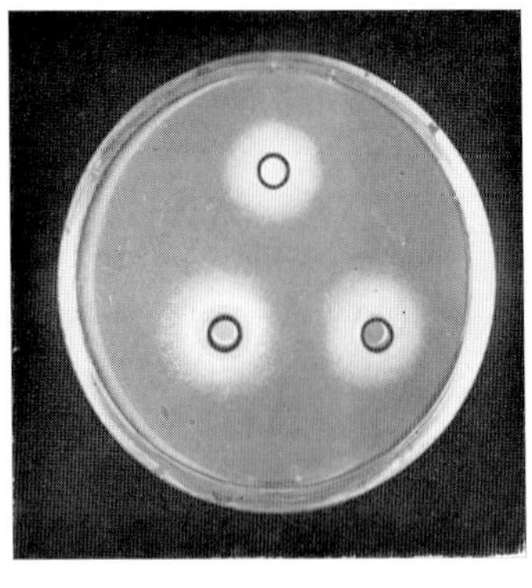

(*a*)

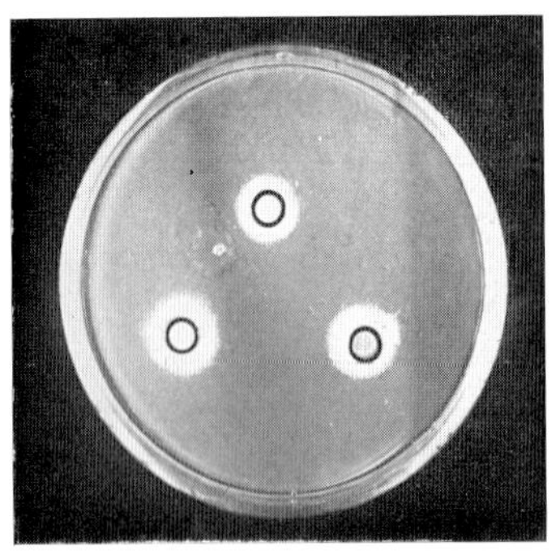

(*b*)

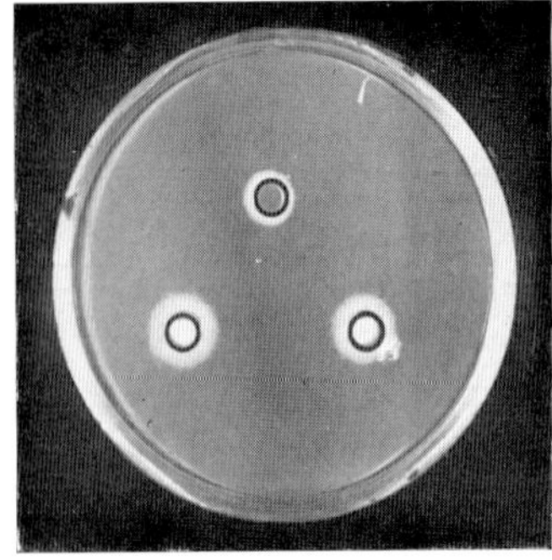

(*c*)

For explanation see p. 216

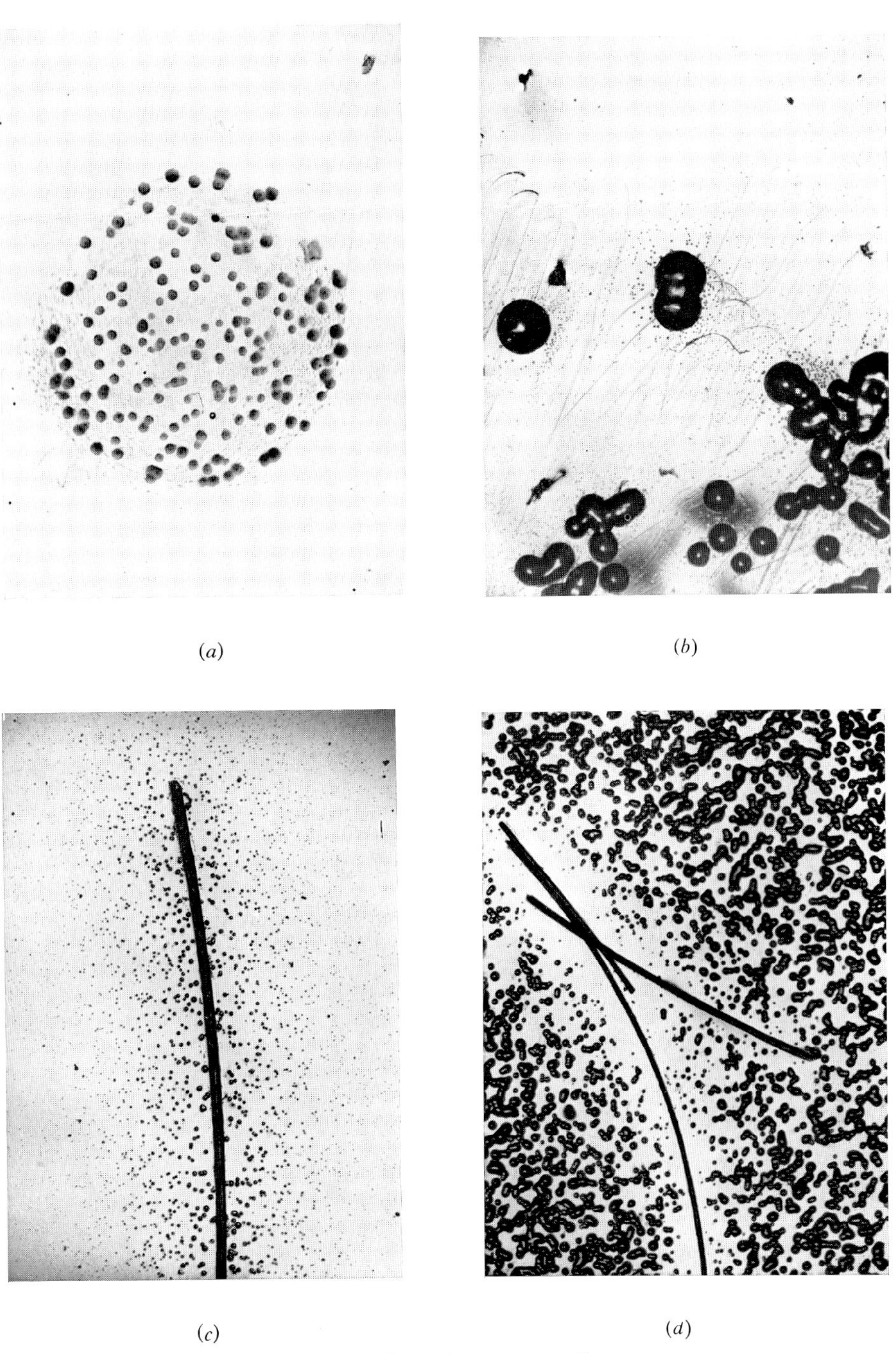

(a) (b)

(c) (d)

For explanation see p. 216

THE EFFECT OF UNSATURATED FATTY ACIDS ON GRAM-POSITIVE BACTERIA

By E. KODICEK

Dunn Nutritional Laboratory, University of Cambridge, and Medical Research Council

I. INTRODUCTION

The wall of the bacterial cell has to be considered not merely as a boundary between the living cytoplasm and the surrounding medium containing potential nutrients, but as an active part of the organism. This view is supported by recent studies on the effect of antibiotic agents which apparently alter the structure and properties of the wall and of the adjacent cytoplasm. It is strikingly demonstrated by studying the effect of various surface-active anionic detergents such as long-chain fatty acids.

While the literature of this subject dates as far back as 1911 (Lamar), this kind of work has received more impetus during recent years. While previous workers have used rather high concentrations of these detergents, I shall deal with experiments employing extremely weak concentrations, since these illustrate more clearly the potent action of these substances.

Two outstanding features of the present investigation may be mentioned here:

(1) Free long-chain fatty acids act as bacteriostatic agents on a number of Gram-positive bacteria. Thus the C_{12-14} series of saturated fatty acids (lauric and myristic acids) show this depressing effect in a concentration of 1 : 100,000. The *cis*-forms of the free unsaturated fatty acids show a more marked effect, the toxicity increasing with increase in the number of double bonds. The toxicity in increasing order may thus be written: oleic acid < linoleic acid < linolenic acid.

(2) The action of the unsaturated fatty acids in these weak concentrations is bacteriostatic and can be reversed by surface-active agents like lecithin, sterols, tocopherol, proteins, etc. The effect of lauric acid, on the other hand, is not reversible. Esters of the unsaturated acids are not bacteriostatic, but in many circumstances are found to be growth-promoting, especially in the presence of proteins or other surface-active agents which may interact with them.

The experiments to be described were performed mainly on *Lactobacillus casei* ϵ, employing a semi-synthetic medium used for riboflavin assay (Kodicek, 1948).

II. BACTERIOSTATIC EFFECT

As the effects of the unsaturated fatty acids were found to be dependent on the number of bacteria present, a standardized inoculum was used throughout. The fatty acids were added to the medium in $CHCl_3$ solution and the $CHCl_3$ was evaporated *in vacuo*.

When saturated fatty acids like palmitic or stearic acids were added in a concentration of 160 μg./10 ml. medium, a slightly increased growth was

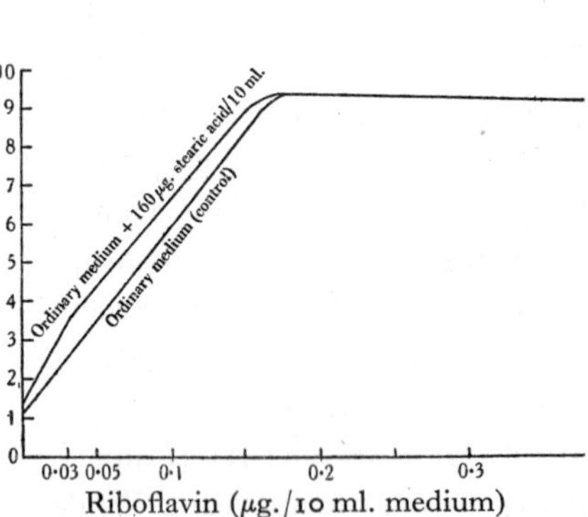

Fig. 1. Effect of stearic acid in ordinary medium (48 hr. incubation).

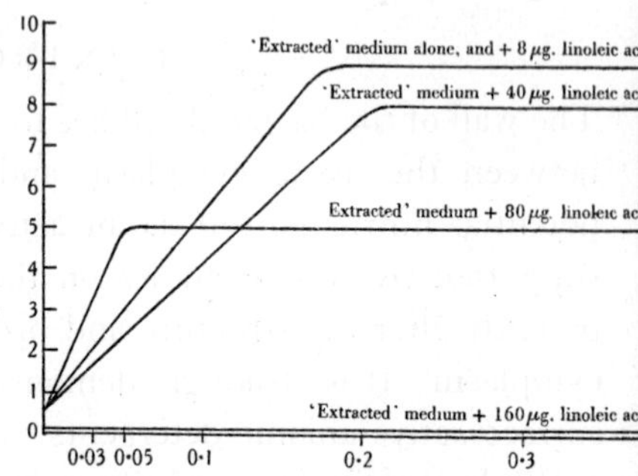

Fig. 2. Effect of different amounts of linoleic acid in chloroform-extracted medium (72 hr. incubation)

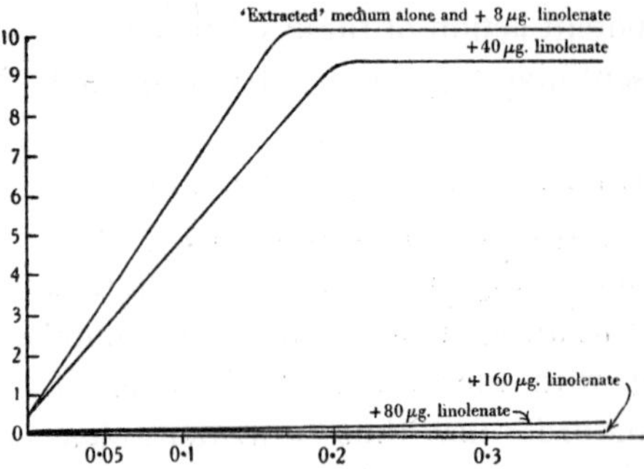

Fig. 3. Effect of different amounts of sodium linolenate in chloroform-extracted medium (48 hr. incubation).

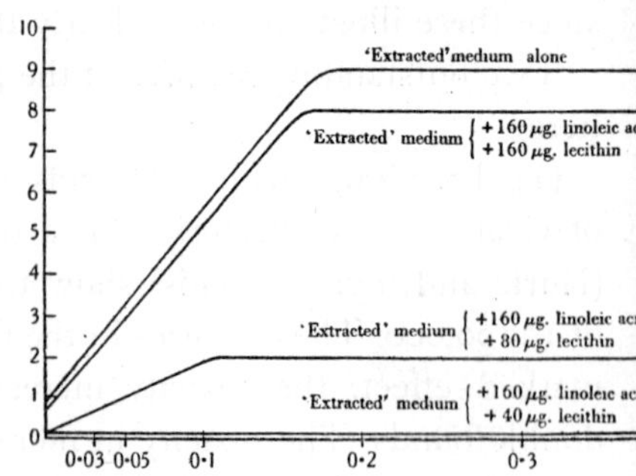

Fig. 4. Effect of different amounts of lecithin in presence of linoleic acid (72 hr. incubation)

observed (Fig. 1), confirming previous reports of Bauernfeind, Sotier & Boruff (1942) and Strong & Carpenter (1942). However, when linoleic acid was added, growth and acid production were depressed over a period of at least 72 hr. This was dependent on the concentrations of linoleic acid (Fig. 2), provided all other factors remained constant. A more pronounced effect was observed with linolenic acid (Fig. 3). Free oleic acid produced a depression in the first 48 hr. of incubation, but at 72 hr. the bacteria started to grow and eventually recovered completely. This exceptional behaviour of oleic acid after prolonged incubation may be explained, as will be mentioned later, by its special connexion with biotin metabolism.

The effect of unsaturated fatty acids was shown to be bacteriostatic, at least in the concentrations used in these experiments. When lecithin or cholesterol or some other surface-active substances were present in the medium, the recovery of the bacteria which occurred depended on the ratio fatty acid to surface-active agent (Fig. 4). This reversal was observed irrespective of whether the antagonist was given from the start of the experiment or after a 48 hr. inhibition by linoleic acid.

Table 1. *Effect of fatty acids on* Lactobacillus casei

Addition, 160 μg./10 ml. media

	Normal or increased growth	Depressed growth	Reversible by sterols, etc.
Caproic acid (C_6)	+	.	.
Caprylic acid (C_8)	+	.	.
Lauric acid (C_{12})	.	±	o
Ethyl laurate	++	.	.
α-Methyl lauric acid	.	++	o
Palmitic acid (C_{16})	++	.	.
Stearic acid (C_{18})	++	.	.
Oleic acid (9-*cis*, C_{18})	.	± (48 hr.)	+
Sodium oleate	.	± (48 hr.)	+
Methyl oleate	+	.	.
Elaidic acid (9-*trans* C_{18})	+	.	.
α-Elaeostearic acid (9, 11, 13-*cis* C_{18})	.	+	+
β-Elaeostearic acid (9, 11, 13-*trans* C_{18})	+	.	.
Linoleic acid (9, 12-*cis* C_{17})	.	++	+
Sodium linoleate	.	++	+
Methyl linoleate	+	.	.
Sodium linoleate (9, 12, 15-*cis* C_{18})	.	++	+
Methyl linolenate	+	.	.

Table 1 shows the effect of various fatty acids on *L. casei* tested either alone or in combination with cholesterol. In the saturated series, the C_{12} compound, lauric acid, depressed growth, but the effect was not reversible by cholesterol; the ethyl ester had no effect. The branched-chain, α-methyl lauric acid was very active even in low concentrations, but here also the inhibition could not be reversed by cholesterol. Of the unsaturated series, the depressing effect increased with the number of double bonds. It was limited to the *cis*-forms of the free acid (Kodicek & Worden, 1946), esterification abolished the inhibition. Cholesterol reversed the effect of these acids in all cases.

More complex compounds were also tested for their bacteriostatic activity (Table 2). Surface-active compounds of the types I, II and III were found to behave differently in lactobacilli than in yeast. Both the acid and ethyl ester of the symmetric di-*n*-heptyl type were strongly inhibiting, and their effect could not be reversed by cholesterol. The straight-chain C_{14} type had no depressing effect. Exactly the opposite was found with yeast cells (Schulman, 1948):

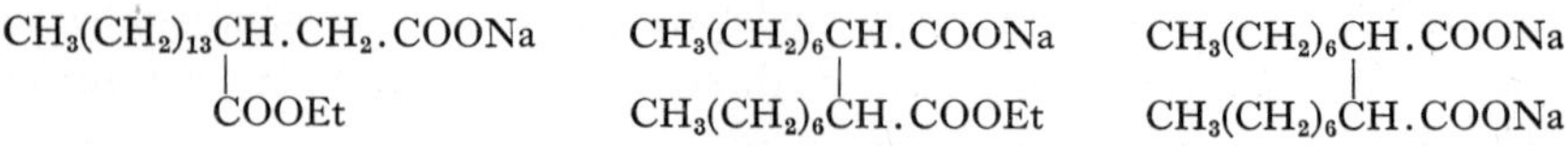

$$CH_3(CH_2)_{13}CH.CH_2.COONa \quad | \quad COOEt$$

Tetra-decanoic succinate ethyl half ester (I)

$$CH_3(CH_2)_6CH.COONa \quad | \quad CH_3(CH_2)_6CH.COOEt$$

Di-*n*-heptyl succinate ethyl half ester (II)

$$CH_3(CH_2)_6CH.COONa \quad | \quad CH_3(CH_2)_6CH.COONa$$

Di-*n*-heptyl succinate (III)

Chalmaugric acid, an unsaturated higher acid of the cyclopentane series (IV),

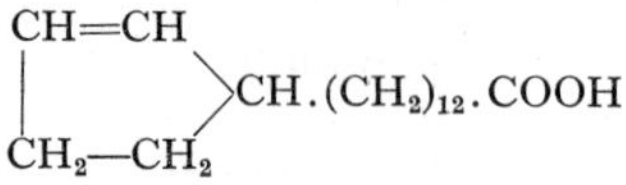

$$\begin{array}{l} CH{=}CH \\ | \qquad\qquad \rangle CH.(CH_2)_{12}.COOH \\ CH_2{-}CH_2 \end{array}$$

d-13-(2-cyclopentenyl) tridecanoic acid (IV),

was moderately inhibiting and its action could be reversed by cholesterol.

Table 2. *Effect of branched fatty acids and other compounds on* Lactobacillus casei

Addition, 160 μg./10 ml. media

	Normal or increased growth	Depressed growth	Reversible by sterols, etc.
Tetra decanoic succinate ethyl-half ester	±	.	.
Di-*n*-heptyl succinate ethyl-half ester	.	++	o
Di-*n*-heptyl succinic acid	.	++	.
Aerosols IB	+	.	.
Aerosols AY	+	.	.
Chalmaugric acid	.	+	+
Vitamin A acid	.	++	o
Vitamin A alcohol	+	.	.
Vitamin A acetate	+	.	.
Vitamin D_2	.	±	(By linoleic acid)
Stilboestrol	.	++	o
Hexoestrol	.	++	o
Dienoestrol	.	++	o
Dienoestrol dipropionate	+	.	.
Dienoestrol dibenzoate	+	.	.
Saponin	+	.	.
Sodium taurocholate	+	.	.
3:4-Benzpyrene	+	.	.
1:2:5:6-Dibenzanthracene	+	.	.
Methylcholanthrene	+	.	.
Butter yellow	+	.	.

Vitamin A compounds, which can be considered as derivatives of β-ionone with a fatty acid chain containing four conjugated double bonds (V),

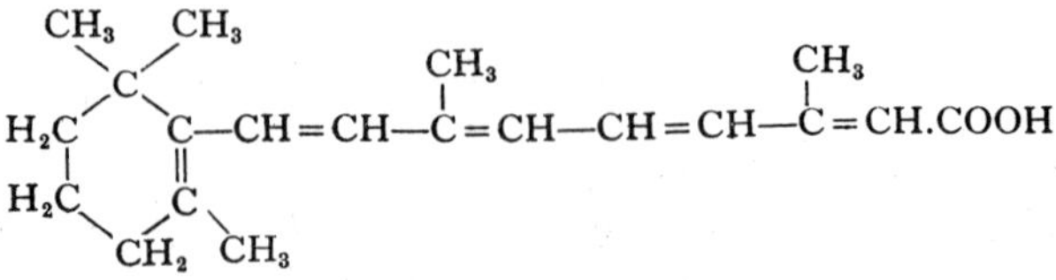

$$\begin{array}{l} \quad CH_3 \quad CH_3 \\ \qquad \diagdown \; \diagup \qquad\qquad\qquad CH_3 \qquad\qquad\qquad\qquad CH_3 \\ \qquad\; C \qquad\qquad\qquad\qquad | \qquad\qquad\qquad\qquad\quad | \\ H_2C \quad\; C{-}CH{=}CH{-}C{=}CH{-}CH{=}CH{-}C{=}CH.COOH \\ \;\; | \qquad\quad \| \\ H_2C \quad\;\; C \\ \qquad CH_2 \quad CH_3 \end{array}$$

Vitamin A acid (V) (Arens & van Dorp, 1946)

when tested for their inhibiting activity, were found to be active only when a free carboxylic group remained in the side-chain. The effect of the vitamin A acid was only observed when the acid was autoclaved together with the medium, while additions of sterile vitamin A acid after autoclaving had no effect. It is therefore highly probable that the molecule underwent further changes or disruption on heat treatment. On the other hand, Aerosols IB ($n=5$) and AY ($n=8$), highly potent detergents, the long chain di-esters of sodium sulphosuccinate, e.g.

$$\begin{array}{l} C_nH_{2n}.COO.CH_2 \\ \qquad\qquad\qquad\;| \\ C_nH_{2n}.COO.CH.SO_3^-...Na^+ \end{array}$$

were inactive, at least in concentrations comparable to those employed in our experiments.

Unsubstituted synthetic oestrogens were bactericidal and could not be reversed by sterols. In confirmation of the observations of Faulkner (1943) and Brownlee, Copp, Duffin & Tonkin (1943) on other Gram-positive bacteria, we found that both diethyl stilboestrol and hexoestrol, in amounts of 160 μg./10 ml., inhibited the growth of *L. casei*. The action of these compounds could not be reversed by the addition of lecithin or cholesterol. Dienoestrol (4:4-dihydroxy-γ:δ-diphenyl-β:δ-hexadiene) was also inhibitory. Its dibenzoate and dipropionate esters did not stop growth, which indicates the importance of the two hydroxyl groups for the inhibition.

Other surface-active agents like saponin, taurocholate and also carcinogens and crystalline B vitamins were inactive (Kodicek & Worden, 1944, 1945). Vitamin D_2 (calciferol) depressed growth, but had a different action when tested together with linoleic acid as will be shown later.

The effect of the compounds tested varies according to the micro-organisms used, which is quite understandable, since there is some evidence that the composition of the cell wall differs from strain to strain. One generalization can be made, that Gram-positive bacteria are more affected than the Gram-negatives (Table 3). The most sensitive organism of the strains tested was *L. casei*. The other Gram-positives needed increasing concentrations of linoleic acid to produce an effect.

Table 3. *Effect of linoleic acid* (160 μ*g.*/10 *ml.*) *on bacteria*

	Normal or increased growth	Depressed growth	Effective concentration (μg./10 ml.) of media
L. casei	.	++	160
L. arabinosus	.	+	640
Str. faecalis	.	±	320
Str. agalactiae	.	++	160
Staph. albus	.	+	320
Listeria monocytogenes	.	++	160
Erysipelothrix rhusiopathiae	.	++	160
E. coli	++	.	160
Proteus vulgaris	++	.	160

III. REVERSAL OF LINOLEIC ACID INHIBITION

The most interesting feature of the inhibition by unsaturated fatty acids is its reversibility by other surface-active agents. Of the other inhibitors only chalmaugric acid had a reversible action; the action of the other inhibitors studied could not be reversed in this manner. Lecithin was one of the most potent of the surface-active agents which reversed the blocking action of the unsaturated fatty acids. Similarly, cholesterol is very effective. A number of sterol compounds were investigated (Table 4); only vitamins D_2 and D_3 were equally effective, while lumisterol and coprostanone showed a slight activity. The effect of the D vitamins could be differentiated from that of cholesterol by treatment with digitonin which forms insoluble complexes with the latter compound.

Table 4. *Reversal effect of lecithin and sterols on* Lactobacillus casei *blocked by linoleic acid*

Addition, 160 μg./10 ml. medium

	Growth in presence of linoleic acid (160 μg.; reversal)	Growth in presence of linoleic acid with digitonin treatment
Lecithin	+	.
Cholesterol	+	0
Cholestanol	0	0
Cholestanone	0	0
7-Dehydrocholesterol	0	0
Dihydrocholesterol	±	0
Lumisterol	±	±
3:5-Dinitro-lumisterol	±	±
Ergosterol	0	0
Vitamin D_2	+	+
Vitamin D_3	+	+
Coprostanone	±	±
Zymosterol	0	0
Sitosterol	0	0

Tocopherol and its acetate effectively reverse the depressing action of unsaturated fatty acids, as does calcium chloride when added in high concentrations (710 μg./10 ml. medium) (Table 5). The effect of various proteins was tried in view of the work of Macheboeuf & Tayeau (1941), Lwoff (1947), Dubos (1946) and Pollock (1947). Horse serum, edestin and haemoglobin effected a reversal, while insulin was inactive. Norite charcoal was active, while other brands like activated charcoal (B.D.H.) and Actibon had no effect. As a point of interest, it was noted that charcoal added without linoleic acid depressed the growth of lactobacilli quite markedly. A similar finding was observed when testing the effect of calciferol alone on *L. casei*.

While ascorbic acid had no effect, ascorbyl palmitate reversed the inhibition to a greater extent than palmitic acid. The slight reversing effect of the short chain fatty acids like oxaloacetic acid, etc., will be discussed below.

Sodium taurocholate, saponin, allantoin, glutathione, methionine, and all the known crystalline B vitamins had no reversing effect.

Table 5. *Reversal of the inhibitory action of linoleic acid on* Lactobacillus casei *by various substances*

Addition, 160 μg./10 ml. medium

	Growth in presence of linoleic acid (160 μg.; reversal)
Calcium chloride (710 μg.)	+
α-Tocopherol acetate	+
α-Tocopherol	+
Horse serum (0·2 ml.)	+
Edestin	+
Haemoglobin	+
Norite charcoal (20 mg.)	+
Activated charcoal (B.D.H.) (20 mg.)	o
Actibon charcoal (20 mg.)	o
Insulin	o
Palmitic acid	±
Ascorbyl palmitate	+
Ascorbic acid	o
Saponin	o
Sodium taurocholate	o
Dihydroxy maleic acid	±
Sodium fumarate	±
Oxaloacetic acid	±
Malic acid	±

IV. PHYSICO-CHEMICAL EXPLANATION

From the results obtained it is assumed that these phenomena can be explained in physico-chemical terms. While former authors (Lamar, 1911; Frobisher, 1926) tried to explain similar findings of bacteriostasis by a changed surface tension, it seems that the action of the fatty acids is far more complex, and it must involve a change in the permeability of the wall, or in the rate of permeation of substances through the cell wall regulated by a mechanism more complex than diffusion (incretion). This effect can be brought about either by blocking the entry of various nutrients into the cell or by allowing the diffusion of essential metabolites into the external environment.

We have compared the surface area of added linoleic acid with the surface area of the bacteria in the medium (Kodicek & Worden, 1945), and concluded that there is enough linoleic acid present even in the lowest effective concentration to cover the whole surface of the bacteria by a monolayer of the acid. Thus these necessarily rough estimates do not exclude such a physico-chemical mechanism. While the C_{12-14} series of saturated fatty acids are also bacteriostatic, acting probably by the same mechanism (Valko, 1946; Hotchkiss, 1946), however, more interesting biologically are the effects of the unsaturated series involving a dynamic equilibrium effect with other surface-active substances occurring in nature such as sterols, vitamins, lecithins and proteins. The more so, as it seems that only the naturally occurring *cis*-forms of these acids are bacteriostatic (Kodicek & Worden, 1946). Similar bacteriostatic effects were observed by Hettche (1934) with *Staph. aureus*, Chaix & Baud (1947) with *Glaucoma piriformis*, Bergström, Theorell & Davide (1946) and Dubos & Davis (1946, 1947) on tubercle bacilli.

The evidence of the action of unsaturated fatty acids on bacteria *in vitro* has been followed by findings of antibiotics in biological material. Humfeld (1947) found in wheat bran an antibiotic fatty acid, and McKee, Dutcher, Groupe & Moore (1947) have found antibacterial lipides in *Tetrahymena geleii*.

The polarity of the carboxyl radical is of importance for the bacteriostatic activity as is shown by our results and those of Dubos (1947). Esterification of the acid changes the effect of the unsaturated fatty acids and they become, if anything growth-promoting, especially when native protein is present to form further molecular associations.

The physico-chemical aspect of the action of unsaturated fatty acids is supported by various evidences of their physico-chemical properties. Thus, McBain (1942) cites Palmers and collaborators as saying: 'apparently the first step in the recruitment into a homogeneous mixed phase of the lipides incompatible in the dry condition is the solvation of the polar groups and perhaps also the unsaturated or partially oxidized double bonds of the fatty acids.' It is thus not only the changed surface activity but also the solvation of polar groups which may be of importance and would change the properties of the attached proteins of the cell wall. It is well known that lecithins, sterols and other surface-active agents can form complexes with unsaturated fatty acids. Dervichian & Pillet (1944) found that the surface area of mixed lecithin and oleic acid is smaller than calculated and suggests a colloidal association of the molecules. Formation of complexes by injecting anionic detergents into a monolayer of cholesterol has been confirmed by Schulman & Stenhagen (1938). Marked differences were obtained on penetration, depending on the degree of unsaturation and the *cis* and *trans* isomerism. These findings are paralleled by our observations in bacteria.

Certain proteins can also form associations with fatty acids more or less loosely bound, possibly by forces of the van der Waals type (Tayeau, 1944). Macheboeuf's lipoprotein isolated from horse serum was found to be a quite stable complex, since only boiling alcohol would completely separate the protein from the lipid. Similar results were obtained with an oleic acid-ovalbumin complex (Gorter & Hermans, 1942). It seems therefore that such surface-active agents with certain polar groupings may (*a*) act as competitors to the lipoproteins (Knaysi, 1938) of the cell wall, and thus remove the unsaturated fatty acids from their surface, or (*b*) change the whole lattice structure by a formation of molecular associations at the site of the former adsorption. Their action and the effect of unsaturated fatty acids will evidently depend on the composition of the cell wall. This is quite clearly shown in the different behaviour of the various species of the Gram-positive micro-organisms. One of the most striking examples is the behaviour of tubercle bacilli, which have been extensively studied by Davis & Dubos (1947).

Though the physico-chemical mechanism seems at present to be the most acceptable, nevertheless a possible metabolic effect of oleic acid cannot, in view of our present inadequate knowledge, be completely excluded. Perhaps isotopic experiments will answer this question, whether oleic acid is metabolized or not.

V. THE REVERSAL EFFECT OF STEROLS

The action of sterols in reversing the bacteriostatic effect of unsaturated fatty acids has been studied in some detail, and it is possible to make some rough generalization. It is, however, clear that on the basis of the rather incomplete knowledge of the complex systems involved any more definite conclusions are premature. The effect of the sterols will have to be interpreted in the last instance in physico-chemical terms. The sterols which effect a reversal of the inhibition need to have a hydroxyl on C_3 in ring I and a double bond at C_5. Their action is potentiated if the second ring is open and three double bonds appear there as in vitamin D. The sterols must form sufficiently small micelles to be able to form complexes, and thus ergosterol is inactive.

VI. ATTEMPT OF A UNIFIED THEORY (BARRIER THEORY)

Baker, Harrison & Miller (1941*a*, *b*, *c*) have studied the action of various detergents on bacteria and noted the antagonistic effect of lecithins. The site of the effect will be in the lipoprotein layer of the cell wall. Franke & Schillinger (1943), and more recently Williams & Fieger (1947), suggest a change in the permeability of the cell membrane: 'If the cell were wetted (by a detergent) just sufficiently to enable it to make better contact with the nutrients of the environment, the cell should grow more rapidly by obtaining its substrate more readily.' If, however, the agent disarranged specific vital structures, then the changed permeability of the membrane may cause loss of vital constituents of the cells into the medium as observed by Hotchkiss (1946). The growth-promoting activity seems to be regulated by various factors, and the term 'wetting' may be too indefinite to describe these conditions. Some of the long-chain fatty acids act as depressors or activators of growth. Oleic and linoleic acids for example support growth in some organisms (Pollock, 1948; Peterson & Peterson, 1945; Williams, Broquist & Snell, 1947; Whitehill, Oleson & Subbarrow, 1947), but for some others they are bacteriostatic agents. For other organisms oleic acid in small concentrations is a growth promoter, while in high concentrations it stops growth. This rather paradoxical behaviour may possibly be unified, if one makes the assumption that its action depends also on the structure of the substrate to which it is adsorbed. As there are sterols, lecithins or other active substances available in the surface layers of the cell wall, it could be assumed that they will act in the same way as lecithins or sterols added to the medium, namely, reverse the bacteriostatic effect of the unsaturated fatty acids. It is obvious, of course, that there would be only a limited amount of these compounds available, and therefore only a small amount of the acid will be bound in this form. Hence, the unsaturated fatty acids in small concentrations would be growth-promoting, but above a certain concentration, with all the available 'reversing agents' 'neutralized', the acid would then act as a bacteriostatic agent. They may thus first enhance the entry of essential nutrients into the cell, or later produce changes in the cell wall which result in loss of nutrients into the external environment, or as the matter may be—block the entry of nutrients into the cell. The same phenomena would, of course, happen if there are other compounds in the medium—apart from the unsaturated fatty acids—which may combine with the acids adsorbed on the cell wall, thus changing its lattice and its physico-chemical properties, or competing with the lipoproteins of the cell wall for the bacteriostatic agents. Tubercle bacilli which contain an appreciable amount of lipides in the cell wall may form an extreme example of this.

VII. OLEIC ACID AND BIOTIN

Oleic acid has been found to replace biotin effectively for *L. casei* (Williams & Fieger, 1945, 1946, 1947), for *L. arabinosus* (Williams, 1945) and for a number of other lactobacilli (Williams *et al.* 1947; Hutchings & Boggiano, 1947). Similarly, vaccinic acid (Axelrod, Hofmann & Daubert, 1947) and some synthetic detergents like Tween 80 (non-ionic oleate) and others (Williams & Fieger, 1947) were equally effective. Shull, Thoma & Peterson (1948) report that oleic, linoleic, ricinoleic, vaccinic acids and Tween 80 can replace biotin for *Clostridium sporogenes*, and suggest that either biotin or oleic acid was a component of concentrates containing the 'sporogenes vitamin' of Knight & Fildes (1933). Snell & Williams (1939) found that biotin can be replaced by a widely distributed organic acid for the growth of *Cl. butylicum*. Trager (1947*a*, *b*, 1948) recently reported the preparation of a fat-soluble fraction, FSF, from blood plasma which could also replace biotin. The activity, however, seems to appear in the non-saponifiable fraction of plasma and thus cannot be oleic acid even in a combined form. FSF replaced biotin also in avian biotin deficiency. On the other hand, Axelrod *et al.* (1947) found a neutral fat-soluble fraction from plasma which will replace biotin for *Lactobacillus casei* and is contained in the saponifiable portion.

Avidin did not prevent replacement of biotin by lipids in experiments with *L. casei* (Williams & Fieger, 1947; Williams *et al.* 1947). However, working with *L. arabinosus*, Broquist & Snell (1948) reported that egg white or purified avidin inactivates the growth-promoting effect of oleic acid. The amounts used have no stoichiometric relation to oleic acid, but are directly proportional to the biotin-binding activity of the preparations. Inactivation of biotin present as an impurity has been ruled out. The close physiological relationship between oleic acid and biotin is thus further emphasized, though the mechanism of the action of avidin is still unknown.

Williams & Fieger (1947) have suggested that biotin may not be a component of an enzyme system, but like oleic acid is rather involved in cell permeability; hence their interaction. On the other hand, Potter & Elvehjem (1948) found that aspartic acid and oleic acid can almost completely replace biotin, but neither substance does it effectively alone. Aspartic acid and oxaloacetic acid were effectively metabolized in the presence of biotin. These findings support the view that biotin is concerned —apart from its interaction with oleic acid—with the metabolism of aspartic acid, and the authors quote their findings as support of the coenzyme activity of biotin. These observations link up with the suggestion that biotin might function as a general CO_2 donor (Burk & Winzler, 1943). More recent evidence of biotin being concerned with CO_2 fixation is supplied by Shive & Rogers (1947), who found that the vitamin is concerned in the biosynthesis of oxaloacetic acid by carboxylation of pyruvic acid and of α-ketoglutaric acid, possibly from oxalosuccinic acid. Stokes, Larsen & Gunness (1947) found a similar involvement of biotin in aspartic acid metabolism. Thus the metabolism of tricarboxylic acids may be linked up with biotin and indirectly with oleic acid.

We have obtained a definite, though small, reversal of the blocking action of linoleic acid by oxaloacetic, fumaric, hydroxymaleic and malic acids when added in very small amounts (160 μg./10 ml.) to the medium. It may be quite possible that these acids, being involved in the bacterial metabolism, have supplied nutrilites which were being constantly lost by the changed permeability as an effect of the unsaturated fatty acids. When the growth-promoting effect of unsaturated fatty acids is observed, and this usually occurs when they are present in non-polar ester form, their action would be reversed. They would prevent the escape of these metabolites and thus effectively spare biotin, which may be involved as a coenzyme in the carboxylation mechanism of tricarboxylic acids. This view would be compatible both with the surface activity of the fatty acids and the coenzyme nature of biotin.

VIII. UNSATURATED FATTY ACIDS IN HIGHER ORGANISMS

The essential nature of unsaturated fatty acids for mammals is well known (see review by Bernhard, 1947). The mechanism of their action, however, has not been elucidated. It is tempting to speculate whether the effects on micro-organisms of naturally occurring fatty acids may have some parallel in higher organisms. This is obviously still very much in the realm of speculation; nevertheless, there are, even in animals, certain interactions between unsaturated fatty acids and compounds like sterols, vitamin E, etc. It has been already mentioned that Trager (1947*a*, *b*) found that biotin could be replaced in chickens by oleic acid.

Vitamin E and linoleic acid. The function of vitamin E in mammals is connected with the metabolism of unsaturated fatty acids (Granados, Mason & Dam, 1947). Feeding of highly unsaturated fatty acids to vitamin E-deficient rats produced marked changes in the appearance of the pathological acid-fast pigment in adipose tissue and in the decolorization of incisors which occurs in such a deficiency. Whether these findings can be explained by the same physico-chemical mechanism as our observations in bacteria will need further investigation.

Immunological reactions, haemolysis and permeability of capillaries. Pollock (1948) is discussing elsewhere the possible relationship between unsaturated fatty acids and the mechanism of combating infections. Lamar (1911) studied the haemolytic and bacteriolytic properties of unsaturated fatty acids and their soaps, and concluded that among the higher acids there was a close relationship between iodine number and haemolytic effect. The greatest activity was shown by potassium linolenate, the lytic properties of which were six times, and those of potassium linoleate four times, those of sodium oleate when tested against virulent pneumococci and red cells. Bull (1937) states that the germicidal properties of ground tissues and of tissue extracts such as liver or brain are attributed to their lipid contents.

The haemolytic systems containing anionic detergents have been extensively studied by Ponder (1945, 1946), who concludes that red cells and bacteria show similar effects. Cholesterol and lecithins inhibit the haemolytic action of the detergents. These findings agree with those of Maizels (1945) and Lee, Puh & Tsai (1945), who report that most lysins are inhibited by cholesterol. Plasma also inhibits the haemolysis (Mazella, 1946). Serum cholesterol esters and proteins are low in the haemolytic disease of the newborn (Rothe-Meyer & Hickmans, 1946). These observations tally with *in vitro* experiments performed by Macheboeuf & Tayeau (1941), and Tayeau (1945) on the interaction of proteins, linoleic acid and other lipids. The isolation of a naturally occurring haemolysin in the blood has been described by Laser & Friedmann (1945). It seems to be a long-chain unsaturated fatty acid (Laser, 1948). The fat-soluble material from plasma having the biological activity of biotin has also been shown to be strongly haemolytic (Trager, 1947*a*).

Only scattered evidence is available on the effect of lipids on the permeability of capillaries. Holman & Swanton (1946) describe a necrotizing arteritis in dogs after kidney damage when cod-liver oil was administered in the diet. The toxic factor may have been the high unsaturated fatty acid content of the oil.

As to immunological reactions and lipids, the antigenicity of synthetic esters of oleic acid has been described by Middlebrook & Dubos (1947). A phospholipid isolated from beef heart, cardiolipin, yielding on hydrolysis fatty acids with a high iodine number and a phosphorylated polysaccharide, acts as an antigen (Pangborn, 1942), thus resembling another phospholipid prepared from tubercle bacilli (Pinner, 1928). Carter (1947) has obtained a lipid fraction, from group O, *Rh*-positive red cells which is antigenic when injected simultaneously with a protein carrier. It inhibits agglutinins present in anti-*Rh* serum and fixes complement in conjunction with human anti-*Rh* serum. The substance is heat-stable and is probably an *Rh*-hapten in impure form. Lipides will also affect the *in vitro* precipitation reactions. Kahn's antigen is strongly activated by certain phospholipides (Wheeler & Brandon, 1947). Kosyakov (1947) describes the lipid nature of the blood group antigen B_2 from human and animal blood.

There is thus increasing evidence that fatty acids and other lipids are concerned in some way with immunological reactions of the body. Our knowledge is necessarily patchy, but will increase with time. It is very likely that these effects have a physico-chemical mechanism similar to that discussed for bacteria.

We have observed in our bacteriological experiments an antagonistic effect between vitamin D, calcium and unsaturated fatty acids. Very little is known about the biochemical function of vitamin D in bones and other cells of the body, and so far no record is available of unsaturated fatty acids being involved in vitamin D metabolism in higher animals. Nevertheless, the striking effects of these substances in unicellular organisms may warrant a search for possible interactions in higher organisms.

IX. SUMMARY

Certain unsaturated fatty acids are bacteriostatic and depress the acid production of Gram-positive bacteria. Their effect increases with increasing unsaturation. It seems that only naturally occurring *cis*-forms have such an activity. The effect is observed even in high dilutions. It is reversed by other surface-active agents, which can form molecular associations with the acids. Lecithin, certain sterols, tocopherol and certain proteins are equally effective. Certain long-chain saturated fatty acids also block the growth of bacteria. However, their action does not seem to be reversed by these substances. A physico-chemical theory of the mechanism has been suggested and discussed. The possibility of a similar mode of action for these lipides has been put forward for higher organisms.

My thanks are due to Drs M. Stephenson, J. F. Danielli, E. Gale and E. M. Cruickshank for allowing me to discuss with them the problems which arose during the work, and for their helpful suggestions; and also to Drs J. H. Schulman, A. E. Alexander and A. I. Macmullen for kindly supplying me with samples of various surface-active agents, Dr J. F. Arens for the sample of vitamin A acid, and Mr A. L. Bacharach, Glaxo Laboratories Ltd., for kindly supplying the various sterols used in this work.

REFERENCES

Arens, J. F. & van Dorp, D. A. (1946). *Nature, Lond.*, **157**, 190.
Axelrod, A. E., Hofmann, K. & Daubert, B. F. (1947). *J. Biol. Chem.* **169**, 761.
Baker, L., Harrison, R. W. & Miller, B. F. (1941*a*). *J. Exp. Med.* **73**, 241.
Baker, L., Harrison, R. W. & Miller, B. F. (1941*b*). *J. Exp. Med.* **74**, 611.
Baker, L., Harrison, R. W. & Miller, B. F. (1941*c*). *J. Exp. Med.* **74**, 621.
Bauernfeind, J. C., Sotier, A. L. & Boruff, C. S. (1942). *Industr. Engng Chem.* (Anal. ed.), **14**, 666.
Bergström, S., Theorell, H. & Davide, H. (1946). *Nature, Lond.*, **157**, 306.
Bernhard, K. (1947). *Z. Vit. Horm. u. Fermentforsch.* **1**, 199.
Broquist, H. P. & Snell, E. E. (1948). *J. Biol. Chem.* **173**, 435.
Brownlee, G., Copp, F. C., Duffin, W. M. & Tonkin, I. M. (1943). *Biochem. J.* **37**, 572.
Bull, H. B. (1937). *The Biochemistry of Lipids.* New York: J. Wiley and Sons Inc.
Burk, D. & Winzler, R. J. (1943). *Science*, **97**, 57.
Carter, B. D. (1947). *Amer. J. Clin. Path.* **17**, 646.
Chaix, P. & Baud, C. (1947). *Arch. Sci. Physiol.* **1**, 3.

DAVIS, B. D. & DUBOS, R. J. (1947). *J. Exp. Med.* **86**, 215.
DERVICHIAN, D. & PILLET, J. (1944). *Bull. Soc. Chim. biol., Paris*, **26**, 454.
DUBOS, R. J. (1946). *Proc. Soc. Exp. Biol., N.Y.*, **63**, 56.
DUBOS, R. J. (1947). *Experientia*, **3**, 45.
DUBOS, R. J. & DAVIS, B. D. (1946). *J. Exp. Med.* **83**, 409.
DUBOS, R. J. & DAVIS, B. D. (1947). *J. Exp. Med.* **85**, 9.
FAULKNER, G. H. (1943). *Lancet*, 245, p. 38.
FRANKE, W. & SCHILLINGER, A. (1943). *Biochem. Z.* **316**, 313.
FROBISHER, M. (1926). *J. Infect. Dis.* **38**, 66.
GORTER, E. & HERMANS, J. J. (1942). *Proc. K. Akad. Wet. Amst.* **45**, 804.
GRANADOS, H., MASON, K. E. & DAM, H. (1947). *Acta path. microbiol. scand.* **24**, 86.
HETTCHE, H. O. (1934). *Z. Immunitätsforsch.* **83**, 506.
HOLMAN, R. L. & SWANTON, M. C. (1946). *Proc. Soc. Exp. Biol., N.Y.*, **63**, 87.
HOTCHKISS, R. D. (1946). *Ann. N.Y. Acad. Sci.* **46**, 479.
HUMFELD, H. (1947). *J. Bact.* **54**, 513.
HUTCHINGS, B. L. & BOGGIANO, E. (1947). *J. Biol. Chem.* **169**, 229.
KNAYSI, G. (1938). *Bot. Rev.* **4**, 83.
KNIGHT, B. C. J. G. & FILDES, P. (1933). *Brit. J. Exp. Path.* **14**, 112.
KODICEK, E. (1948). In the Press.
KODICEK, E. & WORDEN, A. N. (1944). *Nature, Lond.*, **154**, 17.
KODICEK, E. & WORDEN, A. N. (1945). *Biochem. J.* **39**, 78.
KODICEK, E. & WORDEN, A. N. (1946). *Nature, Lond.*, **157**, 587.
KOSYAKOV, P. N. (1947). *Byull. Eksp. Biol. Med.* **23**, 93.
LAMAR, R. V. (1911). *J. Exp. Med.* **13**, 1.
LASER, H. & FRIEDMANN, E. (1945). *Nature, Lond.*, **156**, 507.
LASER, H. (1948). Private communication.
LEE, J. S., PUH, Y. C. & TSAI, C. (1945). *Proc. Chin. Physiol. Soc., Changtu*, **2**, 161.
LWOFF, A. (1947). *Ann. Inst. Pasteur*, **73**, 735.
LYONS, C. G. & RIDEAL, E. K. (1929). *Proc. Roy. Soc.* A, **124**, 333.
MCBAIN, J. W. (1942). *Advances in Colloid Science*, **1**, 126.
MACHEBOEUF, M. & TAYEAU, F. (1941). *Bull. Soc. Chim. biol., Paris*, **23**, 31.
MCKEE, C. M., DUTCHER, J. D., GROUPE, V. & MOORE, M. (1947). *Proc. Soc. Exp. Biol., N.Y.*, **65**, 326.
MAIZELS, M. (1945). *Quart. J. Exp. Physiol.* **33**, 183.
MAZELLA, H. (1946). *Arch. Soc. Biol. Montevideo*, **13**, 227.
MIDDLEBROOK, G. & DUBOS, R. J. (1947). *J. Immunol.* **56**, 301.
PANGBORN, M. C. (1942). *J. Biol. Chem.* **143**, 247.
PETERSON, W. H. & PETERSON, M. S. (1945). *Bact. Rev.* **9**, 49.
PINNER, M. (1928). *Amer. Rev. Tuberc.* **18**, 497.
POLLOCK, M. R. (1947). *Brit. J. Exp. Path.* **28**, 295.
POLLOCK, M. R. (1948). This volume.
PONDER, E. (1945). *J. Gen. Physiol.* **29**, 1.
PONDER, E. (1946). *J. Gen. Physiol.* **30**, 15.
POTTER, R. L. & ELVEHJEM, C. A. (1948). *J. Biol. Chem.* **172**, 531.
ROTHE-MEYER, A. & HICKMANS, E. (1946). *Arch. Dis. Child.* **20**, 160.
SCHULMAN, J. H. (1948). Private communication.
SCHULMAN, J. H. & STENHAGEN, E. (1938). *Proc. Roy. Soc.* B, **126**, 356.
SHIVE, W. & ROGERS, L. L. (1947). *J. Biol. Chem.* **169**, 453.
SHULL, G. H., THOMA, R. W. & PETERSON, W. H. (1948). *Fed. Proc.* **1**, 188.
SNELL, E. E. & WILLIAMS, R. J. (1939). *J. Amer. Chem. Soc.* **61**, 3594.
STOKES, J. L., LARSEN, A. & GUNNESS, M. (1947). *J. Bact.* **54**, 219.
STRONG, F. M. & CARPENTER, L. E. (1942). *Industr. Engng Chem.* (Anal. ed.), **14**, 909.

TAYEAU, F. (1944). *C.R. Soc. Biol., Paris*, **138**, 700.
TAYEAU, F. (1945). *Bull. Trav. Soc. pharm. Bordeaux*, **83**, 62.
TRAGER, W. (1947*a*). *Proc. Soc. Exp. Biol., N.Y.*, **64**, 129.
TRAGER, W. (1947*b*). *J. Exp. Med.* **85**, 663.
TRAGER, W. (1948). *Fed. Proc.* **1**, 282.
VALKO, E. I. (1946). *Ann. N.Y. Acad. Sci.* **46**, 451.
WHEELER, A. H. & BRANDON, E. M. (1947). *Amer. J. Clin. Path.* **17**, 130.
WHITEHILL, A. R., OLESON, T. T. & SUBBAROW, Y. (1947). *Arch. Biochem.* **15**, 31.
WILLIAMS, V. R. (1945). *J. Biol. Chem.* **159**, 237.
WILLIAMS, V. R. & FIEGER, E. A. (1945). *Industr. Engng Chem.* (Anal. ed.), **17**, 127.
WILLIAMS, V. R. & FIEGER, E. A. (1946). *J. Biol. Chem.* **166**, 335.
WILLIAMS, V. R. & FIEGER, E. A. (1947). *J. Biol. Chem.* **170**, 399.
WILLIAMS, W. L., BROQUIST, H. P. & SNELL, E. E. (1947). *J. Biol. Chem.* **170**, 619.

THE ACTION OF PENICILLIN ON THE ASSIMILATION AND UTILIZATION OF AMINO-ACIDS BY GRAM-POSITIVE BACTERIA

By ERNEST FREDERICK GALE

Medical Research Council Unit for Chemical Microbiology, Biochemical Laboratory, University of Cambridge

The growth of any cell involves an increase in its parts. One of the most important constituents of the bacterial cell is protein, and the growth of a bacterial cell involves the synthesis of protein. The proteins of bacteria resemble those of other cells in being built up from some twenty-odd amino-acids. Consequently the synthesis of cell protein must follow on the provision of all these amino-acids in the right relative proportions. Such provision can be made either by synthesis or by assimilation of the ready-formed amino-acids from the environment. Some bacteria, notably those which are usually found living in biological fluids, are unable to synthesize some of their essential amino-acids and are consequently dependent upon assimilation of these amino-acids for their growth purposes. Knight (1936) and Lwoff (1943) have put forward the hypothesis that these organisms have lost the power to synthesize certain amino-acids, since their growth in the complex nutrient media associated with living tissues is accomplished more easily by assimilation than by synthesis, so that the latter ability has been lost by disuse. The synthetic disability may be simple, such as that of *Eberthella typhosa* which requires only tryptophan, or complex such as that of *Streptococcus haemolyticus*, strains of which may require as many as nineteen different amino-acids in a preformed state. There is no obvious distinction between the nutritionally exacting organisms and those which are fully competent at amino-acid synthesis, but many of the organisms with very complex nutritional requirements are Gram-positive.

Penicillin is a selective poison which has no action against the human host, has little action against some groups of bacteria but is highly toxic to other groups. Amongst the least sensitive bacteria are those, such as *Aerobacter aerogenes* and *Escherichia coli*, which are able to synthesize all their amino-acid requirements from ammonia, while the most sensitive organisms include those such as *Streptococcus haemolyticus* and *Staphylococcus aureus* which have wide synthetic disabilities. It has become customary to state that the Gram-positive bacteria are markedly more sensitive to penicillin than are the Gram-negative species, and this is, in general, true. However, the Gram-negative *Neisseria* are amongst the organisms most sensitive to penicillin, so the Gram reaction does not provide a strict rule regarding penicillin sensitivity and, as will be developed below, it is more probable that the dependence of an organism upon assimilation processes rather than upon synthesis determines its penicillin sensitivity.

Studies with Gram-positive bacteria, *Staphylococcus aureus* in particular, have enabled us to study the assimilation process in these organisms, since the Gram-positive species differ from the Gram-negative in that they are able to effect a concentration of certain amino-acids in the free state within the internal environment prior to metabolism (Gale, 1947*a*; Taylor, 1947). The organisms studied cannot synthesize glutamic acid but are able to assimilate it from the external environment. To enable the amino-acid to cross the cell wall and enter the internal environment, the cell has to provide energy by some form of exergonic metabolism such as glycolysis. When such energy is made available, glutamic acid enters the cell and becomes concentrated inside such that the internal concentration is markedly higher than that holding in the external environment at equilibrium (Gale, 1947*a*). The level of free glutamic acid measured within the cell will be determined by the balance between the rate at which the amino-acid is assimilated from the external environment and the rate at which it is metabolized within the cell (Gale & Mitchell, 1947). Thus anything which interferes with the internal metabolism without affecting assimilation will tend to produce a rising level within the cell and, vice versa, anything which interferes with assimilation while not affecting the internal metabolism will lead to a falling level within the cell.

When *Staph. aureus* is grown in an amino-acid-rich medium and cells are removed from culture at various times within the growth period, it is found that free glutamic acid accumulates within the cell and that the level within the cell rises steadily throughout the growth period to a maximum level attained at the end of cell division (Gale, 1947*b*). If an inhibitory concentration of penicillin is added to the medium in which the culture is growing, then the accumulation of glutamic acid within the cell increases normally for approximately 1 hr. after the penicillin addition and then begins to fall abruptly. It would appear that the entry of glutamic acid into the cell is blocked within an hour or so of the addition of penicillin to the medium. This can be checked directly in the following manner: if the cells are grown in a medium containing the minimum free glutamic acid compatible with growth, then the cells, on harvesting, contain a very low internal level of glutamic acid; if these cells are now incubated in a solution of glutamic acid of comparatively high concentration together with glucose as energy source, glutamic acid enters the cells until the internal level has risen to the new equilibrium value and the rise in the internal concentration can be taken as a measure of the assimilation. Cells harvested during the growth period in a normal medium can assimilate 600–700 μl. glutamic acid (1 μmol. = 22·4 μl.) per 100 mg. cells; cells harvested from the same medium 30 min. after the addition of 10 units penicillin/ml. can assimilate only 14% of this, after 60 min. the assimilation ability has fallen to 4% of the control, and after 90 min. the cells can no longer assimilate glutamic acid at all (Gale & Taylor, 1947). The respiration and fermentation activities of these cells are significantly the same as those of the normal control cells, and their ability to assimilate lysine is unimpaired (lysine assimilation differs from that of glutamic acid in being a purely physical process dependent upon the properties of the cell wall (Gale, 1947*a*)). Methods have been worked out for the study of the internal metabolism of glutamic acid in such cells, and it is found that the internal metabolism is the same in the penicillin-treated cells as in the normal cells. However, in the penicillin-treated cells, the internal glutamic acid is rapidly depleted as a result of this metabolism, whereas in the normal cells glutamic acid is assimilated from the external environment as fast as it is metabolized within the cells. It is evident that the action of penicillin has been to block the assimilation of the amino-acid without affecting its metabolism inside the cell. Since the accumulated glutamic acid within the cell acts as a reservoir supplying the anabolic needs of the cell (Gale, 1947*b*; Gale & Mitchell, 1947), it follows that protein synthesis, etc., must cease soon after the action of penicillin has impeded the entry of glutamic acid into the cell.

Not only do different bacterial species vary in their sensitivity towards penicillin, but the sensitivity of a single strain can often be altered by serial subcultivation in the presence of increasing concentrations of penicillin. Demerec (1945) has investigated this process of 'training' and concludes that a culture often contains mutants whose resistance to penicillin is slightly greater than that of the bulk of the cells, and that the training process consists of the suppression of the sensitive cells and selective growth of the more resistant ones. The training process can be pushed to very high levels of penicillin resistance in the case of *Staph. aureus*, although this is not the case with other genera (Bellamy & Klimek, 1948*a*). The action of penicillin on the assimilation of glutamic acid by a strain of *Staph. aureus* trained to resist 60 units penicillin/ml. was tested; the assimilation was unaffected by the addition of 10 units penicillin/ml. to the growing culture but completely blocked by an addition of 100 units/ml. Consequently the increase in resistance was accompanied by a similar increase in the amount of penicillin required to block the assimilation process. A search was therefore made for some property of the assimilation process which could be correlated with the penicillin resistance (Gale, 1947*c*).

The assimilation process can be studied if the internal metabolism is inhibited; this can be effected by working with washed suspensions of cells, in which no protein synthesis takes place (Gale, 1947*b*), treated with crystal violet which inhibits residual internal metabolism of glutamic acid (Gale & Mitchell, 1947; Elliott & Gale, 1948). Using such preparations of cells the

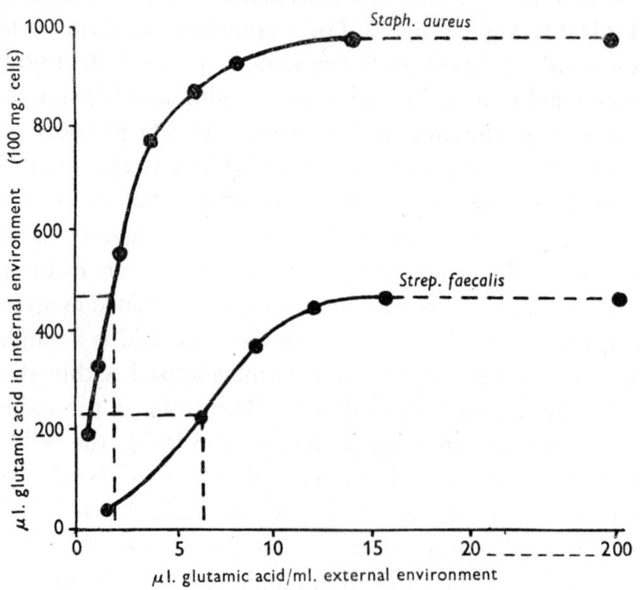

Fig. 1. Relation between internal concentration and external concentration of glutamic acid for *Staphylococcus aureus* and *Streptococcus faecalis*

dependence of the internal concentration of glutamic acid on the concentration holding in the external environment at equilibrium has been investigated. It is found that the internal concentration reaches a maximum which remains constant and independent of external concentration for all except very low values of the latter. The external concentration necessary to bring about saturation of the internal environment varies with the organism, as does the slope of the curve relating internal and external values when the former is below maximum. Some initial studies showed that a penicillin-sensitive *Staph. aureus* gave a concentration curve rising very steeply to the maximum in comparison with the curve obtained for a *Streptococcus faecalis* which was comparatively penicillin-resistant. As a measure of the slope of the concentration curve, the assimilation constant was determined and defined as that external concentration giving rise to an internal concentration equal to half that attained at saturation. The reciprocal of the assimilation constant is a measure of the assimilation 'affinity' or ability of the cell in question to assimilate glutamic acid.

When a survey was made of the assimilation constants of a variety of organisms it was found that the higher the resistance to penicillin, the higher the value of the assimilation constant (Gale, 1947*c*). The correlation was proved by taking a strain of *Staphylococcus aureus*, training it in the usual way to become resistant to penicillin, and determining its assimilation constant at increasing levels of penicillin resistance. As the penicillin resistance increases so does the assimilation constant, and at penicillin levels higher than 1000 units/ml. the assimilation constant increases very rapidly. Since an increasing assimilation constant means a decreasing efficiency of assimilation, it would seem that an organism living at such high penicillin levels would have to adopt a method of amino-acid provision other than assimilation. Bellamy & Klimek (1948*b*) have shown that if *Staph. aureus* is trained to withstand penicillin at a level of 4000–6000 units/ml., then the organism undergoes a profound change. It changes from a Gram-positive coccus to a Gram-negative bacillus, becomes a strict aerobe, and loses many of its fermentative abilities. This change has been confirmed (Gale & Rodwell, 1948) using a different strain of *Staph. aureus* and different penicillin preparations. We have studied the amino-acid metabolism of the Gram-negative organisms and have found that, although their catabolic activities are significantly the same as those of the parent strains, they have the ability to synthesize all their amino-acid requirements from ammonia. The parent strains both require a variety of amino-acids, seven in one case and ten in the other, and also require nicotinamide and thiamine as growth factors. The resistant Gram-negative organisms require thiamine only and will grow normally in a mixture of salts, ammonia, glucose and thiamine; their growth is assisted by cystine but will take place in its absence.

The evidence brought forward so far suggests that the action of penicillin is to block certain assimilation processes, and that development of resistance to penicillin involves the selection of mutants which depend less upon assimilation than upon synthesis for growth purposes until, if the resistance is forced high enough, mutants are selected which can synthesize all their amino-acids. The Gram-negative organisms derived from *Staph. aureus* do not contain any free glutamic acid and so fall into line with the previous finding (Taylor, 1947) that free amino-acids cannot be demonstrated within the internal environment of Gram-negative organisms.

In Table 1 are set out the amino-acid requirements and the penicillin sensitivity of organisms derived from the two parent *Staph. aureus* strains 6773 and 209. The organism 6773(2000) is a Gram-positive *Staph. aureus* derived from 6773 by serial subcultivation in increasing concentrations of penicillin up to the level 2000 units/ml.; its requirements are intermediate between those of the parent 6773 and those of the Gram-negative organism obtained at a level of 6000 units/ml. The organisms PT 1 and PT 2 are both

Table 1. *Relation between synthetic abilities and penicillin resistance in* Staphylococcus aureus

A. *Mutants obtained by training in presence of penicillin*

Organism ...	6773	6773	6773	209	209 PT	209 PT 1 (reversion mutants)	209 PT 2
Penicillin resistance (30 hr. test) (units/ml.) ...	5	2000	6000	0·06	7000	400*	1000*
Gram reaction ...	+	+	−	+	−	+	+
Nutrients							
Nicotinamide	+	+	−	+	−	−	−
Thiamine	+	+	+	+	+	+	+
Proline	(88)	(88)	−	+	−	(88)	(64)
Histidine	+	+	−	+	−	(64)	(64)
Valine	+	+	−	+	−	(88)	(88)
Glycine	+	(88)	−	+	−	(160)	+
Glutamic acid	(40)	−	−	−	−	(64)	−
Aspartic acid	+	(64)	−	+	−	(160)	(64)
Leucine	(40)	(40)	−	+	−	−	−
Cystine	+	+	−	+	−	+	+
Phenylalanine	(40)	(64)	−	−	−	−	−
Arginine	+	+	−	−	−	+	+

* Penicillin resistance doubled after 84 hr. incubation.
+ = presence essential for growth in same time as in complete medium.
− = presence not essential for growth in same time as in complete medium.
(88) = in absence of amino-acid growth equal to that in complete medium took place after a delay of 88 hr.

B. *Mutants obtained by training in depleted media*

Organism ...	6773	6773	209	209
Penicillin resistance (30 hr. test) (units/ml.) ...	5	100	0·06	250
Nutrients				
Nicotinamide	+	+	+	+
Thiamine	+	+	+	+
Proline	+	−	+	−
Histidine	+	+	+	+
Valine	+	−	+	−
Glycine	+	−	+	−
Glutamic acid	+	−	−	−
Aspartic acid	+	−	+	+
Leucine	+	−	+	−
Cystine	+	+	+	+
Phenylalanine	+	−	−	−
Arginine	+	−	−	−

+ = essential for growth. − = non-essential for growth.

Gram-positive *Staph. aureus* isolated from cultures of the Gram-negative 209 PT which have thrown reverse mutations, the method of isolation being that described by Bellamy & Klimek (1948*b*). 209 PT can synthesize all its amino-acids from ammonia, but the reversions cannot synthesize certain amino-acids (glycine, arginine and cystine) and can only adapt to the absence of others after a prolonged period. Two points appear of particular interest: the reverted organisms have retained the ability to synthesize nicotinamide (not possessed by the original parent 209) but are nutritionally exacting towards arginine, whereas the original parent 209 did not require this amino-acid. The penicillin resistance of these reversion mutants is very high, and it was noticed during testing that, just as the organism adapted to the absence of a variety of amino-acids if incubated over a long period, so did its resistance increase as incubation was prolonged.

Gladstone (1937) showed that it is possible to train *Staph. aureus* strains to become non-exacting towards amino-acids by a process of serial subcultivation in media from which the essential amino-acids are progressively withdrawn. Using Gladstone's technique the strains 6773 and 209 have been trained to dispense with most, if not all, of their amino-acid requirements, and Table 1 shows that, as the organism is trained to synthesize its amino-acid requirements, so does its penicillin resistance increase at the same time. The training by nutritional attrition promotes selective growth of non-exacting mutants, and these are found to have high penicillin resistance; training to increased penicillin resistance promotes selective growth of resistant mutants, and these are found to have high synthetic abilities towards amino-acids. When these results are taken in conjunction with those previously described on the blocking of amino-acid assimilation by penicillin and on the decreasing assimilation ability accompanying increase in penicillin resistance, it becomes clear that penicillin is a selective agent which prevents the growth of organisms dependent upon assimilation processes for their protein synthesis (Gale & Rodwell, 1949).

It is improbable that assimilation is a property specific to Gram-positive organisms; the fact that organisms such as *Eberthella typhosa* and the *Neisseria* are nutritionally exacting towards certain amino-acids indicates that they must be able to assimilate these essential amino-acids. The Gram-negative organisms, however, do not possess the ability to concentrate amino-acids in the internal environment after assimilation, and it may be that this ability in the Gram-positive organism represents a form of evolution in that the internal concentration mechanism compensates to some extent for the loss of synthetic ability. The *Neisseria* may possibly represent some intermediate stage of this evolution, as they may, by a change of staining conditions, be made to react as Gram-positive although they cannot be classed as such (Verhoeff, 1940). It is clear that further work is required on the metabolism of these organisms before we can understand why they should be as sensitive to the action of penicillin as the highly exacting Gram-positive cocci. Within the Gram-positive group, the positive reaction to the stain has been investigated by Henry & Stacey (1943) and by Bartholomew & Umbreit (1945), who have shown that the staining substrate is a protein lying in the surface layers of the cells and which has a magnesium ribonucleotide complex as prosthetic group. The presence of this Gram-complex is certainly linked with the assimilation-concentration mechanism (Taylor, 1947) and, possibly indirectly, with the penicillin sensitivity of the cell. As shown above, if the ability of the cell to synthesize amino-acids and to resist penicillin is pushed to the limit, then the cell becomes Gram-negative, while Dufrenoy & Pratt (1947) have shown that *Staphylococcus aureus* becomes Gram-negative if it is grown in bacteriostatic concentrations of penicillin. Frieden & Frazier (1947) have also found that reducing the magnesium content of the growth medium increases the penicillin resistance of the culture, and it may be that such treatment promotes the growth of mutants which are less dependent upon the Gram-complex assimilation mechanism.

It seems clear that the normal growth of the Gram-positive cell is dependent upon its ability to assimilate and concentrate those amino-acids which it cannot synthesize; in turn this is connected in some way with the presence of the Gram-complex which must, consequently, be synthesized as the cell divides in order that the daughter cells may be viable in turn. Penicillin acts only on growing cultures, and it is beginning to look as though its fundamental action must lie somewhere on these synthetic processes occurring during cell division. Krampitz & Werkman (1947) showed that high concentrations of penicillin will inhibit the dissimilation of exogenous or endogenous ribonucleate by washed suspensions of *Staph. aureus*; no effect could be shown on the metabolism of desoxyribonucleate. The amounts of penicillin used were of the order 400–2000 units/ml., which were approximately 10,000 times the quantity required to prevent the growth of an inoculum; the authors point out that the amounts of organism used in their experiments were also many times those used in an inoculum. This argument may or may not be valid, but there is, at any rate, an indication here that penicillin can disorganize ribonucleic acid metabolism in the cell, and the Gram-complex is a ribonucleotide complex.

Table 2 shows results obtained for the nucleic acid content of a penicillin-sensitive *Staph. aureus* before and after the addition of penicillin to the growing culture. Organisms were harvested at the time of the penicillin addition, and at 30 and 60 min. after the addition. In all cases the ribonucleic-P and desoxyribonucleic-P were estimated. In the normal culture there is a steady drop of both desoxy- and ribonucleic acid content with the age of the culture, much as described by Malmgren & Heden (1947). The addition of penicillin has been followed by a significant decrease in the

Table 2. *Effect of penicillin on the nucleotide content of* Staphylococcus aureus

Staph. aureus D grown at 37° C. in casein-digest-glucose medium and penicillin 5 units/ml. added to growing cultures as indicated below. All quantities expressed as μg. P/mg. dry weight of cells.

Age of culture ...	4 hr.	4½ hr.		6 hr.		Gram-negative resistant organism
Time for which penicillin was present before harvesting (min.)	—	—	30	—	120	—
Total P content	32·6	30·2	31·7	27·6	26·2	15·2
Ribonucleic acid P	20·2	18·8	17·7	15·5	13·9	8·38
Desoxyribonucleic acid P	2·30	2·26	2·71	1·58	2·5	3·06
Ratio $\frac{\text{Ribonucleic P}}{\text{Desoxyribonucleic P}}$	8·8	8·3	6·5	9·8	5·55	2·7
Dry weight of cells in culture (mg./ml.)	0·121	0·190	0·154	—	—	—
Ribonucleic P/ml. culture	2·45	3·47	2·72	—	—	—
Desoxyribonucleic P/ml.	0·25	0·43	0·42	—	—	—

ribonucleic-P, 6% below the control in the 30 min. penicillin-treated culture and 10% in the 120 min. culture; this is accompanied by a marked increase in the desoxyribonucleic-P which is 20% greater than in the control after 30 min. contact with penicillin and 58% after 120 min. The ratio of ribonucleic-P to desoxyribonucleic-P averages 8·7 for the normal culture and has fallen to 6·5 for the 30 min. penicillin culture and to 5·55 for the 120 min. penicillin culture. There is definite evidence here of a disorganization of nucleic acid metabolism in the presence of penicillin. If we calculate the total amount of the nucleic acids per ml. of culture we can see that the synthesis of desoxyribonucleic acid is the same in the presence of penicillin as in its absence, but that, whereas the ribonucleic acid increases by 42% during the 30 min. period in the normal culture, in the penicillin culture it increases by only 10%; this is further evidence that the disorganization of nucleic acid metabolism relates to the synthesis of the ribonucleic acid fraction. If the original suggestion of Brachet (1933) is correct, that ribonucleic acid gives rise to desoxyribonucleic acid, then these results could be interpreted as meaning that the penicillin has inhibited the formation of ribonucleic acid but not its transformation into desoxyribonucleic acid. Analyses carried out on the highly resistant Gram-negative organism derived from *Staph. aureus* 6773 show that this has the very low ratio of 2·7, while its desoxyribonucleic-P content is higher than that of the Gram-positive organisms investigated. The action of penicillin on the Gram-positive organism has thus had the effect of altering its nucleic acid composition in the direction of that of the Gram-negative organism.

The present position of our knowledge concerning the action of penicillin can be summarized thus:

Penicillin appears to act as a selective poison suppressing the growth of bacteria which are dependent upon assimilation rather than upon synthesis for their growth requirements; the ability to concentrate amino-acids within the internal environment after assimilation is confined to Gram-positive species and can be correlated with the presence of the ribonucleotide complex which is responsible for the positive Gram reaction; there is evidence that penicillin exerts a direct effect upon nucleic acid metabolism within the growing cell, and it is possible that its action in blocking assimilatory processes is a consequence of the disturbance of the nucleic acid synthesis.

REFERENCES

Bartholomew, J. W. & Umbreit, W. W. (1945). *J. Bact.* **48**, 567.
Bellamy, W. D. & Klimek, J. W. (1948*a*). *J. Bact.* **55**, 147.
Bellamy, W. D. & Klimek, J. W. (1948*b*). *J. Bact.* **55**, 153.
Brachet, J. (1933). *Arch. Biol.* **44**, 519.
Demerec, M. (1945). *Ann. Mo. Bot. Gdn*, **32**, 131.
Dufrenoy, J. & Pratt, R. (1947). *J. Bact.* **54**, 283.
Elliott, W. H. & Gale, E. F. (1948). *Nature, Lond.*, **161**, 129.
Frieden, E. H. & Frazier, C. N. (1947). *Arch. Biochem.* **15**, 265.
Gale, E. F. (1947*a*). *J. Gen. Microbiol.* **1**, 53.
Gale, E. F. (1947*b*). *J. Gen. Microbiol.* **1**, 327.
Gale, E. F. (1947*c*). *Nature, Lond.*, **160**, 407.
Gale, E. F. & Mitchell, P. D. (1947). *J. Gen. Microbiol.* **1**, 299.
Gale, E. F. & Rodwell, A. (1948). *J. Bact.* **55**, 161.
Gale, E. F. & Rodwell, A. (1949). *J. Gen. Microbiol.* **3**.
Gale, E. F. & Taylor, E. S. (1947). *J. Gen. Microbiol.* **1**, 314.
Gladstone, G. P. (1937). *Brit. J. Exp. Path.* **18**, 322.
Henry, H. & Stacey, M. (1943). *Nature, Lond.*, **151**, 671.
Knight, B. C. J. G. (1936). *Bacterial Nutrition.* M.R.C. Special Report Series. H.M. Stationery Office.
Krampitz, L. O. & Werkman, C. H. (1947). *Arch. Biochem.* **12**, 517.
Lwoff, A. (1943). *l'Évolution physiologique.* Hermann et Cie, Paris.
Malmgren, B. & Heden, C.-G. (1947). *Act. Path. Microbiol. Skand.* **24**, 417.
Taylor, E. S. (1947). *J. Gen. Microbiol.* **1**, 86.
Verhoeff, F. H. (1940). *J. Amer. Med. Ass.* **115**, 1545.

ADAPTATION OF BACTERIA TO RESIST DRUG ACTION (WITH SPECIAL REFERENCE TO *BACT. LACTIS AEROGENES*)

By C. N. HINSHELWOOD

Department of Physical Chemistry, Oxford

This paper will be concerned principally with the adaptation of bacterial cells to resist the action of inhibitory substances. The theme is closely related to the question of selective toxicity because, according to one of the simplest hypotheses, adaptation depends upon changes in the material constitution of the cell, which are caused directly by the specific influence of the inhibitory drug upon various rates of synthesis.

The selectivity appears at the very outset of the study. Certain types of cell adapt themselves rapidly to ignore the presence of one drug, but seem quite incapable of acquiring resistance to another. For example, *Bact. lactis aerogenes*, which has been subjected to rather close examination, can be serially subcultured in presence of enough phenol to reduce the growth rate to about half the normal. After as many as 100 subcultures, each involving about eight successive cell divisions, there is not the slightest detectable recovery towards a normal growth rate. In sharp contrast with this lack of response to training we find that comparatively few cell divisions in presence of acridine derivatives or sulphonamides confer a considerable degree of adaptation (Davies & Hinshelwood, 1943; Davies, Hinshelwood & Pryce, 1944, 1945; Pryce, Davies & Hinshelwood, 1945; Pryce & Hinshelwood, 1947). The bacterium referred to is soon trained to grow optimally in presence of a concentration of 2:8 diaminoacridine (proflavine), 100-fold that which would, initially, have inhibited all growth.

The way in which a selective interference with relative reaction rates could provoke an automatic adaptive response is shown in a qualitative and schematic manner in Fig. 1. *a* and *b* are two essential parts of the cell, *a* furnishing an intermediate necessary to the reproduction of *b*. The concentration of this intermediate is represented by the density of the hatched lines in the figure. The drug, without impeding the reproduction of *a*, cuts down the output of the intermediate (as shown by the lighter shading in the second line of the diagram) so that the synthesis of *b* is slowed down. *b*, however, is essential to the cell and division must wait until some more or less standard critical amount of it has been built up. By this time the amount of *a* has become greater than normal, and the concentration of intermediate which it now supplies will have risen (right-hand side of second line of diagram). When the cell has at length divided, the ratio of *a* to *b* remains abnormal, and since *a* now produces a greater concentration of intermediate than originally, the influence of the drug will be antagonized (as shown in the last line of the figure).

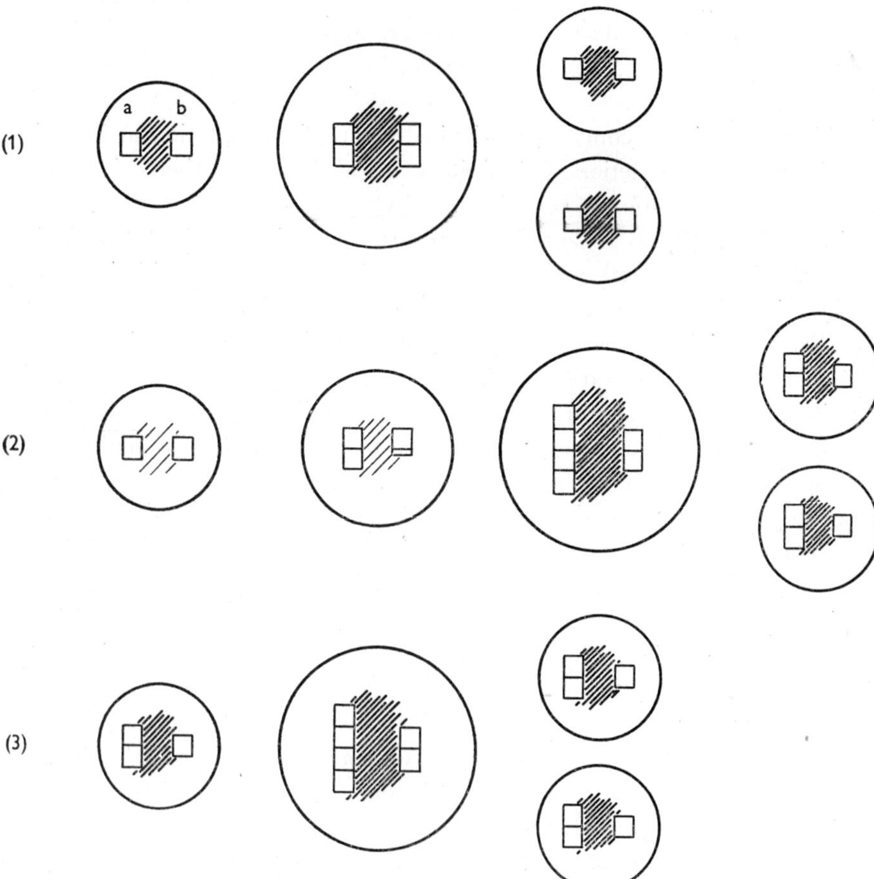

Fig. 1. (1) normal state; (2) inhibition by drug; (3) adapted state, drug present.

Examination of the differential equations of this linked enzyme model shows that the adaptive response will be complete as soon as the amount of new material synthesized outweighs the original, and that the overall growth rate will be restored to normal.

The hypothesis that drug adaptation occurs in some such way as this is of interest in that (1) it predicts specificity, since there would be no adaptation to a drug which reduced all the rates equally; (2) different drugs are envisaged as intervening at different stages of the sequence of synthetic reactions; (3) the response expected is automatic, rapid and common to all the cells; (4) the adapted cells might be expected to show somewhat modified enzymatic properties. There is evidence that all these conditions are in fact fulfilled.

There is, moreover, as would be expected, a quantitative relation between the concentrations of drug at which the cells have been trained and the properties which the strain shows when tested at still higher concentrations (Davies *et al.* 1945; Pryce & Hinshelwood, 1947). This is

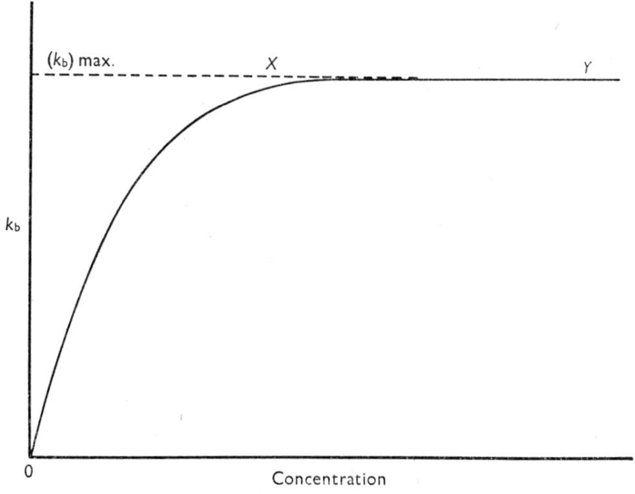

Fig. 2.

well shown if one plots the concentration of the drug against the lag which it induces. The 'lag-concentration' curve for a given bacterial strain and a given drug is of a characteristic form, and the curves for various trained strains constitute a family with spacings quantitatively related to the 'training concentrations'. The mathematical relationships observed in these families have been studied in some detail by my colleagues Drs Davies, Pryce and James.

A somewhat complex problem is presented in the interpretation of the permanence and stability of drug adaptation. Casual observation would suggest that sometimes it is relatively permanent and sometimes easily lost on further subculture in the absence of the drug which provoked it. The equations of the model referred to above in their simplest form predict reversibility of the adaptation. But this may be modified by a special circumstance.

The relation between the growth rate of the cell substance b and the concentration of the intermediate which is employed in the process will in general be of the Langmuir form shown in Fig. 2. If we are operating on the part OX of the curve, the expansion of a relative to b, as envisaged in the hypothesis about the adaptive process, will be reversible, because, when the drug is removed the rate of synthesis of b rises again relatively to that of a. But if we are operating on the section XY, removal of the drug allows an increased concentration of the intermediate to appear without any corresponding increase in the rate of synthesis of b. With trained cells in presence of the drug we may easily be working on OX, and with the trained strains in absence of drug, along XY.

The rate of growth of a is given by

$$x_a = (x_a)_0 \, e^{k_1 t},$$

where x_a is the amount at time t and $(x_a)_0$ the initial amount. For untrained cells, or for trained cells in the presence of the drug at the training concentration, the growth of b is given by

$$dx_b/dt = k_2 c x_b = k_b x_b,$$

where c is the intermediate concentration, and, since a and b are in equilibrium proportions,

$$k_b = k_1.$$

There is reason to believe that in normal cells c normally corresponds to a point near X in Fig. 2, so that k_b approaches $(k_b)_{\text{max.}}$ and thus $(k_b)_{\text{max.}} = k_1$.

The increase in the concentration, c, which occurs when the trained cells are returned to the drug-free medium gives no further increase in $(k_b)_{\text{max.}}$ so that we still have

$$x_a = (x_a)_0 \, e^{k_1 t} \quad \text{and} \quad x_b = (x_b)_0 \, e^{(k_b)_{\text{max.}} t} = (x_b)_0 \, e^{k_1 t}.$$

Thus

$$x_a/x_b = (x_a)_0/(x_b)_0.$$

$(x_a)_0/(x_b)_0$ has the value established during training, and this now shows no rapid change. Now the concentration c is determined by the ratio and thus remains higher than in untrained cells. Consequently if the bacteria are once more exposed to the drug they are able to function normally unless the concentration of drug exceeds that at which they were trained.

According to this view the stability is not absolute. If some other disturbance of conditions changes the ratio x_a/x_b back towards that characteristic of an untrained cell it stays there unless a fresh process of training occurs.

Induced loss of drug resistance is in fact observable, and here again a considerable element of specificity and selectivity is encountered. A few examples may be quoted. *Bact. lactis aerogenes* trained to sulphanilamide may be made to lose its adaptation by culture in presence of proflavine, though the process is unavailing if the cells have previously been trained to proflavine itself (James & Hinshelwood, 1947). Proflavine adaptation may be destroyed by subculture in presence of cresol at suitable concentrations (Davies *et al.* 1945).

The facts about spontaneous loss of adaptation are at first sight puzzling, but seem on detailed study to conform to a general pattern which may be described as follows. When the cells have acquired the power to grow normally by a few divisions in presence of the drug, they easily revert on subculture in their normal medium. When the training process has been longer continued, the reversion is slower or postponed for several subcultures. When the training has been very thorough, the adaptation shows a very high degree of stability, though the cells are still subject to the kind of induced reversion referred to above. A similar pattern is shown by the phenomena of adaptation to use new sources of carbon.

It may be said at once that the behaviour is in many respects just what would be encountered in the selection of a resistant strain from a mixed population, the difficulty of reversion being simply a measure of the thoroughness with which the pure resistant strain has been isolated. But it must also be remarked that the selection view has grave difficulties which will be considered at a later stage.

At the present moment the most satisfactory interpretation of the training-reversion pattern would seem to be somewhat as follows. Relative stability (or metastability) is the normal condition. The cell, however, is a complex interlocking economy, and in the process of training various secondary adjustments are rendered necessary by the major adaptive change. In the early stages these have not had an opportunity to establish themselves, and the resulting disharmony exerts a disturbance which leads to an induced reversion, comparable in some respects with that brought about by other drugs. Actual mechanical distortions consequent upon changed enzyme proportions are not excluded. The whole question of adaptive stability seems to present problems well worth further investigation.

An interesting topic of investigation is provided by the enzyme changes accompanying drug adaptation. Training to sulphanilamide causes some increase in the catalase activity of *Bact. lactis aerogenes* (Cole & Hinshelwood, 1947), while training to proflavine causes a certain loss. Secondary changes in dehydrogenase activity accompany the adaptation to high concentrations of crystal violet (Davies & Hinshelwood, 1947). This field is obviously a very large one and will require considerable time for its exploration.

What has been said so far is based essentially upon the hypothesis that the relative proportions of enzymatic materials change. From some points of view it is also helpful to envisage actual modifications in the texture of the material itself. These could well include minor changes in the composition of complex polysaccharides or proteins, and also changes in the modes of packing and folding of chains. This latter factor would naturally influence the spacing of specific active groups in interfaces. It is specially useful to consider qualitative modifications of this kind in studying the adaptation of cells to utilize new sources of growth material, but similar ideas could be applied to the development of drug resistance. This form of hypothesis is, on closer inspection, seen to be not so very different from that of changed enzyme proportions. The spacings and foldings of the macromolecular material will probably show statistical variations from one element of cell volume to another, and we may suppose that some regions will be less liable to interference by drug absorption than others. During growth in presence of the drug there will be a relative expansion of these comparatively immune sites at the expense of the others (in a way which can be shown by calculation to lead to optimum growth rate). Once the favoured type of site has expanded (by what is virtually a sort of internal selection) the new configuration will tend to persist (though not absolutely) because during autosynthesis the new material will tend to be laid down in a way conforming to the arrangement of the old. In one sense the expanded sites constitute 'new enzymes', and the possibility that has been outlined reconciles in some degree views about qualitative and quantitative changes and also about selection as against adaptation.

The development of drug-immune enzyme sites might more easily occur if the cell utilized a sequence of growth reactions other than its normal one. There is, in fact, evidence that in some examples of drug adaptation (as well as in adaptation to new substrates), an 'alternative mechanism' comes into play, for example, in the training of *Bact. lactis aerogenes* to sulphonamides (Davies & Hinshelwood, 1943; Davies, Hinshelwood & James, 1947).

Quite apart from any hypotheses it is significant that in presence of sulphonamides the growth curve (logarithm of bacterial count plotted against time) is, for untrained cells, of a composite form, a slower mode of growth being superseded at a certain point by a more rapid one. The transition from one to the other occurs progressively earlier as training proceeds. The view has been expressed that in such examples a less efficient combination of reactions is replaced by a more efficient one when the requisite time has been allowed for the overcoming of the lag.

One of the respects in which selective action appears most prominently is in connexion with the phenomenon of reciprocal or 'cross' adaptation. Of this there are numerous examples, e.g. cells of *Bact. lactis aerogenes* trained to proflavine also show a considerable degree of training to methylene blue and vice versa. Where the adaptation is reciprocal it suggests that the stage in the sequence of synthetic reactions at which the two drugs act is the same. It must, however, be borne in mind that many drugs probably have multiple effects and that only one of these may be common to two drugs showing the cross-training effect. In such circumstances the reciprocity will only be partial, as, indeed, it appears to be with *Bact. lactis aerogenes*, proflavine and methylene blue.

In the study of the adaptation of *Bact. lactis aerogenes* to the utilization of new sources of carbon a rather close relation appears between growth rate and the dehydrogenase activity towards the carbon substrate. It is interesting, therefore, to inquire into the effects of growth-inhibiting drugs on the dehydrogenases. This problem has been investigated by Dr A. M. James, who finds a certain general parallelism between the inhibition of growth and the inhibition of dehydrogenase activity. It is, however, far from quantitative, and the situation is complicated by the fact that the action of drugs such as proflavine on the enzyme activity shows a marked time effect. Nevertheless, training of the cells to grow in presence of a given drug leads to a strain whose dehydrogenase activity is also much less than normally influenced by that drug. Furthermore, well-marked reciprocal relations appear, 'cross-training' of dehydrogenase activity following the same general lines as the cross-adaptation of growth.

Another very striking manifestation of selectivity is the power of various drugs to inhibit the cell-division function of certain bacteria while leaving their rate of elongation relatively unaffected. This results in very pronounced morphological changes. It looks as though the synthesis of certain specific factors is a necessary prelude to division, while the formation of an independent set precedes the elongation of the cell. Many curious and interesting observations can be correlated in the light of this view (Hinshelwood & Lodge, 1944). For example, *m*-cresol (Spray & Lodge, 1943) and proflavine (Davies *et al.* 1944) greatly lengthen the lag which precedes any increase in the cell substance of *Bact. lactis aerogenes*. They have a still more adverse effect on cell division, so that long filaments tend to be formed. Addition of small amounts of salt to the medium shortens the drug-induced lag, but does not affect the division, so that elongation and division are thrown still more out of balance and the filament formation is much enhanced. Larger additions of salt increase the lag again and so allow division once more to keep pace with elongation with reduction of the filamentation (James & Hinshelwood, 1947).

It should be added that the power of the drug to interfere with division in many cases rapidly wanes as an adaptive response develops.

An interesting question arises in connexion with the actual uptake of drugs and inhibitory agents by cells. It has sometimes been thought that the development of drug resistance depends upon the emergence of a type of cell impermeable to the drug. As far as *Bact. lactis aerogenes* is concerned this hypothesis does not offer much help in explaining the development of resistance to proflavine or to methylene blue.

Two lines of evidence speak against it. A. R. Peacocke has measured the actual uptake of proflavine and of methylene blue by the cells as a function of concentration (expressing the results in the form of 'absorption isotherms'). Now if training depended upon decrease in permeability, one might expect trained strains to take up less drug under comparable conditions than untrained strains. This is by no means found to be true, and with proflavine the trained strains actually take up more than the normal cells.

Another piece of evidence depends upon observations (S. Jackson) of the time required for proflavine to exert its inhibitory effect when added to an actively growing culture. The results show that this time is independent of the concentration gradient between the inside and the outside of the cell, and that, therefore, rate of penetration is unlikely to be a significant factor.

Another aspect of the question of selectivity appears in the results on the uptake of proflavine by the cells. The inhibitory action of proflavine is antagonized by hydrogen ions, as has been shown for the aminoacridines generally by Albert, Rubbo, Goldacre, Davey & Stone (1945). The actual uptake of the drug, however, changes little over a pH range in which the lag would vary widely. The inhibitory action would seem, therefore, to be exerted at specialized sites in the cell and not at every site where it can be absorbed.

Since it is of fundamental importance in the whole discussion of adaptive phenomena, something should be said about the role of selection in training. This powerful mechanism must of course operate to accentuate differences which have already given an advantage in respect of growth or survival to any members of a population, but the question is whether in the development of drug resistance it is of primary or of secondary importance.

The views which may be held are classifiable approximately as follows: (*a*) The original bacterial population is mixed and contains inherently resistant strains in small proportion. These are selected during the training process. (*b*) Chance mutations, independent of the drug, produce resistant strains which are selected. (*c*) Adaptation occurs by a direct response in each cell to the action of the drug. (*a*) will be discarded. Enough work has been done with pure strains to make it unlikely, and, the number of original strains that would have to be assumed to account for the manifold adaptations which can occur approaches the absurd.

With regard to (*b*) and (*c*) the situation is more complex. There are two issues. The first is whether the change occurring during the development of the resistance originates in a few special cells or in the whole population, and the second, which is quite distinct, is whether the initial change is provoked by the drug or due to chance.

Drug resistance being transmissible to daughter cells, the view that it may be an active response to the drug is sometimes stigmatized as 'Lamarckian', on the ground that the resistance is an 'acquired character'. The famous stretch acquired in the giraffe's neck is not, of course, transmitted, but, the fact that it leaves the reproductive material unchanged seems to be a good and sufficient reason. Irradiation of this material with gamma-rays would almost certainly give some transmissible changes, and a bacterial cell, as has often enough been pointed out, constitutes its own reproductive material. But, we may be told, the transmissible change must be a chance mutation and not an adaptation. The logical interpretation of this seems a little unsatisfactory. Chance means the operation of an unknown cause, while drug action represents a known cause. What the 'Lamarckian' objection amounts to is an assertion that changes due to unknown causes are transmissible while those due to known causes are not.

As to whether causes unconnected with the drug are more or less likely to lead to resistance than the action of the drug itself, the following considerations are perhaps relevant:

(*a*) Adaptation occurs not only to numerous highly specific and selective drugs but to appropriate combinations of these. The role which would have to be attributed to random changes seems surprisingly great.

(*b*) The sort of automatic response discussed earlier in this paper would, on the view of random changes, have to be shown not merely wrong in form (which it may very well be) but impossible in principle.

An extremely interesting argument has been advanced by Luria & Delbrück (1943), according to which the statistical variations of resistance in different tests are too great to be accounted for except by the assumption that the earliest stages of development depend upon chance mutations too rare for the appearance of uniform statistical averages. This argument is satisfactory in so far as it is applied to the choice between the two possibilities actually formulated by Luria & Delbrück, which are, on the one hand, the case where each cell changes under the influence of the drug and, on the other, the case where only rare mutants are thrown off. But other sets of circumstances exist, notably the case, quite definitely encountered in practice, where the cells are rapidly dying under the influence of the drug. Here, if an adaptive change occurs, it must be complete before the whole population dies. Now according to one view (for which good arguments exist) certain types of adaptation involve a long lag phase. Under drastic conditions the only cells to survive to the end of this will be a few exceptional ones. Death-rate curves are approximately exponential in form, with, however, a tail where randomness will become more and more pronounced the nearer one comes to its end. Thus, if there is a race between adaptation and dying, the statistical variations will become greater the more keenly the race is contested, as in the case treated by Luria & Delbrück. While one may accept their conclusion that the spread of results indicates a stage in the process where relatively few individuals are concerned, I do not feel able to agree that it necessarily points to the operation of chance mutations. The element of chance may well be in the survival rather than in the development of the resistance. What constitutes the element of chance leading to the exponential death-rate curve is another question, and a much more general one than that of drug resistance.

As to the general issue whether changes are initiated in exceptional cells or in the bulk of the population, it would not be surprising if it turned out in the end that all types of behaviour from one extreme to the other were encountered, according to the bacterial species and the drug. But with *Bact. lactis aerogenes* and proflavine there is evidence that the changes affect the bulk of the population. The argument is as follows. In a certain state of training the cells lose their resistance on cultivation in the drug-free medium. If we accept the theory that this represents a re-selection of non-resistant cells, then we must assume the population which changes to be a mixture of two types. Now stably resistant cells can be obtained by further training (i.e. on the selection view, further purification of resistant strain). These can be mixed with ordinary non-resistant cells, and any change in the balance of the population on cultivation in the drug-free medium can be studied. The proportions do not, in fact, change (Pryce & Hinshelwood, 1947). Yet loss of resistance does, in fact, occur with partially trained populations. The loss must, therefore, be located in the individual cells, since it is not due to changes in proportion of resistant and non-resistant. Conversely, it is natural to conclude, the training affects the individuals.

REFERENCES

ALBERT, A., RUBBO, S. D., GOLDACRE, R. J., DAVEY, M. E. & STONE, J. D. (1945). *Brit. J. Exp. Path.* **26**, 160.

COLE, E. H. & HINSHELWOOD, C. N. (1947). *Trans. Faraday Soc.* **43**, 266.

DAVIES, D. S. & HINSHELWOOD, C. N. (1943). *Trans. Faraday Soc.* **39**, 431.

DAVIES, D. S. & HINSHELWOOD, C. N. (1947). *Trans. Faraday Soc.* **43**, 257.

DAVIES, D. S., HINSHELWOOD, C. N. & JAMES, A. M. (1947). *Trans. Faraday Soc.* **43**, 138.

DAVIES, D. S., HINSHELWOOD, C. N. & PRYCE, J. M. G. (1944). *Trans. Faraday Soc.* **40**, 397.

DAVIES, D. S., HINSHELWOOD, C. N. & PRYCE, J. M. G. (1945). *Trans. Faraday Soc.* **41**, 163, 778.

HINSHELWOOD, C. N. & LODGE, R. M. (1944). *Proc. Roy. Soc.* B, **132**, 47.

JAMES, A. M. & HINSHELWOOD, C. N. (1947). *Trans. Faraday Soc.* **43**, 274.

LURIA, S. E. & DELBRÜCK, M. (1943). *Genetics*, **28**, 491.

PRYCE, J. M. G., DAVIES, D. S. & HINSHELWOOD, C. N. (1945). *Trans. Faraday Soc.* **41**, 465.

PRYCE, J. M. G. & HINSHELWOOD, C. N. (1947). *Trans. Faraday Soc.* **43**, 742, 752.

SPRAY, G. H. & LODGE, R. M. (1943). *Trans. Faraday Soc.* **39**, 424.

THE SIGNIFICANCE OF HYDROGEN-ION CONCENTRATION IN THE STUDY OF TOXICITY

By E. W. SIMON and G. E. BLACKMAN
Department of Agriculture, University of Oxford

I. INTRODUCTION

It has long been known that while the toxicity of a weak acid or a weak base varies with the hydrogen-ion concentration of the medium in which it is applied to the organism, yet pH has little or no effect on the toxicity of a substance which does not dissociate in solution (Labes, 1922; Clowes & Keltch, 1931). Hence it is imperative when recording the toxicity of a weak acid or a weak base to state the pH at which the measurement is made, or to determine the toxicity over a range of pH levels (Fleischer & Amster, 1923; Gershenfeld & Milanick, 1941; Baker, Harrison & Miller, 1941).

This paper is concerned with problems that arise in evaluating the comparative toxicity of a group of compounds one or more of which is a weak acid or a weak base. A study of the way in which the toxicity of different substances varies with pH shows that determinations of toxicity which are made at only one pH level may give misleading results. Fig. 1 *a* (taken from the data of Hoffman, Schweitzer & Dalby, 1940, 1941) shows the effect of pH on the fungistatic activity of phenol and of acetic acid. Acetic acid is the more toxic, but the relative toxicity varies with pH. At pH 2·5, acetic acid is about 4·5 times more active than phenol, but at pH 5·5 it is 16 times as active. It is interesting to note that the standard techniques for comparing the activity of a disinfectant with that of phenol (Rideal-Walker test, B.S.S. no. 541 and Chick-Martin test, B.S.S. no. 808) specify that the tests must be carried out at a pH of 7·3–7·5. Fig. 1 *b*, which is taken from our own data, shows the influence of pH on the fungistatic activity of *o*-nitro-phenol and of 2:4-dinitro-phenol to *Trichoderma viride*. 2:4-dinitro-phenol is the more toxic of the two compounds at a low pH, but at a high pH it is less toxic than *o*-nitro-phenol. Other workers who compared the toxicity of these two compounds to fungi have recorded data which are apparently contradictory, but which fall into line with our observations on the effect of pH on the level of toxicity. Woodward, Kingery & Williams (1934) measured the toxicity of 2:4-dinitro-phenol made up in 1% sodium carbonate and found it less active than *o*-nitro-phenol to *Monilia tropicalis*, but Plantefol (1922) who used an acid medium, observed that 2:4-dinitro-phenol was the more toxic of the two compounds to *Sterigmatocystis nigra*. Similar variations in relative toxity will also arise when the toxicity of an acid substance is to be compared with that of a base or a non-electrolyte.

In many instances it is necessary to compare the suitability of different toxic compounds for a particular purpose. Such a comparison is best made at the pH at which the compounds will be used in practice, for instance, that of the blood stream or of a spray solution. But if the object of toxicity measurements is to arrive at an understanding of the differences in toxicity in terms of the physical and chemical properties of the substances, more attention must be paid to the pH at which the tests are made.

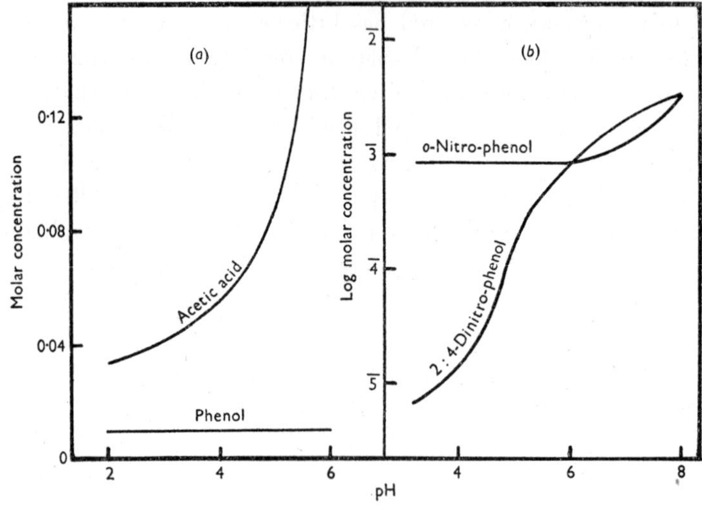

Fig. 1. *a*. The effect of pH on the concentrations of phenol and of acetic acid required to prevent the growth of 'common moulds'. (Data from Hoffman *et al.* 1940, 1941). *b*. The effect of pH on the concentrations of *o*-nitro-phenol and of 2:4-dinitro-phenol required to halve the growth rate of *Trichoderma viride*.

Such a problem arose during a study of the comparative toxicity of substances allied to 3:5-dinitro-*o*-cresol, which involved the measurement of the fungistatic activity of aromatic nitro and hydroxy compounds, some of which are non-electrolytes (nitro derivatives of anisole and phenetole), and some of which are acids (nitro derivatives of phenol).

II. EXPERIMENTAL METHODS

Reduction in the growth rate of a fungus has been used as the criterion of physiological response in this study of the influence of pH on toxic action. The procedure followed was that described by Brian (1945). *Trichoderma viride* was grown on modified Czapek Dox medium with 2·5 g./l. of dextrose. After autoclaving, the medium was brought to the desired pH by the addition of N/10-NaOH or HCl; pH was measured colorimetrically. The toxic material was then added as a suspension in water, and three replicate Petri dishes were poured for each treatment. Inoculations were made from the region just inside the circumference of a vigorously growing 4-day-old plate of *T. viride* by transferring a small cylinder of agar and mycelium on to the centre of a freshly poured plate. The plates were incubated at 25° C. The diameter of the mycelium in each plate was

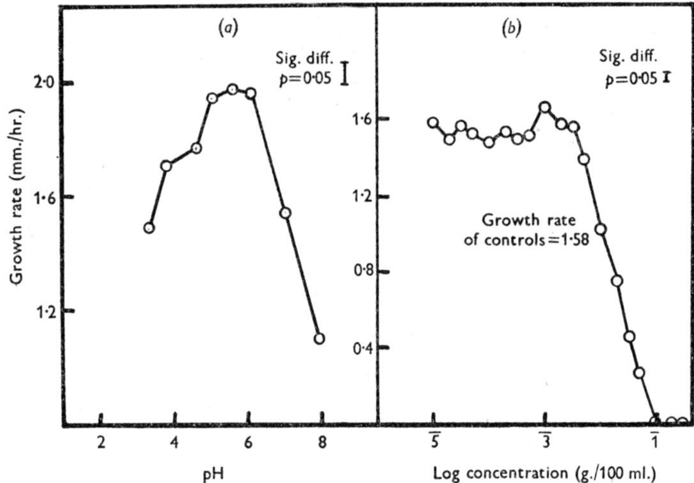

Fig. 2. *a*. The effect of pH on the growth rate of *Trichoderma viride*. *b*. The effect of increasing concentrations of 3:5-dinitro-*o*-cresol on the growth rate of *T. viride* at pH 7·0.

recorded as the mean of measurements taken in two directions at right angles. The first measurement was usually taken after about 40 hr., and three subsequent readings were made at intervals of 6 or 12 hr. The rate of radial growth was constant over this period, and for each plate it was computed as the coefficient of the linear regression equation. The growth rates in the replicate plates of each treatment agreed closely, the average coefficient of variation being only 4%.

The growth rate of the fungus in control plates to which no toxic material had been added showed some variation with pH (Fig. 2*a*). Experiments were conducted at pH levels from 3 to 8, and over this range the highest growth rate was double the lowest. Fig. 2*b* shows how the growth rate was influenced by the addition of a range of concentrations of 3:5-dinitro-*o*-cresol to media buffered at pH 7·0. Concentrations below 0·003 g./100 ml. have no consistent effect, and the growth rate does not depart from that of control plates; more than this amount causes a decrease in the growth rate, and concentrations of the order of 0·1 g./100 ml. stop growth altogether. From such a graph, the total concentration of 3:5-dinitro-*o*-cresol required to halve the growth rate can be read off. At pH 7 the value is 0·013 g./100 ml.

The concentration necessary to produce a 50% reduction in growth rate has been used as the criterion of toxicity in all the present experiments. Values for other pH levels were obtained in the same way in each series of experiments, and the results for the various toxic materials tested are expressed graphically by plotting the logarithms of the molar concentrations required to halve the growth rate of the fungus against pH. From the total concentrations of acid substances the corresponding concentrations of undissociated molecules have been computed. The dissociation constant for 3:5-dinitro-*o*-cresol is given by Krahl & Clowes (1938); values for the other compounds tested are listed in *International Critical Tables* (1929). The phosphate present in the medium (1 g. KH_2PO_4/l.) gave adequate buffering, since closely similar results were obtained when the whole experiment with 3:5-dinitro-*o*-cresol was repeated with ten times this concentration of phosphate.

III. EXPERIMENTAL RESULTS

Fig. 3 shows the effect of pH on the toxicity of five non-acid substances (*m*-nitro-anisole, *p*-nitro-anisole, 2:4-dinitro-anisole, *p*-nitro-phenetole and 2:4-dinitro-phenetole). It is seen that the toxicity of the phenetole derivatives varies slightly with pH, while that of the anisole derivatives scarcely varies at all.

The toxicity of a group of weak acids at various pH levels is shown in Figs. 4 and 5. Phenol (Fig. 4*a*) is an extremely weak acid, with a pK value (10·0) lying above the pH range over which experiments were possible, and here pH has no effect on toxicity. *o*-, *m*- and *p*-nitro-phenols (Figs. 4*b*, *c* and *d*) have pK values of 7·3, 8·3 and 7·2 respectively, and at pH levels well below the pK value, variation in pH has no effect on the toxicity of any of the compounds. At higher pH levels the toxicity decreases; in each instance larger total concentrations are necessary to reduce the growth rate by 50%, but it should be noted that the corresponding concentrations of undissociated molecule decrease as pH is raised.

2:4-dinitro-phenol, 2:6-dinitro-phenol and 3:5-dinitro-*o*-cresol (Figs. 5*a*, *b* and *c*) have lower pK values—4·0, 3·6 and 4·4 respectively—and the toxicity of these compounds decreases rapidly as pH is raised. The corresponding concentrations of undissociated molecules show a slight rise at pH levels around the pK value, followed by a fall at higher pH levels.

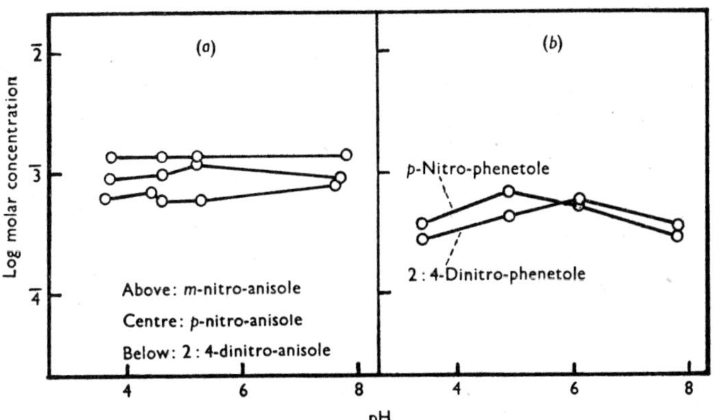

Fig. 3. The effect of pH on the concentrations of (*a*) anisole derivatives and (*b*) phenetole derivatives required to halve the growth rate of *Trichoderma viride*.

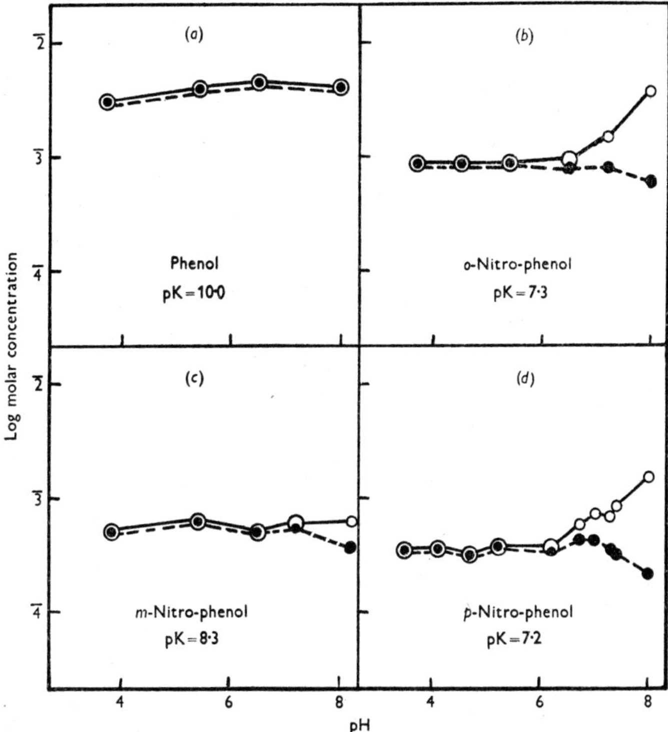

Fig. 4. The effect of pH on the concentrations of phenol and its mononitro derivatives required to halve the growth rate of *Trichoderma viride*. The full lines show the total concentrations and the broken lines the corresponding concentrations of undissociated molecules.

IV. DISCUSSION

Effect of pH *on the toxicity of weak acids and bases*

It is seen from Fig. 3 that the toxicity of non-acid compounds such as the anisole and phenetole derivatives appears to be independent of pH. At pH levels 2 or more pH units less than the pK value of the substance, pH has no effect on the toxicity of a compound with an acid group (Fig. 4). It is assumed that at lower pH levels than could be tested, the toxicity of the three dinitro derivatives (Fig. 5) would also be independent of pH. This assumption is supported by the work of Cruess & Richert (1929), who measured the fungicidal activity of benzoic acid from pH 2·5 to 7, and found that at pH levels below the value of pK there was no variation of toxicity with pH (Fig. 5*d*). Comparison of these results shows that the effect of pH on toxicity does not depend on pH alone, but it is also a function of pK. The total concentrations of toxic material necessary to produce the same responses from the organism rise at the higher pH levels, but the point along the pH scale at which this rise commences is governed by pK. Hence the effect of pH on toxicity can be related directly to the degree of dissociation of the toxic substances.

The present experiments have been concerned with the effect of pH on the toxicity to one fungus of a group of closely related aromatic substances, but the results are of a much wider significance. Table 1 lists references to work on acid substances in which the data presented are sufficiently detailed to show the variation of activity with pH. Only those acids are listed for which values of pK have been found in the literature. Examination by us of these results shows that when the logarithms of the equi-effective total concentrations and of the corresponding concentrations of undissociated acid molecules are plotted against pH, the curves resemble those seen in Figs. 3–5. These investigations cover a wide range of acids with pK values varying from 3 to 5·5 (apart from phenol, pK = 10·0). It is of interest to note that this relationship between pH and the biological activity of an acid is not confined to instances of toxic action. The data of Hoffman, Schweitzer & Dalby (1942) show that the toxic action of *p*-aminobenzoic acid against the spores of common moulds falls off at pH levels above the pK value, while Wyss, Lilly & Leonian (1944), working with a mutant of the fungus *Neurospora crassa* for which *p*-aminobenzoic acid is an essential growth factor, demonstrated that the acid becomes less active in promoting growth at pH levels above pK.

Table 2 lists comparable data for basic substances. At low pH levels the total concentration of a base required to produce a given effect rises and the corresponding concentration of undissociated base molecules falls.

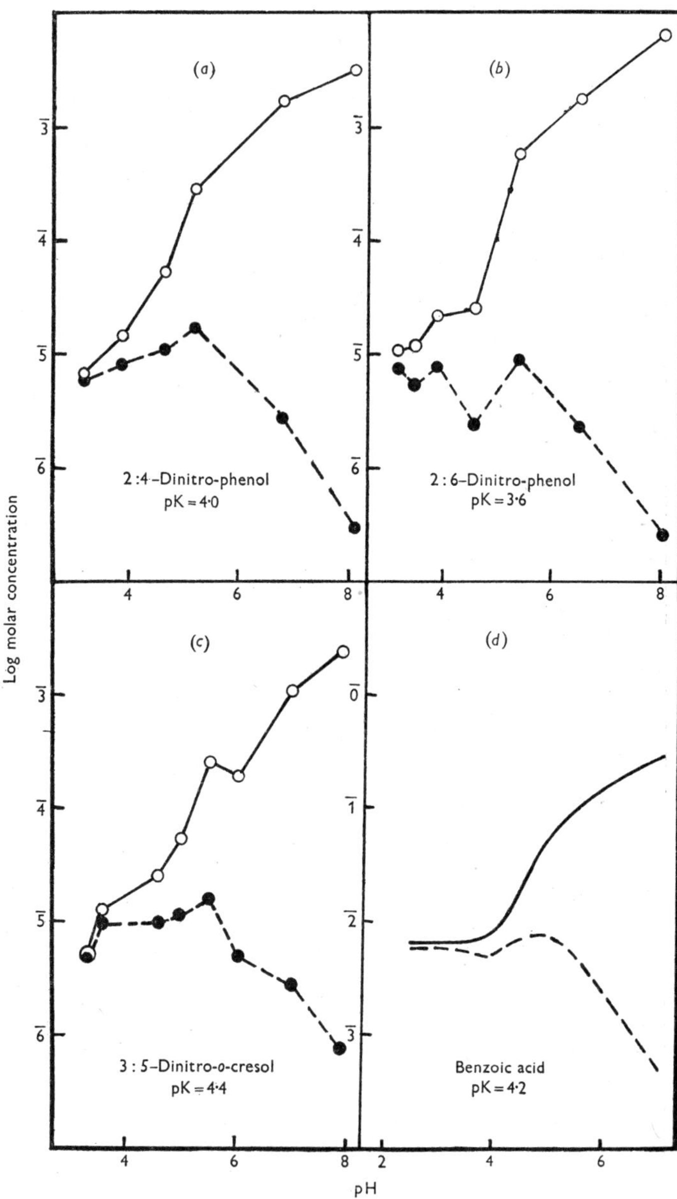

Fig. 5. The effect of pH on the concentrations required to halve the growth rate of *Trichoderma viride*. *a* and *b*. Dinitro derivatives of phenol. *c*. 3:5-Dinitro-*o*-cresol. *d*. The effect of pH on the concentrations of benzoic acid required to prevent the growth of *Mucor* (data from Cruess & Richert, 1929.) In each figure the full lines show the total concentrations, and the broken lines the corresponding concentrations of undissociated molecules.

Table 1

Author	Criterion of physiological response	Acids
Brian (1945)*	50 % reduction in growth rates of fungi *Trichoderma viride* and *Fusarium graminearum*	3:5-Dinitro-*o*-cresol
Smith, Walker & Hooker (1946)	50 % germination of spores of fungus *Colletotrichum circinans*	Acetic and benzoic acids
Huntington & Rahn (1945)	Inhibition of growth of a wine yeast	Acetic, propionic, butyric, monochloracetic and bromopropionic acids
Rahn & Conn (1944)	Do.	Benzoic and salicylic acids
Hoffman, Schweitzer & Dalby (1939)	Inhibition of growth of 'common moulds'	Normal fatty acids from formic to pelargonic. Three isomers of valeric acid
Hoffman, Schweitzer & Dalby (1940)	Do.	Acetic, monochloracetic, propionic, chlor-propionic and iodopropionic acids
Hoffman, Schweitzer & Dalby (1941)	Do.	Benzoic and salicylic acids, phenol
Hoffman, Schweitzer & Dalby (1942)	Do.	*p*-Aminobenzoic acid
Wyss, Lilly & Leonian (1944)*	Production of 20 mg. average dry weight of mycelium in 72 hr. by fungus *Neurospora crassa*	*p*-Aminobenzoic acid
Wyss, Ludwig & Joiner (1945)	Inhibition of growth of fungi *Aspergillus niger* and *Trichophyton interdigitale*	Propionic, caproic and caprylic acids
Foley, Herrmann & Lee (1947)	Inhibition of growth of fungus *Trichophyton gypseum*	Propionic and caprylic acids
Goshorn & Degering (1938)	Inhibition of growth of *Bacterium coli* and of *Staphylococcus aureus*	Benzoic and phenylacetic acids
Cruess & Richert (1929)	Inhibition of growth of various fungi and bacteria	Benzoic acid
Eagle (1945)	Immobilization of 55 % of populations of trypanosomes and of spirochaetes	Acid-substituted phenyl arsenoxides
Dierick (1943)	100 % mortality of eggs of insects *Ephestia Kühniella* and *Rhopalosiphum crataegellum*	3:5-Dinitro-*o*-cresol
Cowles (1942)	Inhibition of growth of *B. coli*	Sulphonamides
Brueckner (1943)	Inhibition of growth of *Staphylococcus aureus*	Sulphonamides

* These authors record the response of the organism to all of the concentrations of toxic agent used at each pH tested. To analyse the data, it has been necessary to select a suitable response and read off the appropriate concentration for each pH.

Table 2

Author	Criterion of physiological response	Bases
Crane (1921)*	Death of 'most' of a population of *Paramoecium caudatum* in 120 min.	Strychnine and quinine
Walker (1928)	Toxicity to protozoan *Colpidium*	Quinine
Trevan & Boock (1927)	Anaesthesia of cornea of rabbit's eye within 10 min.	Cocaine, novocaine and other anaesthetics
Clowes & Keltch (1931)	Anaesthesia of larvae of annelid *Arenicola* within 5 min.	Cocaine
Silverman & Evans (1944)*	Visible growth of *B. coli* after about 10 hr.	Mepacrine
Albert, Rubbo, Goldacre, Davey & Stone (1945)	Inhibition of growth of various bacteria	Two mono-amino-acridines and four di-amino-acridines

* These authors record the response of the organism to all of the concentrations of toxic agent used at each pH tested. To analyse the data, it has been necessary to select a suitable response and read off the appropriate concentration for each pH.

Table 3

Author	Criterion of physiological response	Toxic substances
Tyler & Horowitz (1937)	Reduction by 95–100 % of rate of cell division in eggs of *Strongylocentrotus purpuratus*	2:4-Dinitro-phenol
Smith (1925)	Inhibition of cell division in fertilized eggs of *Echinarachnius parma*	Benzoic and salicylic acids
Krahl & Clowes (1938)	Reduction by 50 % of rate of cell division in fertilized eggs of *Arbacia punctulata*	Chlor- and nitro-phenols
Clowes, Keltch & Krahl (1940)	Do.	Barbituric acid derivatives
Krahl, Keltch & Clowes (1940)	Do.	Local anaesthetic bases

The investigations listed in Tables 1 and 2 were made with a wide variety of test organisms, and yet they have all yielded similar results. On the other hand, anomalous results have been obtained in experiments in which reduction in the rate of cell division of fertilized Echinoderm eggs was the criterion of physiological response. Table 3 lists the references to such experiments which we have analysed graphically. This examination has shown that while the total concentrations of the acids or bases required to produce the standard toxic effect change with pH, the corresponding concentrations of undissociated acid or base do not vary with pH (or show no consistent variation), even at pH levels at which there is much dissociation.

In the present experiments it has been found that an appropriate concentration of toxic material reduced the growth rate of *Trichoderma viride* by 50%, and that the growth rate remained at this level for periods of several days, so long as the fungus was still in contact with the toxic substance. If the fungus was transferred to a medium containing no toxic material, the normal growth rate was rapidly attained. Similar observations have often been made in the study of toxicity, and it is concluded that in such cases of reversible action an equilibrium exists between the concentrations of the substance in the external medium and at the site of action within the cell. Change of pH may influence both the rate at which equilibrium is attained, and the final equilibrium conditions. Analysis of the influence of pH on toxicity will be considerably simplified by eliminating the former, and only taking measurements when equilibrium conditions have been reached. Experiments on the effect of pH on the fungistatic or bacteriostatic power of a toxic material may last for several days, and the organism is in contact with toxic material for the whole of this period, so that it is reasonable to suppose that a state of equilibrium has been reached. In some of the experiments listed in Tables 1–3, the period of contact is very short (Clowes & Keltch, 1931, 5 min.; Trevan & Boock, 1927, 10 min.), and yet the experiments all yield results of the same general nature. It would therefore appear that equilibrium conditions are reached quickly and were in fact attained in all the experiments under consideration.

Davson & Danielli (1943) have reviewed the work on the rate of penetration of weak acids and bases from solutions at different pH levels and have found that the rate of penetration is proportional to the concentration gradient of undissociated molecules. This relationship shows that only the undissociated molecules are penetrating the cell; when equilibrium is established the internal and external concentrations of undissociated molecules are either equal or are related by a partition coefficient.

Toxicity is measured by the dose of toxic substance that must be applied to the organism to produce a standard response. The response must be brought about by a particular concentration of toxic material at the site of action within the cell, and it is assumed that this concentration is independent of the pH conditions outside the cell. In the Echinoderm egg experiments listed in Table 3, the equi-effective external, and hence internal, concentrations of undissociated molecules do not vary with pH. This observation led Krahl & Clowes (1931) to adopt the view that toxicity is only exerted by the undissociated molecules of a weak acid or base. However, it must be stressed that the behaviour of Echinoderm eggs in the experiments listed in Table 3 is exceptional, and that in the general case the equi-effective concentration of undissociated acid or base molecules is not constant; at pH levels at which there is much dissociation, the concentration decreases. Some workers have overlooked this decrease because they only measured toxicity at pH levels close to the pK values of the substances under test, e.g. Huntington & Rahn (1945). Many others measured the equi-effective total concentrations and did not compute the corresponding concentrations of undissociated molecules. Hence it is not surprising that the view should be widely held that the undissociated molecule is the only toxic form of a weak acid or base.

In the general case it must be supposed that the ion is also exerting a toxic effect. Since the undissociated molecules are in equilibrium across the membrane, there will be two ion concentrations, an internal and an external depending upon the respective pH levels. Of these two, it is probable that the internal ion concentration alone exerts a toxic action. There is indirect evidence that changes in the pH of the external medium bring about changes in the internal hydrogen-ion concentration (Gale, 1947; Wolf & Wolf, 1947). The two pH levels may not be equal but changes in the external pH are likely to be accompanied by proportionate changes in the internal pH. On this basis a general interpretation of the present observations and those of other workers can be put forward (Simon, 1948). A shift in the internal pH will lead to changes in the ratio of the internal concentrations of ion and molecule. If both ion and molecule have a toxic action within the cell, then the effect of pH can be interpreted in terms of the differing toxicity of ion and molecule and their respective concentrations at the site of action. The influence of change of pH on toxicity is a result of the consequent change in the proportion of ion to molecule at the site of action.

Effect of pH *on the toxicity of moderately strong acids and bases*

The influence of pH on the toxicity of moderately strong acids and bases is similar to its effect on weak acids and bases, but the observations must be interpreted in a different way. Gershenfeld & Milanick (1941) measured the bactericidal powers of a sulphonic acid and of a quaternary ammonium chloride and found that the former was more active in acid solutions, and the latter in alkaline. Similarly, we have found that 2:4:6-trinitro-phenol (picric acid) is most toxic to *Trichoderma viride* in acid solutions.

All these substances are completely dissociated at the pH levels used in the experiments and they must act as anions or cations respectively. In this connexion the view has been held (Valko & DuBois, 1944) that the uptake of surface-active materials is preceded by attachment to an oppositely charged centre on the cell surface. Thus there is a competition between cations and hydrogen ions for negative centres, and between anions and hydroxyl ions for positive centres. In each case, bacterial growth is arrested when the ratio of ions reaches a particular value. Therefore changes of pH influence toxicity as a direct result of the changes in hydrogen- and hydroxyl-ion concentrations.

CONCLUSIONS AND SUMMARY

Toxicity measurements which are conducted with no thought beyond their immediate practical value are best made at the pH at which the compounds are likely to be used. When the tests are part of a theoretical investigation of differences in toxicity between compounds, more attention must be paid to the pH levels of the solutions in which the compounds are applied to the organism.

If none of the compounds dissociate in solution then it is immaterial at what pH the tests are made. Weak acids and weak bases should be tested at pH levels at which there is very little dissociation. The toxicity of a weak acid decreases in the presence of increasing concentrations of the anion; that of a weak base in the presence of the cation. At a pH level two or more pH units below the pK value of the substance (or above pK_b for a base) toxicity is uninfluenced by ionization. In the absence of any knowledge of pK, the toxicity of an acid substance should be measured at a low pH, that of an alkaline one at a high pH.

To obtain an adequate measure for the toxicity of moderately strong acids or bases such as synthetic detergents or picric acid, a pH level must be chosen arbitrarily and standardized for all tests. pH 7 is a convenient standard to choose.

The authors wish to acknowledge that these investigations have been undertaken with the aid of grants from the Agricultural Research Council.

REFERENCES

ALBERT, A., RUBBO, S. D., GOLDACRE, R. J., DAVEY, M. E. & STONE, J. D. (1945). *Brit. J. Exp. Path.* **26**, 160.

BAKER, Z., HARRISON, R. W. & MILLER, B. F. (1941). *J. Exp. Med.* **73**, 249.

BRIAN, P. W. (1945). *J. Soc. Chem. Ind., Lond.*, **64**, 315.

British Standard Specification no. 541 (1934). *Determining the Rideal-Walker Coefficient of Disinfectants.*

British Standard Specification no. 808 (1938). *Modified Technique for the Chick-Martin Test for Disinfectants.*

BRUECKNER, A. H. (1943). *Yale J. Biol. Med.* **15**, 813.

CLOWES, G. H. A. & KELTCH, A. K. (1931). *Proc. Soc. Exp. Biol., N.Y.*, **29**, 312.

CLOWES, G. H. A., KELTCH, A. K. & KRAHL, M. E. (1940). *J. Pharmacol.* **68**, 312.

COWLES, P. B. (1942). *Yale J. Biol. Med.* **14**, 599.

CRANE, M. M. (1921). *J. Pharmacol.* **18**, 319.

CRUESS, W. V. & RICHERT, P. H. (1929). *J. Bact.* **17**, 363.

DAVSON, H. & DANIELLI, J. F. (1943). *The Permeability of Natural Membranes.* Cambridge University Press.

DIERICK, G. F. E. M. (1943). *Tijdschr. PlZiekt.* **49**, 22.

EAGLE, H. (1945). *J. Pharmacol.* **85**, 265.

FLEISCHER, L. & AMSTER, S. (1923). *Z. ImmunForsch.* **37**, 327.

FOLEY, E. J., HERRMANN, F. & LEE, S. W. (1947). *J. Invest. Dermatol.* **8**, 1.

GALE, E. F. (1947). *The Chemical Activities of Bacteria*, pp. 60–5. London: University Tutorial Press.

GERSHENFELD, L. & MILANICK, V. E. (1941). *Amer. J. Pharm.* **113**, 306.

GOSHORN, R. H. & DEGERING, E. F. (1938). *J. Amer. Pharm. Ass.* **27**, 865.

HOFFMAN, C., SCHWEITZER, T. R. & DALBY, G. (1939). *Food Res.* **4**, 539.

HOFFMAN, C., SCHWEITZER, T. R. & DALBY, G. (1940). *J. Amer. Chem. Soc.* **62**, 988.

HOFFMAN, C., SCHWEITZER, T. R. & DALBY, G. (1941). *Industr. Engng Chem.* **33**, 749.

HOFFMAN, C., SCHWEITZER, T. R. & DALBY, G. (1942). *J. Amer. Pharm. Ass.* **31**, 97.

HUNTINGTON, G. I. & RAHN, O. (1945). *J. Bact.* **50**, 655.

International Critical Tables of Numerical Data, Physics, Chemistry and Technology (1929). **6**, 259ff.

KRAHL, M. E. & CLOWES, G. H. A. (1938). *J. Cell. Comp. Physiol.* **11**, 1, 21.

KRAHL, M. E., KELTCH, A. K. & CLOWES, G. H. A. (1940). *J. Pharmacol.* **68**, 330.

LABES, R. (1922). *Biochem. Z.* **130**, 14.

PLANTEFOL, M. L. (1922). *C.R. Acad. Sci., Paris*, **172**, 123.

RAHN, O. & CONN, J. E. (1944). *Industr. Engng Chem.* **36**, 185.

SILVERMAN, M. & EVANS, E. A. (1944). *J. Biol. Chem.* **154**, 521.

SIMON, E. W. (1948). D.Phil. Thesis. Oxford University.

SMITH, H. W. (1925). *Amer. J. Physiol.* **72**, 347.

SMITH, F. G., WALKER, J. C. & HOOKER, W. J. (1946). *Amer. J. Bot.* **33**, 351.

TREVAN, J. W. & BOOCK, E. (1927). *Brit. J. Exp. Path.* **8**, 307.

TYLER, A. & HOROWITZ, N. H. (1937). *Proc. Nat. Acad. Sci., Wash.*, **23**, 369.

VALKO, E. I. & DUBOIS, A. S. (1944). *J. Bact.* **47**, 15.

WALKER, E. (1928). *Biochem. J.* **22**, 292.

WOLF, A. W. & WOLF, T. W. (1947). *The Fungi*, **2**, chap. 7. New York: John Wiley and Sons, Inc.

WOODWARD, G. J., KINGERY, L. B. & WILLIAMS, R. J. (1934). *J. Lab. Clin. Med.* **19**, 1216.

WYSS, O., LILLY, V. G. & LEONIAN, L. H. (1944). *Science*, **99**, 18.

WYSS, O., LUDWIG, B. J. & JOINER, R. R. (1945). *Arch. Biochem.* **7**, 415.

THE EFFECTS OF PHENYLCARBAMATES ON THE GROWTH OF HIGHER PLANTS

By G. W. IVENS and G. E. BLACKMAN
Department of Agriculture, University of Oxford

I. INTRODUCTION

The general formula of the carbamates can be expressed as NH_2COOR or $R^1NHCOOR$, where R and R^1 may be either alkyl or phenyl groups. In the literature some confusion exists as to their nomenclature; many workers, particularly on the continent, use the words 'carbamate' and 'urethane' synonymously, while others restrict the term 'urethane' to ethyl carbamates of general formula $RNHCOOC_2H_5$. It is in the latter restricted sense that the term 'urethane' is used in this paper.

The inhibitory properties of carbamates have been recognized for a long time, and they have been shown to affect various functions of a wide range of organisms. Ethyl carbamate (urethane, $NH_2COOC_2H_5$) has been widely used as an inhibitor. It inhibits respiratory activity and suppresses division in the ciliate, *Glaucoma pyriformis* (Lwoff, 1934), arrests division of *Arbacia* eggs (Loeb & Wasteneys, 1912), and inhibits respiration and division of bacteria such as *Bact. coli*, *Staphylococcus aureus* (Weinstein & McDonald, 1945) and luminous bacteria (Taylor, 1935). Boell & Taylor (1933) observed that a number of homologous carbamates depressed the E.M.F. of frog skin, while cancer research workers have found that such compounds produce mitotic abnormalities and inhibit cell division in various tissues (Dustin, 1929, 1947; Ludford, 1936; Haddow & Sexton, 1946). The action of urethane in inhibiting some enzyme processes has been studied by Michaelis & Quastel (1941) and Quastel (1943).

The effects of carbamates on the growth of plants have been studied by a number of investigators. Friesen (1929) working with the compound $NH_2COOC_6H_5$ (which he called phenylurethane but which should be referred to as phenylcarbamate) showed that a concentration of 0·002% delayed germination and reduced the growth rate of oats and wheat, eventually killing a proportion of the seedlings. Ethyl phenylcarbamate or phenylurethane ($C_6H_5NHCOOC_2H_5$) has been found to inhibit the growth and respiration of sections of *Avena* coleoptile grown in culture solution (Bonner, 1933, 1936). Later Lefevre (1939) pointed out that the action of phenylurethane on cell division is similar to that of colchicine. He observed that the extension growth of roots and coleoptiles of wheat seedlings was arrested by a saturated solution of phenylurethane diluted to one-twentieth of its original strength. Cytological investigation of the root showed that mitosis was blocked at metaphase, and that this led to the formation of polyploid nuclei and other abnormalities. Since then a number of workers have observed similar effects with phenylurethane and other carbamates (and a large number of unrelated compounds, both organic and inorganic) on a wide range of plants (Kostoff, 1938; Simonet & Guinochet, 1939; Favorsky, 1939; Levan, 1940, 1945, 1946; Levan & Levring, 1942; Levan & Östergren, 1943; Mangenot & Carpentier, 1944; Gavaudan, 1943; Östergren, 1944; and, for a complete bibliography up to 1942, Krythe & Wellensiek, 1942). As these changes show so many similarities to those produced by colchicine treatment, Levan proposes the use of the term '*c*-mitosis' for all such mitotic abnormalities, and the term in this sense has been adopted throughout the present paper. The use of this term, however, does not necessarily imply that the effects and mode of action of ethyl phenylcarbamate are the same as those of colchicine.

Templeman & Sexton (1945) first suggested that carbamates might be of value as selective herbicides. They found that concentrations of ethyl phenylcarbamate and *iso*propyl phenylcarbamate (hereafter abbreviated to EPC and iso-PPC) which killed oats and other cereals were without effect on some dicotyledonous plants, including mangolds, sugar beet, flax, rape and yellow charlock (*Sinapis arvensis*). Of the large number of phenylcarbamates and related compounds subsequently tested, iso-PPC was observed to be the most active, being approximately three times as effective as EPC in killing grasses (Templeman & Sexton, 1946). Similar results were obtained in the U.S.A., where grasses including *Zea*, *Oryza*, *Festuca*, *Lolium*, *Agropyron* and *Phleum* species were found to be relatively susceptible to EPC and iso-PPC, while other grasses, including *Setaria*, *Sorghum* and *Milium* species, were much more resistant. This resistance was shared by many dicotyledonous crops, such as soya bean, sunflower, radish and turnip. Buckwheat was one of the few dicotyledonous plants of a similar susceptibility to the less resistant grasses (Allard, Ennis, De Rose & Weaver, 1946; Ennis, 1947; Mitchell & Marth, 1947).

The objects of the present investigation have been twofold. First, to gain some understanding of the factors underlying the differential susceptibility of plants to poisons of the phenylurethane type; secondly, to estimate their value as selective herbicides, with especial reference to the control of graminaceous weeds.

II. EXPERIMENTAL RESULTS

Effects on extension growth and cell size

Seeds of oats and barley will germinate in a range of solutions of EPC, but at the higher concentrations elongation of the roots ceases after a length of about 3–5 mm. has been attained. Even with the highest concentration used—128 p.p.m.—some growth always takes place. With lower concentrations, growth is slowed down but continues for several days; with less than 1·0 p.p.m. the growth rate is unaffected no matter how long treatment is continued.

Friesen (1929), describing the effects of phenyl carbamate on oats and wheat, stated that root-hair production is stimulated. In the present experiments, EPC, although it affects seedlings of oats and barley in a similar manner, does not appear to stimulate the production of root hairs. In a treated root, the growth of which has been arrested by EPC, hairs are formed to the very tip of the root, whereas in a normal root their production is restricted to the zone of cell differentiation several millimetres behind the tip. It appears therefore that, although meristematic activity is inhibited by EPC, differentiation as shown by root-hair formation is unaffected and continues to the root apex.

Deysson (1944) noted that in roots of *Allium cepa* treated with 100 p.p.m. EPC, swelling of the root tips occurred, and Lefevre (1939) observed the same effect in wheat. In the present investigation it is found that high concentrations cause some swelling of the roots of barley and oats so that the tip of a root which has stopped growing does not retain the dimensions of a normally growing tip, but increases to a thickness which equals, or slightly exceeds, that of fully differentiated roots of control plants. There is, however, nothing corresponding to the large tumours noted by Levan (1942, 1946) and Levan & Östergren (1943) in their work on colchicine and other substances. They found that so-called *c*-tumours were produced on the roots of a number of plants, including oats and barley, and that the formation of tumours occurred independently of *c*-mitosis. Thus with colchicine, the threshold concentration for *c*-mitosis in *A. cepa* was between 200 and 250×10^{-6} mol./l., while *c*-tumours were produced by 150×10^{-6} mol./l. On the other hand, with acenaphthene, *c*-mitosis was caused by 5 to 6×10^{-6} mol./l., while 32×10^{-6} mol./l. were required for the formation of tumours. Again, with *Pisum sativum*, both colchicine and acenaphthene caused *c*-mitosis though acenaphthene did not produce tumours.

Shoot extension, although less sensitive to EPC, is inhibited in much the same way as root growth (see Text-fig. 1). With higher concentrations, 32 p.p.m. and over, the coleoptiles of oats and barley only reach a length of about 10 mm. and the first leaf fails to emerge.

Shoots of seedlings treated with EPC also swell to varying extents if the concentration used is sufficient to stop growth. The coleoptile swellings do not reach any considerable size in either of the cereals used, but with linseed the whole of the hypocotyl swells and may reach a thickness of 3–4 times that of control seedlings. If seeds of oats and barley are sown in soil sprayed with EPC, the first leaf frequently reaches a length of several centimetres before growth is finally arrested. In this case the leaf itself is thicker than normal. If the treatment has not been too drastic, subsequent leaves may appear and grow normally, while the first leaf remains in the same condition for several weeks before dying.

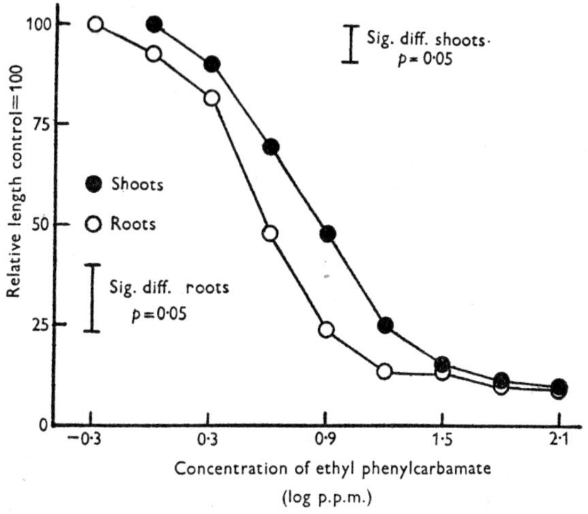

Text-fig. 1. The inhibitory effects of a range of concentrations of ethyl phenylcarbamate on the length of roots and shoots of oat seedlings after 120 hr. treatment.

The production of these abnormal growths is not accompanied by any increase in the number of cells in the region of swelling, because the concentrations which are sufficient to cause swelling also stop cell division. Abnormal growth is caused by vacuolation and swelling of the meristematic cells, accompanied by a more or less pronounced change in the direction of cell extension.

Levan (1942), in a discussion of tumour formation by the action of colchicine on plants, pointed out that cell extension is not due to the increased number of chromosomes present, as, even if mitotic activity is stopped by the action of X-rays, subsequent colchicine treatment leads to the formation of tumours. He suggested that colchicine causes tumours by interfering with the action of growth substances. In this connexion, Havas (1938) found that, whereas colchicine applied to tomato leaves caused an epinastic curvature, EPC had no effect.

Although EPC does not induce the production of large root tumours in barley and oats, it causes considerable increase in cell size. Root tips of barley seedlings, which had been treated with a range of concentrations of EPC, were fixed, sectioned and stained and counts made of the numbers of cells in the fields visible under a $\frac{1}{12}$ in. objective. In Text-fig. 2 the total number of cells visible under twenty fields, expressed as a percentage of the control values, is plotted against log concentration. The method does not permit a very high standard of accuracy (the significant difference between treatment means being 22·2% at the 5% probability level), but marked changes are evident. Treatment for 12 hr. has not produced a statistically

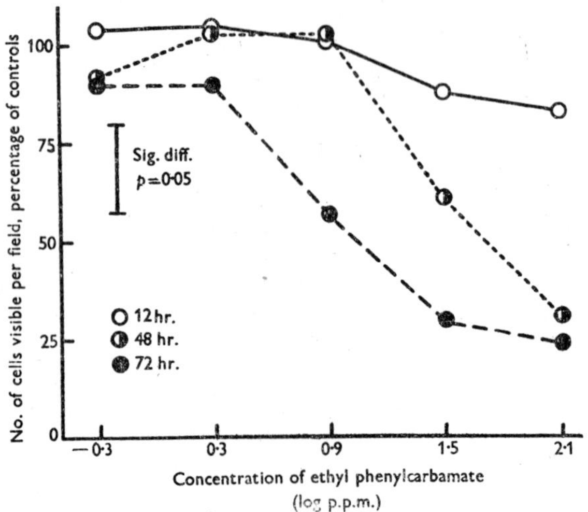

Text-fig. 2. The effects of varying concentrations of ethyl phenylcarbamate on the size of cells in the root-tip meristem of barley after 12, 48 and 72 hr. treatment.

significant effect, though there is a tendency for the size of cells to increase with the highest concentration. This trend is more marked after 48 hr. treatment with concentrations above 8 p.p.m. (see Pl. 1). After 96 hr. treatment, even concentrations over 2 p.p.m. give an increase in cell size. The increase in cell size reaches a maximum with 128 p.p.m. when the cross-section area of the cells is 4–5 times as great as those seen in the controls. Assuming that the cells extend isodiametrically, there is an 8–12-fold increase in volume.

This increase in cell size may well account for the initial extension of the root which occurs even with the highest concentrations. The extended cells are vacuolated and have the general appearance of mature cortical cells, although they do not undergo the longitudinal extension that usually occurs in the maturation of normal cortex. Differentiation of vascular tissue also proceeds much nearer to the root apex than in a normally growing tip. Mangenot (1939) also noted the differentiation of the apical region and the formation of vascular tissue in the roots of *Allium cepa* treated with colchicine. These observations, together with the formation of root hairs previously mentioned, suggest that the final arrest of growth after treatment with EPC is due to the exhaustion of the supply of meristematic cells which are not replaced owing to the failure of cell division.

A series of experiments has been performed to determine the inter-relationship between 'presentation time' and the concentration of EPC on the reduction in root growth of oats. Germinating seeds were selected at a stage when the primary roots had emerged from the seed, but were not above 1 mm. in length. They were then placed on filter-paper in Petri dishes and treated with a range of concentrations for the appropriate time, washed in tap water and replaced in clean dishes containing filter-paper moistened with distilled water. At each concentration there were three replicates of ten seeds per dish, and all experiments were carried out at 25° C. The longest root of each seedling was selected; its length was measured at the end of the presentation time (or after 48 hr. if the presentation time was less than this). A second measurement was made 48 hr. after the first.

The mean growth, made during the second period, expressed as a percentage of the control growth, plotted against log concentration, for each treatment time, gives a sigmoid curve which can be treated statistically by means of the probit transformation (Finney, 1947). A series of probit lines expressing the degree of inhibition produced by various times of treatment with a range of concentrations of EPC is shown in Text-fig. 3. A linear regression coefficient common to all the series of points may be calculated. If the probit value of the percentage reduction in growth rate caused by any combination of treatment time and concentration is plotted against the value of log concentration + log time ($\log C + \log t$), the results seen in Text-fig. 4 are obtained. The regression coefficient of the linear equation of the closest fit to all these points is approximately the same as the coefficient calculated for the series of points in Text-fig. 3. In each case the deviations from the regression line of closest fit are not statistically significant.

Thus it follows that for any degree of inhibition, log concentration + log treatment time is a constant, i.e. $\log C + \log t = k$ or $Ct = k$. This equation is one form of a general equation $C^n t = k$ (n and k being constants) which has frequently been found to relate the concentration and time of treatment required to produce a given toxic effect (Bliss, 1940). Such equations, however, only hold good over a more or less limited range of time and concentration as they take no account of the threshold values; for instance, in this series of experiments, concentrations below approximately 2 p.p.m. and treatments lasting less than about 1 hr. do not reduce the growth rate significantly.

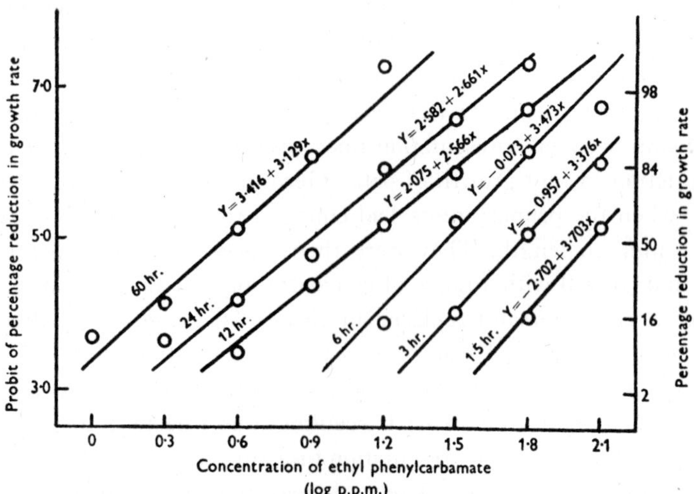

Text-fig. 3. Probit lines showing the relationships between the concentration of ethyl phenylcarbamate and the reduction in root-extension growth of oat seedlings, after times of treatment varying from 1·5 to 60 hr.

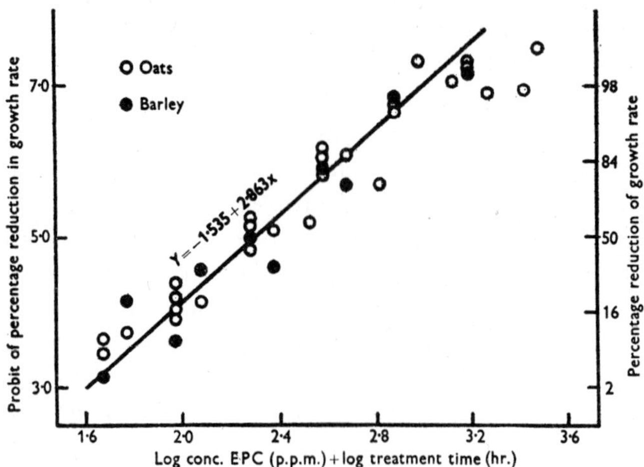

Text-fig. 4. The relationship between the probit value of the percentage reduction in extension growth of oat and barley roots and the sum of the logarithm of the concentration of ethyl phenylcarbamate and the logarithm of the duration of treatment.

Clark (1933) has pointed out that a number of formulae will approximately fit the results obtained by time/concentration experiments, but that none of them expresses more than a portion of the results. He also stressed that even where the effect measured involves a far smaller number of factors than is the case in the present series of experiments, such formulae give little information about the mode of action of the toxic agent. It appears, therefore, that the relationships described above are not due to any especial mode of action of EPC, but are probably related to the differential susceptibility of individual cells, which can be expected in any population, and to varying rates of penetration from the external medium.

The concentration/treatment time relationships of the inhibition of barley root growth are almost identical with those of oat, as is seen in Text-fig. 4. The concentrations required to halve the growth rate, i.e. the '50% points' for barley and oats under the conditions of the experiment, are, respectively, 34 and 29 p.p.m. for 6 hr. treatment, while for 60 hr. treatment the concentration is the same for both species, 3·6 p.p.m.

The inhibition of growth by EPC under the experimental conditions is reversible in its early stages; if the seedlings are transferred to distilled water meristematic activity in the root is again initiated. When the growth rate has been reduced below about 30% of the control value the inhibition is irreversible. This critical point corresponds approximately to the point at which swelling of the meristematic cells begins (Text-figs. 2, 3). Once the cells have started to extend they have lost all power of further division. The forces responsible for this extension and vacuolation may be the same as those which cause the elongation of cells formed by a normal meristem, but there is a change in the direction of extension.

Experiments with the field pea (*Pisum sativum*) show that it is much less susceptible to EPC than either oats or barley, but that the curves obtained by plotting log concentration against root growth are of very similar form. After 60 hr. treatment, the 50% points for oats and pea respectively are 3·6 and 60 p.p.m. The results at present available indicate that the treatment time/concentration curve will be similar to that in Text-fig. 4, but with the threshold values at a considerably higher level.

A series of observations made on the increase in cell size of the root meristem, after treatment with a range of concentrations for 117 hr., shows that, in *Pisum* also, meristematic activity is irreversibly stopped at the point where the cells start to extend. There is a statistically significant increase in the size of meristematic cells with 128 p.p.m., and with the highest concentration used (256 p.p.m.) there is a threefold increase in volume.

Kostoff (1938) found that *Vicia*, *Lathyrus* and *Medicago* species were less susceptible to the action of acenaphthene than were oats, barley and some other graminaceous species. Levan & Östergren (1943) also noted that leguminous plants were more resistant to naphthalene derivatives than some Gramineae. Until the action of a number of other substances has been tested on legumes and grasses under comparable conditions, however, it would be unwise to attach too much significance to these differences.

Experiments with seeds of a variety of plants, treated in Petri dishes at room temperature, showed that wide variations in susceptibility to EPC exist between species. Approximately 50% inhibition of root growth, after about 10 days' treatment, was produced by the following concentrations: 2 p.p.m. for *Avena fatua* and *Lolium italicum*, 5 p.p.m. *Dactylis glomerata* and *Linum usitatissimum*, 8 p.p.m. *Allium cepa*, 20 p.p.m. *Trifolium repens*, 50 p.p.m. *Zea mais*, *Sorghum vulgare* and *Setaria italica*. These results are similar to those obtained by other workers using soil treated with EPC and iso-PPC, in field and greenhouse experiments (Templeman & Sexton, 1946; Mitchell & Marth, 1947).

The results obtained so far indicate that the characteristics of plants which result in their relative resistance or susceptibility to EPC are not restricted to any one family or group of families; among the Gramineae for instance, some species are extremely sensitive to its action, e.g. oats, while others are highly resistant, e.g. *Sorghum*. Nevertheless, some of the commoner graminaceous weeds are very susceptible to EPC, while many crop plants are relatively resistant—a fact which may be of considerable practical value though the principles underlying this differential resistance are, as yet, unknown.

The effects on cell division

Most of the mitotic abnormalities in the meristematic tissues of plants which are caused by EPC arise either from inhibition of development of the spindle or from its disorganization. Thus in a treated barley root, the chromosomes of a cell undergoing mitosis remain scattered at metaphase instead of being arranged on the equatorial plate. In those cells that have produced a spindle before being affected, the spindle breaks down and chromosomes undergo little further movement. Thus in the first stages of *c*-mitosis the chromosomes may be seen either scattered, or in arrested metaphase or anaphase positions. Chromosomes so affected remain stationary for some time and the chromatids slowly separate. Some degree of contraction, greater than that which normally occurs at metaphase, takes place at this stage. This contraction does not appear to be due to prolongation of 'metaphase' as it is frequently evident in prophase also. The pairs of contracted chromatids, lying either parallel or with their arms diverging, are very characteristic of *c*-mitosis and have been noted by a number of workers (Levan, 1938; Gavaudan & Gavaudan, 1937). The chromosomes then slowly pass into the interphase condition. An affected cell may contain either a single tetraploid nucleus or several nuclei with varying chromosome numbers, depending on the distribution of chromosomes at 'metaphase'. Mitoses arrested at anaphase may give rise to cells containing two normal diploid nuclei through failure to develop a complete dividing wall. In the early stages of EPC treatment, longitudinal splitting of chromosomes (apparent from very early stages in barley) proceeds normally, so that tetraploid interphase nuclei can go through further cycles of doubling, with the formation of high polyploids containing hundreds of greatly contracted chromosomes which are scattered, or clumped in irregular groups. The process of doubling of chromosome number appears to continue until more general toxic effects damage the nuclei. With longer treatments, involving several *c*-mitoses, interphase nuclei become irregular in shape, and they may be either very large or split into several fragments. Eventually, after about 12 hr. treatment with concentrations of 32 p.p.m. or over, nuclei become pycnotic. In this connexion the nuclear abnormalities produced after 48 hr. are seen in Pl. 1.

The majority of the substances causing *c*-mitosis have the same primary effect, i.e. an action on the spindle of dividing cells (Favorsky, 1939; Gavaudan & Gavaudan, 1937; Levan, 1938). Their secondary effects, however, which do not result from spindle inhibition, vary considerably. Thus the *c*-mitotic effects of substances which exert a strong toxic action may be modified in various ways.

The susceptibility of individual cells varies considerably, so that after a few hours' treatment, the proportion of abnormal nuclei is small. Later, cells going through the cycle for the first time are tetraploid, while those undergoing their second or third *c*-mitosis are octoploid or higher polyploids. The degree of chromosome contraction seems to depend on the number of *c*-mitoses undergone, so that at any one time chromosomes in all stages of contraction may be seen, ranging from those slightly affected in tetraploid nuclei to minute rods in the high polyploids (see Pl. 1). Östergren (1944) has estimated the *c*-mitotic activity of a number of compounds on the roots of *Allium cepa* by measurement of the degree of chromosome contraction they caused, ten chromosomes being measured per observation. He claimed that chromosome contraction was a very sensitive criterion, and, with some compounds, was able to detect a measurable degree of contraction before there was any apparent effect on the spindle.

With barley, this technique does not seem to be a very satisfactory method of measuring the effect of EPC, since, as mentioned above, chromosomes in all stages of contraction are present simultaneously. Thus, unless a very large number of chromosomes is measured for each observation, no reliable estimate can be obtained.

The duration of the arrested metaphase condition is considerably longer than the normal metaphase, and thus in a treated root there are more nuclei in 'metaphase' than in a control root. The proportion of cells affected rises with concentration and time of treatment, leading to a corresponding increase in the number of arrested metaphases. This increase has been studied in an attempt to estimate the *c*-mitotic activity of EPC.

Root tips of barley, after various treatments, were fixed in Craf fluid (Randolph, 1935), embedded in paraffin wax, cut longitudinally at 10μ, and stained by Feulgen's method. For each treatment, counts were made

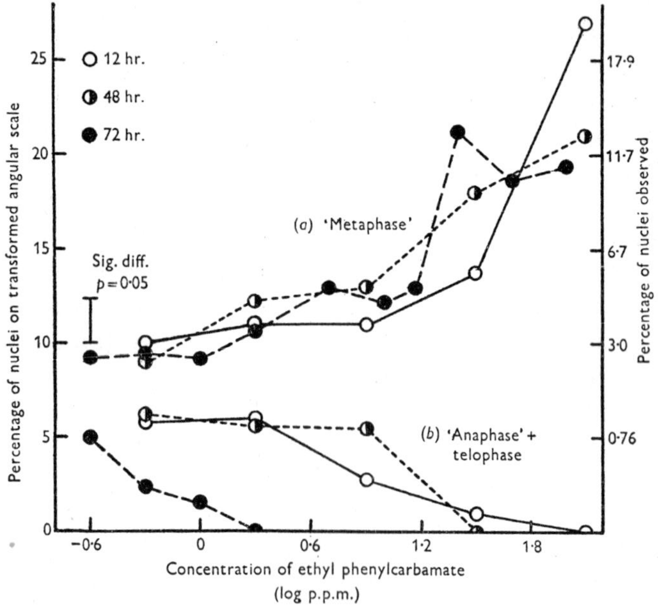

Text-fig. 5. The effects of a range of concentrations of ethyl phenylcarbamate on the percentage of nuclei in the root-tip meristem of barley which are in (*a*) 'metaphase', (*b*) anaphase plus telophase, after varying periods of treatment.

of the nuclei in 'metaphase' visible in the field of a $\frac{1}{12}$ in. objective, four fields being observed in the tip of each of five sections through the central part of the root.

Results typical of those obtained by this method are shown in Text-fig. 5. The numbers of 'metaphases', expressed as percentages of the total numbers of nuclei visible, have been transformed to angular values (the arc sin $\sqrt{}$(percentage transformation) (Bliss, 1937)) in order that the figures may be treated statistically, and these values are plotted against the logarithm of the concentration. The numbers of anaphases and telophases together are also shown. In the preliminary experiments so far performed all treatments have not been replicated and the only estimate of variance between roots is provided by the controls. However, unless the variance between treated roots is greatly different from that between the controls, which does not appear to be the case, the significant difference obtained will not lead to any false conclusions.

The number of unaffected divisions (anaphase and telophase) falls off increasingly with time so that, while after 12 hr. treatment with a concentration of 32 p.p.m. there are still a few normal divisions, after 72 hr. with 2 p.p.m. there are none. The number of arrested metaphases rises with concentration, and after 12 hr. treatment with 128 p.p.m. there are approximately 5 times as many nuclei in this stage than there are metaphases in the controls. The effect of duration of treatment is not clearly demonstrated by these results, but other experiments have shown that at any concentration the number of 'metaphases' increases with lengthening treatment, and, with the higher concentrations, reaches a maximum and then falls, as nuclear disintegration takes place.

In comparing these results with those obtained by measuring root growth, it will be recalled that in the latter a certain degree of recovery is involved. Thus cytological study may show that a certain treatment has stopped cell division, but growth, as it is shown in Text-fig. 3, may have taken place after the same treatment. However, the point at which the percentage of 'metaphases' begins to rise significantly above the control value, and the number of unaffected divisions begins to fall, is in general the point at which a significant depression in growth rate occurs.

As yet no observations have been made on the fate of abnormal nuclei when root tips resume growth after they have been transferred from EPC solution to water. In this connexion, Levan (1938) stated that inactivation of the spindle by colchicine might be reversed if the period of treatment was less than about an hour. He also stated that the more resistant cells, which had remained unaffected by the treatment, divided more rapidly than the polyploid cells after the influence of the drug had been removed, so that the root tended to regain its normal cytological state.

Results of a similar investigation on the numbers of 'metaphases' in the root meristem of *Pisum sativum*, illustrating the effects produced by an exposure of 117 hr. to a range of concentrations of EPC, are shown in Text-fig. 6. Inhibition of the spindle occurs as in barley, but, as might be expected from the greater resistance of *Pisum* to EPC in respect of root growth, this inhibition requires a much higher concentration. The concentration of 64 p.p.m. which causes anaphases and telophases to fall significantly from the control level also brings about a reduction in root growth of approximately 50%.

As yet, only a single series of observations has been made on *Pisum* roots, but the indications are that the proportion of 'metaphases' falls rapidly with concentrations over 64 p.p.m. There is no sign of the increase in number which occurs in barley. This difference may be due to injury of the nuclei as a result of the long duration of treatment in this experiment, though there is no sign of the nuclear degeneration which occurs in barley after a comparably severe treatment. It is also possible that in *Pisum*, interphase nuclei are affected in such a way that, after arrested metaphase chromosomes have passed into the resting stage, they are prevented from undergoing further *c*-mitoses. In support of this, no high polyploid nuclei have yet been observed in *Pisum*. As such polyploids are dependent on the

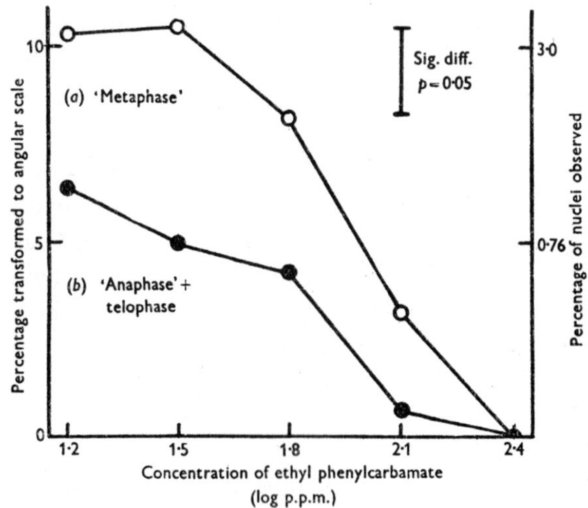

Text-fig. 6. The effects of varying concentrations of ethyl phenylcarbamate on the percentage of nuclei in the root-tip meristem of *Pisum sativum* which are in (*a*) 'metaphase', (*b*) anaphase plus telophase, after 117 hr. treatment.

completion of several *c*-mitoses, their absence would indicate that some factor prevents the occurrence of a chain of abnormal divisions. Until further results are available, however, it would be unwise to speculate on the apparent difference in the mode of action of EPC on barley and *Pisum*.

III. DISCUSSION

The results obtained in this investigation demonstrate that the effects of ethyl phenylcarbamate on root extension share the characteristics of many toxic substances. The sigmoid curve relating root growth to the logarithm of concentration (Text-fig. 1), which can be treated by means of probit analysis (Text-fig. 3), has been observed in other investigations (not yet published) to hold for the effects of chlorinated phenoxyacetic acids and pentachlorophenol on the root growth of maize. Similarly, other experiments (still in progress) have demonstrated that the percentage suppression of germination of various seeds, caused by increasing concentrations of iso-propyl and ethyl phenylcarbamates, exhibits a sigmoid relationship similar to that produced by copper salts and aryl-nitro compounds. It has also been shown that the same type of curve links the effects of varying concentrations of aryl-nitro compounds with (*a*) the reduction in growth rate of the mould *Trichoderma viride* (Simon & Blackman, p. 256), (*b*) the reduction in rate of frond formation of *Lemna minor* and (*c*) the percentage mortality of seedlings of many species (see Blackman, Holly & Roberts, p. 283).

The present investigation has shown that within limits the effect of concentration is inversely related to the duration of treatment, since, as is clear from Text-fig. 4, the product of the concentration and time of treatment required to produce any degree of inhibition of either oat or barley roots is a constant. Relationships of similar form have been recorded for other toxic compounds on various organisms, but it is pointed out that no special significance can be attached to the relationship in this case as little is known of the factors involved.

In attempting to interpret the effects of EPC on the inhibition of root growth, the cytological studies must be regarded as preliminary observations and far from complete. It is, however, of interest to note that the concentrations which inhibit root extension appear to be the same as those which interfere with cell division in the meristematic zone and give rise to nuclear abnormalities. There is evidence that the cytological changes in the pea, which from the point of view of the inhibition of root growth may be classed as a 'resistant' plant, differ from those in the 'susceptible' barley.

Although the nuclear abnormalities have been grouped under the general heading of *c*-mitosis, it is again stressed that this does not imply that the changes produced are precisely similar to those brought about by colchicine. Nevertheless, it has been noted by numerous workers that many substances affecting mitosis appear to act primarily through interference with the spindle, while the differences in their effects, e.g. different degrees of tumour formation, seem to result from secondary actions not immediately connected with the spindle.

At the present juncture, any explanation of the effects of EPC on cell division must be largely speculative. According to previous workers, EPC can be regarded as a typical 'indifferent narcotic'. The theories of narcotic action are many, and reference should be made to the reviews that have been made from time to time (Winterstein, 1926; Henderson, 1930; Clark, 1937; McElroy, 1947). It is, however, generally held that narcotic activity is related to physical properties, and Gavaudan & Poussel (1944) and Östergren (1944) consider that these relationships also hold for compounds causing *c*-mitosis. In support of this view Levan & Östergren (Levan & Östergren, 1943; Östergren, 1944) record that the introduction of a hydrophilic radicle into the molecule of a *c*-mitotic substance (thus decreasing lipoid solubility and increasing solubility in water) reduces the *c*-mitotic activity; they have suggested that the inhibition of spindle formation is brought about by some action on lipoid material. They postulate that a *c*-mitotic compound associates with the lipophil side-chains of the proteins in the spindle, and that above a certain concentration an intramolecular precipitation takes place, leading to folding of the protein chains and disintegration of the spindle. Gavaudan, Dodé & Poussel (1946), on the basis of Dervichian's (1941) views of protein structure, have proposed a somewhat similar theory to account for narcotic action in general. Finally, Ehrenberg (1946) has recorded that weak concentrations of lipophilic substances cause an increased curvature of the spindles in root-tip meristems of *Salix fragilis* × *alba* and suggests that these effects may be due to the contraction of long chain proteins.

Much more evidence is clearly required before such theories can be accepted as offering an explanation of the action of EPC on cell division. All that can be said at present is that the contraction of chromosomes which has been observed might be explained in terms of a folding of protein chains.

Apart, however, from the question of the mechanism of spindle disintegration, it is necessary to consider whether EPC might operate through its effects on cell metabolism. Bonner (1933, 1936), working with oat coleoptiles, observed that both cell extension induced by growth substances and respiration were inhibited by approximately the same concentrations of EPC. Fisher and his co-workers (Fisher & Henry, 1940; Fisher & Stern, 1942), investigating the action of inhibitors on sea-urchin eggs, and Burt (1945), studying the ciliate *Colpoda steinii*, conclude that cell division is associated with some part of the respiratory system which is more susceptible to the action of carbamates than the respiration connected with other aspects of cell activity. How far these conclusions are applicable to higher plants remains unknown, but if such factors are also involved then in the present state of knowledge any interpretation of the cytological effects of EPC will be further complicated.

The authors wish to acknowledge that this investigation, as part of the research programme on phytocidal action, has been supported by grants from the Agricultural Research Council.

REFERENCES

ALLARD, R. W., ENNIS, W. B., DE ROSE, H. R. & WEAVER, R. J. (1946). *Bot. Gaz.* **107**, 589.

BLACKMAN, G. E., HOLLY, K. & ROBERTS, H. A. (1948).

BLISS, C. I. (1937). *Bull. Pl. Prot. Leningrad*, **12**, 67.

BLISS, C. I. (1940). *Ann. Ent. Soc. Amer.* **33**, 721.

BOELL, E. J. & TAYLOR, A. B. (1933). *J. Cell. Comp. Physiol.* **3**, 355.

BONNER, J. (1933). *Proc. Nat. Acad. Sci., Wash.*, **19**, 717.

BONNER, J. (1936). *J. Gen. Physiol.* **20**, 1.

BURT, R. L. (1945). *Biol. Bull. Woods Hole*, **88**, 12.

CLARK, A. J. (1933). *The Mode of Action of Drugs on Cells.* London: E. Arnold.

CLARK, A. J. (1937). *Trans. Faraday Soc.* **33**, 1057.

DERVICHIAN, D. G. (1941). *J. Chim. phys.* **38**, 59.

DEYSSON, G. (1944). *C.R. Acad. Sci., Paris*, **219**, 366.

DUSTIN, A. P. (1929). *Arch. Anat. micr.* **25**, 37.

DUSTIN, P. (1947). *Nature, Lond.*, **159**, 794.

EHRENBERG, L. (1946). *Hereditas, Lund*, **32**, 15.

ENNIS, W. B. (1947). *Science*, **105**, 95.

FAVORSKY, M. V. (1939). *C.R. Acad. Sci. U.R.S.S.* **25**, 71.

FINNEY, D. J. (1947). *Probit Analysis.* Cambridge Univ. Press.

FISHER, K. C. & HENRY, R. J. (1940). *Biol. Bull. Woods Hole*, **79**, 371 (Abstract).

FISHER, K. C. & STERN, J. R. (1942). *J. Cell. Comp. Physiol.* **19**, 109.

FRIESEN, G. (1929). *Planta*, **8**, 666.

GAVAUDAN, P. (1943). *C.R. Soc. Biol., Paris*, **137**, 281, 342, 571.

GAVAUDAN, P. & GAVAUDAN, N. (1937). *C.R. Soc. Biol., Paris*, **126**, 985.

GAVAUDAN, P. & POUSSEL, H. (1944). *C.R. Soc. Biol., Paris*, **138**, 246.

GAVAUDAN, P., DODÉ, M. & POUSSEL, H. (1946). *C.R. Acad. Sci., Paris*, **223**, 521, 591.

HADDOW, A. & SEXTON, W. A. (1946). *Nature, Lond.*, **157**, 500.

HAVAS, L. J. (1938). *Growth*, **2**, 257.

HENDERSON, V. E. (1930). *Physiol. Rev.* **10**, 171.

KOSTOFF, D. (1938). *C.R. Acad. Sci. U.R.S.S.* **19**, 197.

KOSTOFF, D. (1938). *Nature, Lond.*, **141**, 1144.

KRYTHE, J. M. & WELLENSIEK, S. J. (1942). *Bibliogr. genet.* **14**, 1.

LEFEVRE, J. (1939). *C.R. Acad. Sci., Paris*, **208**, 301.

LEVAN, A. (1938). *Hereditas, Lund*, **24**, 471.

LEVAN, A. (1940). *Hereditas, Lund*, **26**, 262.

LEVAN, A. (1942). *Hereditas, Lund*, **28**, 244.

LEVAN, A. (1945). *Nature, Lond.*, **156**, 751.

LEVAN, A. (1946). *Hereditas, Lund*, **32**, 294.

LEVAN, A. & LEVRING, T. (1942). *Hereditas, Lund*, **28**, 400.

LEVAN, A. & ÖSTERGREN, G. (1943). *Hereditas, Lund.* **29**, 381.

LOEB, J. & WASTENEYS, H. (1912). *J. biol. Chem.* **14**, 517.

LUDFORD, R. J. (1936). *Arch. exp. Zellforsch.* **18**, 411.

LWOFF, M. (1934). *C.R. Soc. Biol., Paris*, **115**, 237.

MCELROY, W. D. (1947). *Quart. Rev. Biol.* **22**, 25.

MANGENOT, G. (1939). *C.R. Acad. Sci., Paris*, **208**, 222.

MANGENOT, G. & CARPENTIER, S. (1944). *C.R. Soc. Biol., Paris*, **138**, 105.

MICHAELIS, H. & QUASTEL, J. H. (1941). *Biochem. J.* **35**, 518.

MITCHELL, J. W. & MARTH, P. C. (1947). *Science*, **106**, 15.

ÖSTERGREN, G. (1944). *Hereditas, Lund*, **30**, 429.

QUASTEL, J. H. (1943). *Trans. Faraday Soc.* **39**, 348.

RANDOLPH, L. F. (1935). *Stain Tech.* **10**, 95.
SIMON, E. W. & BLACKMAN, G. E. (1948).
SIMONET, M. & GUINOCHET, M. (1939). *C.R. Soc. Biol., Paris*, **130**, 1057.
TAYLOR, G. W. (1935). *J. Cell. Comp. Physiol.* **7**, 409.
TEMPLEMAN, W. G. & SEXTON, W. A. (1945). *Nature, Lond.*, **156**, 630.
TEMPLEMAN, W. G. & SEXTON, W. A. (1946). *Proc. Roy. Soc.* B, **133**, 480.
WEINSTEIN, L. & MCDONALD, A. (1945). *Science*, **101**, 44.
WINTERSTEIN, H. (1926). *Die Narkose*, 2nd ed. Berlin: Julius Springer.

EXPLANATION OF PLATE 1

(*a*) Longitudinal section of barley root tip showing various stages of normal mitosis. ×550. (*b*) Longitudinal section of root tip after 48 hr. treatment with 128 p.p.m. ethyl phenylcarbamate. Note (i) increase in size of cells, (ii) polyploid nuclei, (iii) varying degrees of chromosome contraction, (iv) increased number of nuclei in 'metaphase', (v) multinucleate cells. ×550.

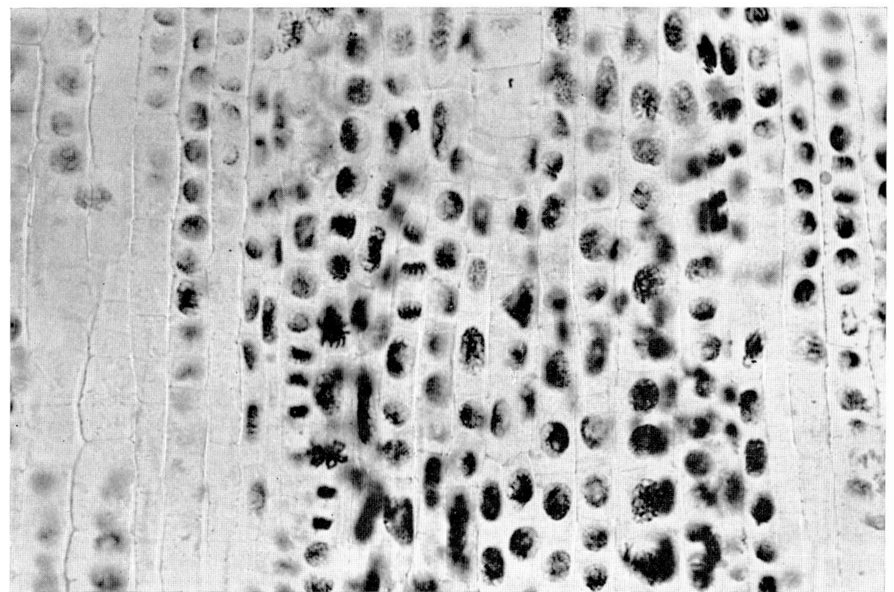

(*a*)

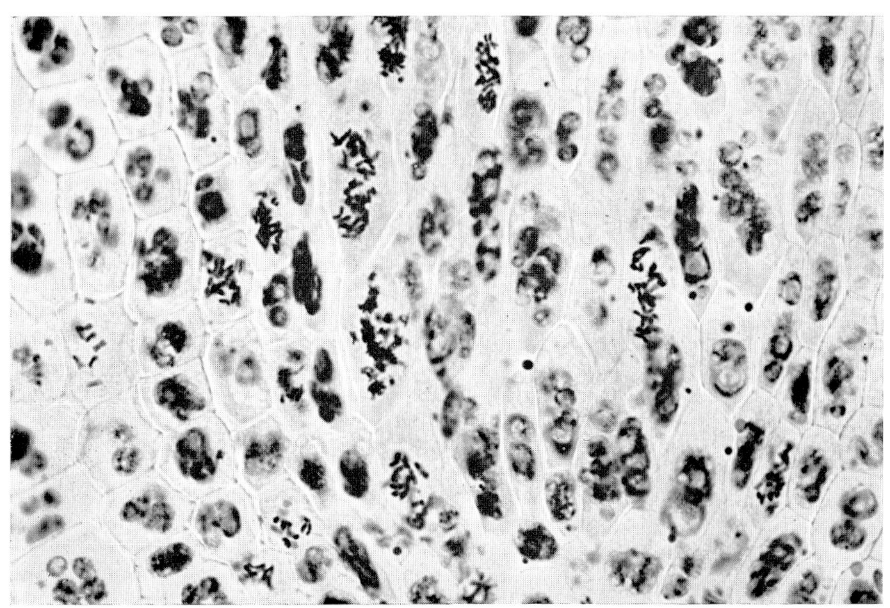

(*b*)

For explanation see p. 282

THE COMPARATIVE TOXICITY OF PHYTOCIDAL SUBSTANCES

By G. E. BLACKMAN, K. HOLLY and H. A. ROBERTS
Department of Agriculture, University of Oxford

I. INTRODUCTION

The development of copper fungicides foreshadowed the discovery that cupric sulphate is differentially toxic to higher plants. In 1895–7 it was demonstrated independently by Bonnet in France, by Bolley in America, and by Schultz in Germany, that a dilute solution of cupric sulphate, when sprayed on cereal crops, killed the ubiquitous annual *Sinapis arvensis* but did not injure the cereal. The subsequent progress can be divided into two phases, that of inorganic toxic substances, and a second period following on the discovery of organic compounds with selective phytocidal properties.

The inorganic substances which have been used include ferrous sulphate, cupric nitrate, ammonium sulphate, sodium nitrate, calcium cyanamide, potassium salts, sulphuric acid and nitric acid. These compounds were applied as dusts or aqueous sprays for the selective destruction of annual dicotyledonous weeds in cereal crops. During the same period developments occurred in the search for substances of general toxicity to higher plants, particularly perennial plants, and by 1930 carbon disulphide, chlorates, thiocyanates and arsenic and boron compounds had been employed for the eradication of vegetation.

The first organic compounds to be used as selective herbicides were the sodium salts of 2:4-dinitro-phenol and 3:5-dinitro-*o*-cresol. In 1932 Truffaut & Pastac (1932, 1944) took out a patent which claimed that these compounds were toxic to many dicotyledonous annuals at concentrations which were not injurious to cereals. When a paste containing sodium dinitro-*o*-cresylate became commercially available in the United States, Westgate & Raynor (1940) and Harris & Hyslop (1942) confirmed that while cereals in the seedling stage were not injured, many annual dicotyledons were killed. Moreover, Harris & Hyslop demonstrated that the efficiency of the spray solution was enhanced by the addition of ammonium sulphate as an activator.

The next major development resulted from the observation made by Templeman in 1940 that in some pots of oats sprayed with α-naphthylacetic acid, there were chance seedlings of *Sinapis arvensis*, which alone were killed. Acting upon this indication that synthetic 'growth substances' or 'growth regulators' could be differentially toxic, Slade, Templeman and Sexton had by the end of 1942 synthesized and tested in pot and laboratory trials a large number of allied compounds. Amongst the most active were 2-methyl-4-chloro-phenoxyacetic and 2:4-dichloro-phenoxyacetic acids. In the meantime, Zimmerman & Hitchcock (1942) had shown that 2:4-dichloro-phenoxyacetic acid was highly toxic, but had not demonstrated its selective action.

On the other hand, Nutman, Thornton & Quastel, studying the part played by growth substances in the mechanism of the entry of nodule organisms into the root hairs of legumes, reached the conclusion in 1942 that 2:4-dichloro-phenoxyacetic acid was selectively toxic to plants. Early in 1943, the results obtained by Slade *et al.* and by Nutman *et al.* were passed to a team led by the first author, so that these substances could be included in comparative field trials with other herbicides then being independently developed. Owing to wartime restrictions of publication in Great Britain, summaries of the work of the three teams were not published until 1945 (Slade *et al.* 1945; Nutman *et al.* 1945; Blackman, 1945*a*). Since then, further accounts have been given by Templeman & Sexton (1946) and by Blackman (1945*b*, 1946).

In the United States, the use of 2:4-dichloro-phenoxyacetic acid as a herbicide was first suggested by Mitchell & Hamner (1944). The first experimental details were given by Hamner & Tukey (1944), who applied the substance to the deep-rooted perennial *Convolvulus arvensis*. This paper, published within five weeks, only describes observations of the toxic effect within the first 10 days after spraying. For a bibliography up to 1946 of the many subsequent American papers, the review by Gilbert (1946) should be consulted. Recent publications include a large number by members of The Special Projects Division, Chemical Warfare Service (Norman, 1946).

Two further types of compound are worthy of mention. Templeman & Sexton (1945) claimed that the alkyl phenyl carbamates are especially toxic to graminaceous plants in the early stages of development. Finally, Chabrolin (1940, 1942) showed that sodium pentachlorophenate is toxic to a large range of dicotyledonous plants, and is suitable for use as a selective herbicide in cereal crops.

The pattern of development, compared with the fields of insecticides, fungicides and bactericides, has been largely divergent. The herbicide investigations have been concerned almost exclusively with field observations; with a few exceptions, the use of statistics either in the design of the experiment or in the analysis of the data has been conspicuous by its absence. Far too high a proportion of papers have contained *ex cathedra* statements with little supporting data. Moreover, individual investigations have in general been restricted to one type of compound, and direct and *simultaneous* comparisons of a range of compounds is seldom available. Finally, where supporting experiments have been carried out in the laboratory and greenhouse they have largely been 'screening' or sorting tests, often of doubtful validity, and have not related to a precise analysis of the differences observed. In consequence, in spite of half a century, there is as yet little critical basis on which to formulate any principles of phytocidal action.

The present series of investigations was started in 1941, and during the war years had as its primary object the development of new techniques of weed control in a variety of crops. The wide scope of the investigations, the use of statistical methods and factorial experiments for the simultaneous comparison of a range of compounds at different concentrations, has provided an array of facts from which some general conclusions can be drawn. Moreover, it has been possible during the last two years to begin research on some of the more fundamental problems revealed by the initial field studies.

II. EXPERIMENTAL TECHNIQUE

Design of experiments

The data recorded in this paper have been obtained almost exclusively from factorial field experiments of randomized block design. The individual treatments were always applied as sprays at a standard volume of a hundred gallons per acre.

Where the effects on resistant crop plants were being studied, the individual plots were harvested and threshed to determine the seed yield. In many instances supporting data on the changes in plant density, tiller and fertile shoot production, seed number per inflorescence or capsule were estimated by random sampling of each plot.

To assess the effect on susceptible species the criterion of percentage kill has been employed. The technique adopted was to estimate by random quadrats, at an appropriate interval after spraying, the relative densities of plants on the treated and control plots.

In the case of perennial plants where true estimates of kill cannot be made unless final counts are made a year later, the seasonal change in plant density on the control and treated plots must be taken into account. Thus, density estimates have to be made on each plot both before the experiment is started and again when it is completed, so that the proper correction factors can be applied to the final data.

Statistical analysis of results

The data for each trial involved dealing with a large number of figures, since where the counts included investigations on crop development the total number of observations amounted to as many as 7000 per experiment.

For some of the observations the methods of analysis of variance or covariance are applicable without transformation of the data. Treatment without transformation is not, however, appropriate to the percentage kill estimates, since the original data consists of plant numbers and the distribution is discontinuous. It is therefore necessary to analyse statistically the square root of the density, or if many of the estimates contain less than ten plants, the square root of the density plus a half. In some experiments the treatments have given such widely divergent degrees of kill that all the data cannot on mathematical grounds be included in a single analysis of variance. Under these circumstances the treatments giving either the highest or lowest degree of control have been omitted from the analysis, for example, in Tables 1, 2 and 3. Because significant differences derived from such analyses are only applicable to the transformed data, the transformed plant densities are included with the percentage kill results in many of the tables.

Where pre-spraying counts have been made in addition to the final counts, as in the perennial weed experiments, the results are expressed as percentage kills based on percentage change in density on each plot. These percentages have been transformed to the angular scale of Bliss (1937), viz. arc sin $\sqrt{(\% \text{ kill})}$, before the analysis of variance was made.

III. THE INTERRELATIONSHIP BETWEEN SPECIES AND THE RELATIVE TOXICITY OF DIFFERENT GROUPS OF COMPOUNDS

It has already been stated in the introduction that the four main groups of compounds employed for selective weed control in cereals and other crops are strong acids, copper salts, aryl nitro compounds and growth substances. To illustrate the main differences in toxicity between these groups, sulphuric acid, cupric chloride, 3:5-dinitro-*o*-cresol (either as a suspension or as the ammonium salt) and sodium 2-methyl-4-chloro-phenoxyacetate have been selected as typical examples of each group. To bring out the interdependence between relative toxicity and species, selected data covering sixteen species of annuals are included in Tables 1–4.

In Table 1 it is to be noted that while with the first four species the compounds exhibit little difference in the degree of kill attained, this is not so in the case of *Atriplex patula*. In this instance there is a descending order of toxic action from dinitro-*o*-cresol to copper chloride with the growth substance and sulphuric acid intermediate.

Table 1. *The interrelationship between species and the toxic effects of different compounds. Exp. II*, 1944

Compound and % concentration of spray (w/v)		Percentage kill in seedling stage					
		Galeopsis tetrahit (% kill)	*Sinapis arvensis* (% kill)	*Erysimum cheiranthoides* (% kill)	*Polygonum convolvulus* (% kill)	*Atriplex patula* (% kill)	*Atriplex patula* √density
Sulphuric acid*	13·8	98·8	84·9	89·1	93·7	76·7	9·0
	18·4	98·8	82·6	87·4	97·9	66·7	11·0
	Mean	98·8	83·7	88·2	95·8	71·7	10·0
Cupric chloride	2·0	95·6	97·7	98·1	97·6	30·9	15·8
	4·0	100·0	100·0	100·0	97·4	35·4	15·4
	Mean	97·8	98·8	99·1	97·5	33·2	15·6
Dinitro-*o*-cresol	0·4	100·0	100·0	99·8	97·1	99·1	—
	0·8	98·5	100·0	97·9	98·7	99·8	—
	Mean	99·4	100·0	98·9	97·9	99·5	—
Sodium methyl chloro-phenoxy-acetate	0·2	99·4	97·7	99·4	95·5	78·3	8·8
	0·4	100·0	99·8	99·2	96·8	86·1	7·0
	Mean	99·7	98·3	99·3	96·2	83·2	7·9
Sig. diff. ($P=0·05$):							
Between treatments							3·8
Between compounds							2·7

* Spray contained a wetting agent of the sulphonated oil type.

Table 2. *The interrelationship between species and the toxic effects of different compounds. Exp. IV*, 1944

Compound and % concentration of spray (w/v)		Percentage kill in seedling stage					
		Scandix pecten-veneris (% kill)	*Papaver rhoeas* (% kill)	*Veronica hederifolia* (% kill)	*Veronica hederifolia* √density	*Fumaria officinalis* (% kill)	*Fumaria officinalis* √density
Sulphuric acid*	9·2	16·0	10·5	33·8	8·57	21·5	6·10
	13·8	6·2	58·2	63·8	6·33	68·1	3·80
	18·4	18·6	52·1	81·7	4·47	57·6	4·43
Cupric chloride	0·5	32·9	0·0	0·0	10·70	0·0	7·13
	1·0	5·5	0·0	18·0	9·57	0·0	6·96
	2·0	0·0	19·4	57·5	6·87	29·2	5·83
Ammonium dinitro-*o*-cresylate	0·275	0·0	56·3	7·5	10·10	81·2	—
	0·55	2·1	90·0	23·4	9·10	96·5	—
	1·10	47·1	95·7	49·4	7·50	99·3	—
Sodium methyl chloro-phenoxy-acetate	0·05	95·6	63·2	80·8	—	93·7	—
	0·15	100·0	86·0	91·0	—	96·5	—
	0·45	100·0	98·1	98·5	—	99·3	—
Sig. diff. ($P=0·05$) between treatments			—		1·62		1·81

* Spray contained a wetting agent of the sulphonated oil type.

This interdependence of the relative phytocidal effects and species is also brought out in Table 2. Whereas sulphuric acid and copper chloride have only given a partial kill of the four species, ammonium dinitro-*o*-

cresylate is highly toxic to *Papaver rhoeas* and *Fumaria officinalis*, but not to *Scandix pecten-veneris* or *Veronica hederifolia*. This partial resistance of the last two species does not apply to the phenoxyacetic acid derivative, since an almost complete kill of all four species has been achieved by the highest concentration.

While the data of Table 2 stress the relative high toxicity level of methyl-chloro-phenoxyacetic acid, this emphasis is altered by the results of Table 3, since the three species cited, especially *Galium aparine*, are largely resistant to this compound. This resistance does not, however, hold for the other herbicides; ammonium dinitro-*o*-cresylate is highly toxic to *Alchemilla arvensis* and *Matricaria chamomilla*. In the case of *Galium aparine*, sulphuric acid is significantly superior to dinitro-*o*-cresol, but this position is reversed for two out of the remaining three sets of observations. Moreover, the reaction of the three species to cupric chloride is individually different.

Lastly, it should be observed that while the two separate experiments on *Matricaria chamomilla* show the same trend, the percentage kills obtained, except for dinitro-*o*-cresol, differ considerably. Such differences are likely to arise because of variations in weather conditions and the stage of growth at the time of spraying; they will be discussed in more detail in later sections.

The results set out in Tables 1–3 are representative of the variable behaviour of annual dicotyledonous plants and could be supplemented by equivalent data covering a further thirty or more species. Such information would, however, only confirm the same general conclusion that individual species are a law unto themselves. Broadly it can be stated that dinitro-*o*-cresol is toxic to the widest range of species, followed by methyl-chloro-phenoxyacetic acid and sulphuric acid and—a long way behind—by cupric chloride. It is significant that no annual dicotyledon so far investigated is resistant in the seedling stage to all four groups of compounds; the nearest to approach this is *Polygonum aviculare*.

This susceptibility of annual dicotyledons to one or more of the four compounds contrasts with the reactions of the Gramineae which, once they have reached the stage of development when active tillering is taking place, are largely resistant. The effects of spraying cereals at this or later stages, as measured by the final grain production, can be seen for wheat and oats in Table 4 and for barley in Table 6. Although at the concentrations employed few of the differences are significant, yet the overall indications are that sulphuric acid may produce some check, followed by cupric chloride. On the other hand, there is little sign of injury after spraying with dinitro-*o*-cresol or methyl-chloro-phenoxyacetic acid.

Table 3. *The interrelationship between species and the toxic effects of different compounds*

Compound and % concentration of spray (w/v)		Percentage kill in seedling stage									
		Exp. III, 1944				% conc.	Exp. I, 1944 *Matricaria chamomilla*		Exp. XIV, 1945 *Galium aparine*	
		Alchemilla arvensis		*Matricaria chamomilla*						
		% kill	√density	% kill	√density		% kill	√density	% kill	√density
Sulphuric acid	*9·2	45·0	10·1	15·7	12·4	9·2	68·0	9·3	49·7	8·9
	*13·8	69·1	7·4	32·3	11·1	13·8	85·5	6·3	94·0	3·0
	*18·4	78·4	6·1	54·9	9·2	18·4	91·3	4·9	94·8	2·6
Cupric chloride	2·0	0·0	14·1	25·3	11·6	1·0	38·7	13·1	25·6	11·0
	4·0	23·9	11·6	35·5	10·6	2·0	55·6	11·1	66·0	7·4
	6·0	7·2	13·1	40·5	10·4	4·0	67·2	9·7	83·3	5·2
Ammonium dinitro-*o*-cresylate	0·3	99·6	—	75·6	6·3	0·2	55·0	10·9	9·9†	11·6
	0·6	99·6	—	84·1	5·4	0·4	77·8	8·0	74·0†	6·3
	0·9	99·6	—	86·3	4·5	0·8	88·3	5·7	78·6†	5·7
Sodium methyl-chloro-phenoxy-acetate	0·2	22·8	11·9	37·0	10·7	0·05	43·0	12·4	0·0‡	—
	0·4	15·8	12·5	25·7	11·5	0·15	37·0	13·0	0·0‡	—
	0·6	29·1	11·5	45·6	10·0	0·45	68·1	9·1	0·0‡	—
Sig. diff. ($P=0·05$): Between treatments			3·1		3·3			3·9		4·0

* Spray contained a wetting agent.
† In this experiment the concentrations of ammonium dinitro-*o*-cresylate were 0·3, 0·6 and 0·9%.
‡ In this experiment the concentrations of sodium methyl-chloro-phenoxyacetate were 0·1, 0·2 and 0·4%.

The same general trends have been observed in many other cereal experiments carried out between 1943 and 1946. Where reductions in yield have been recorded they have largely occurred in spring cereals especially in dry seasons. There is also evidence of concentration effects. Concentrations in excess of 9·2% for sulphuric acid and 2·0% for cupric chloride are liable to reduce the yield level. By contrast, in none of the thirty-eight experiments concerned with the effects of dinitro-*o*-cresol or its salts on cereals have depressions in yields been obtained at the concentrations which destroy dicotyledonous annuals. On the other hand, with sodium methyl-chloro-phenoxyacetate there is a slight risk that there may be some loss in yield once a concentration of 0·2% is exceeded, i.e. more than 2 lb. applied per acre.

Table 4. *The effects of compound on the seed production of cereals*

Compound and concentration of spray	Wheat, Exp. III, 1944		Oats			
			Exp. I, 1944		Exp. III, 1944	
	% conc.	cwt./acre	% conc.	cwt./acre	% conc.	cwt./acre
Sulphuric acid	9·2	23·5	—	—	—	—
	13·8	22·0	—	—	13·8	28·1
	18·4	21·3	—	—	18·4	27·8
	Mean	22·3	—	—	—	28·0
Cupric chloride	0·5	21·1	0·5	12·8	—	—
	1·0	24·2	1·0	13·8	2·0	29·4
	2·0	24·9	1·5	13·1	4·0	27·9
	Mean	23·4	—	13·2	—	28·7
Ammonium dinitro-*o*-cresylate	0·28	25·1	0·3	10·8	—	—
	0·55	24·3	0·5	12·0	0·4*	34·0
	1·10	27·6	0·7	10·9	0·8*	33·3
	Mean	25·6	—	11·2	—	33·6
Sodium methyl-chloro-phenoxyacetate	0·05	26·7	0·025	12·9	—	—
	0·15	25·6	0·05	12·8	0·2	32·5
	0·45	22·9	0·10	12·8	0·4	31·6
	Mean	25·1	—	12·9	—	32·0
Control	—	23·9	—	11·4	—	33·1
Sig. diff. ($P=0·05$):						
Between treatments		n.s.		n.s.		7·4
Between means		n.s.		1·8		5·3

* Wetting agent added to spray solution.

Supporting evidence that there is a similarity in the reactions of other grasses is available from investigations on *Dactylis glomerata*, *Lolium perenne*, *Phleum pratense*, *Phalaris canariensis* and *Festuca rubra*. The main differences between these species, *once they are established*, and the cereals are their greater sensitivity to cupric chloride and the elimination of the differences between the effects of sulphuric acid, dinitro-*o*-cresol and methyl-chloro-phenoxyacetic acid. On the other hand, *Zea mais* is less tolerant to all four types of compound than are the other grasses investigated.

So far, in discussing the toxic action of the several groups of compounds, the measure of phytocidal action has been the degree of kill or the effect on seed production. The experiments on flax, however, bring out that while there may be no external signs of injury, yet there may be marked internal effects. This disparity is clearly brought out in Table 5, where the consequences of spraying flax, when 15 cm. high, on the final weight of seed or straw contrast startlingly with the recovery of fibre after processing.

Table 5. *The interrelationship between compound and concentration on their effects on* Linum usitatissimum (*flax*). *Exp. V*, 1944

Compound and % concentration of spray		Seed production (cwt./acre)	Straw production (cwt./acre)	Fibre production (cwt./acre)
Control		4·6	28·4	4·6
Cupric chloride	0·5	4·8	24·0	4·3
	1·0	4·5	25·7	4·3
	2·0	4·6	25·9	4·3
	Mean	4·6	25·2	4·3
Cupric nitrate	0·5	4·1	25·9	4·3
	1·0	3·7	25·8	3·8
	2·0	4·1	25·5	3·7
	Mean	4·0	25·7	3·9
Dinitro-*o*-cresol	0·3	4·3	23·6	3·6
	0·6	3·8	25·8	3·6
	0·9	4·0	20·7	2·6
	Mean	4·0	23·4	3·3
Sodium dinitro-*o*-cresylate	0·3	3·5	25·5	4·2
	0·6	4·1	27·7	5·0
	0·9	5·0	26·7	3·6
	Mean	4·2	26·6	4·3
Sodium methyl-chloro-phenoxy-acetate	0·025	4·3	26·1	4·6
	0·05	4·9	25·5	4·3
	0·10	4·9	25·5	3·7
	Mean	4·7	25·7	4·2
Sig. diff. ($P=0·05$):				
Between treatments		n.s.	3·4	0·9
Between compounds		n.s.	2·0	0·5

In the first place it is seen that none of the treatments has produced a significant effect on seed production. On the other hand, all the five compounds have on average decreased the yield of straw. There are, in addition, differences between compounds; dinitro-cresol, applied as a suspension, is more toxic than its sodium salt, the suspension is also more damaging than cupric nitrate.

These effects on straw production are, however, small compared with those on fibre production. Whereas the decrease in straw weight brought about by the highest concentration of dinitro-*o*-cresol amounts to 27%, the corresponding figure for fibre is 43%. It is also clear that the disparity between dinitro-cresol and its sodium salt is even more marked when considered on the basis of the fibre data. The two copper salts and the growth substance show the same trends in their influence on straw and fibre production, though for the last there is a greater influence of concentration.

IV. RELATIVE TOXICITY AND THE STAGE OF DEVELOPMENT

No complete picture of the differences in the level of toxicity between compounds can be obtained unless account is taken of the varying influence of the stage of development on the susceptibility of individual species. This difficulty does not arise in the case of the data in Tables 1 and 2, where the species were of the same age, since they all germinated together, but the question does arise in Tables 3 and 4, where the comparisons are not simultaneous but refer to several experiments. This interrelationship between stage of development and resistance or susceptibility has not in the past been sufficiently realized, and without a critical appreciation of this factor comparisons may be invalidated.

Table 6. *Effect of stage of development on the relative toxicity of different compounds. Exps. I and II (barley), 1944*

Compound and % concentration of spray		*Thlaspi arvense* (% kill)		Barley. Seed production cwt./acre	
		Seedling stage	Flowering stage		
Sulphuric acid	4·6	—	81	—	21·9
	9·2	91	96	29·2	20·7
	13·8	97	99	30·2	20·9
	18·4	99	—	30·1	—
	Mean	96	92	30·0	21·2
Cupric chloride	1·0	75	69	31·3	25·4
	2·0	86	88	29·4	24·7
	4·0	95	97	28·8	22·7
	Mean	85	85	29·8	24·3
Ammonium dinitro-*o*-cresylate	0·3	100	19	28·8	26·6
	0·5	100	45	31·1	25·9
	0·7	100	22	31·5	25·4
	Mean	100	29	30·5	26·0
Sodium methyl-chloro-phenoxyacetic acid	0·05	—	100	—	27·5
	0·10	100	100	33·2	27·4
	0·20	100	100	33·4	27·0
	0·40	100	—	29·5	—
	Mean	100	100	31·9	27·3
Control		—	—	25·6	27·4
Sig. diff. ($P=0·05$):					
Between treatments		—	—	4·5	3·3
Between compounds		—	—	2·6	1·9

Some indication of the complexity of the factors involved can be seen from the data set out in Table 6, where the effects are shown of spraying barley infested with *Thlaspi arvense* when the latter was either in the seedling stage or had reached the flowering period. With early spraying all four compounds have completely killed *T. arvense*, and none of the treatments has had any differential effect on barley. In contrast, by delaying spraying the relative toxicity of the compounds is completely changed. As far as barley is concerned, sulphuric acid is now the most damaging followed by copper chloride, while in the case of *Thlaspi* ammonium dinitro-*o*-cresylate is greatly inferior to the other compounds.

Further evidence of the interdependence of phytocidal action on the stage of growth is shown in Table 7. While cupric chloride is equally toxic to *Sinapis arvensis* whether the plants are seedlings or in flower, there is a progressive decrease in the phytocidal action of methyl-chloro-phenoxy-acetic acid.

Table 7. *The effect of stage of development on the relative toxicity of different compounds. Exp. VII*, 1945

Compound and % concentration of spray		*Sinapis arvensis* (% kill) Stage of development		
		Seedling	Pre-flowering	Flowering
Sodium-methyl-chloro-phenoxy-acetate	0·0125	56	79	25
	0·025	90	84	25
	0·05	100	99	39
	Mean	82	87	29
Cupric chloride	0·5	93	74	73
	1·0	99	98	99
	1·5	100	100	100
	Mean	97	90	91

The present investigation and past work on the mode of action of sulphuric acid (Åslander, 1927; Blackman, 1934, 1942; Blackman & Templeman, 1936) provide a sufficient body of results on which to formulate general inclusions as to the operation of the age factor. Åslander has pointed out that the resistance or susceptibility of a plant to damage by acids is largely dependent upon differences in morphological structure and habits of growth. Resistant plants, such as the Gramineae and the onion, have more or less upright shoots with waxy leaves, and the basal meristematic tissues are protected by the leaf sheaths. Dicotyledonous plants, on the other hand, possess initially a single apical growing point, while the leaves, often with a thin cuticle, are set in a horizontal plane. In consequence while the spray droplets are caught by the horizontal leaves they tend to run off upright leaves. Moreover, if the cuticle is waxy, penetration of the acid is retarded, and finally while the basal meristematic tissue of monocotyledons is shielded from the spray droplets the dicotyledonous apical bud is exposed to direct contact. From this it follows that dicotyledonous plants are most susceptible to acid sprays once the cotyledons have expanded, and that with subsequent increasing differentiation of the shoot there will be a falling off in susceptibility. The rate of increasing resistance will be primarily dependent on the rapidity with which axillary buds are formed and the development of the cuticle. *Urtica urens* can be cited as an example of a plant which produces axillary buds at a very early stage and *Chrysanthemum segetum* and *Fumaria officinalis* as plants with waxy cuticles. In other species such as *Scandix pecten-veneris*, *Ranunculus arvensis* and *Papaver rhoeas*, the main growing points are in part protected and so to a certain extent these species share the resistance of the Gramineae. In all these groups, an increased degree of kill can be achieved by adding a wetting agent to the spray solution.

The same trend of decreasing susceptibility with increasing age also holds for dinitro-*o*-cresol, but the gradient during the period of active growth from the seedling to the pre-flowering stage may often be less pronounced. It is often possible to maintain the same degree of kill at varying stages of development by increasing the concentration. On the other hand, with sulphuric acid, raising the concentration will not prevent a sharp fall in the kill obtained as the plant passes from the seedling to the pre-flowering stage, e.g. *Matricaria inodora*, *Polygonum aviculare*, *Papaver rhoeas*.

The relation between the toxic effect of cupric chloride and the stage of growth is dependent on the species. It has already been demonstrated (Tables 6 and 7) that up to flowering there is no falling off in effectiveness for both *Thlaspi arvense* and *Sinapis arvensis*. On the other hand, in cases where only a partial kill is achieved in the seedling stage, e.g. *Raphanus raphanistrum*, there is a steady rise in resistance as the plant becomes older.

So far the discussion on the interdependence of toxic action and stage of development has centred on those plants which can be grouped as susceptible annual species and has not taken into account resistant species, such as the grasses. Here the trend is somewhat different. In the early seedling stage the plants are to some extent susceptible, reach a maximum resistance at the time of active tillering, while the liability to damage again increases as the plants grow older. The influence of stage of development once the tillering stage has been passed is greatest in the case of sulphuric acid and copper chloride (see, for example, the barley data in Table 6).

Further evidence of the interdependence between phytocidal effects and stage of development is illustrated by the results of a trial in which flax was sprayed at heights of 2·5–5 cm. (early) 10–15 cm. (medium) and 60 cm.

(late) (see Table 8). On the basis of the data already discussed (Table 5), it is not unexpected that the effects on total plant weight differ very markedly from those on fibre production. Taking first the main effects, it is seen that whereas cupric chloride has reduced total weight, sodium methyl-chloro-phenoxyacetate has not. Overall there is no significant difference between spraying dates. On the other hand, there is a significant interaction between the increasing damage caused by rising concentration for cupric chloride as against the insignificant change for the growth substance. There is, too, a significant interaction between occasion and concentration, since at the earliest spraying the positive effect of increasing concentration is greatest.

Considering next the fibre data it is clear that all the observed effects on plant weight are magnified. The depression due to cupric chloride is greater, and so the contrast between the two compounds is accentuated. There is now a significant effect of occasion, with the greatest injury caused by early spraying. The contrasting effects of changing from the low to the intermediate concentration between the first and second occasion should be noted. Finally, while at the highest concentration of cupric chloride injury is reduced with increasing age, at the lowest concentration of the growth substance the reverse is the case.

The interdependence of stage of development and the toxicity of growth substances are of the greatest complexity, since the reaction of species is extremely variable. It is, however, the rule, though the exceptions may be many, e.g. *Thlaspi*, that between the cotyledon and flowering stages there is a lessening effect of any given concentration. For example, it is seen in Fig. 1*a* that in the cotyledon stage a complete kill of *Chenopodium album* has been obtained with a 0·05% concentration, while in Fig. 1*b* complete suppression of older plants in other experiments is only achieved when the concentration is at least five times as great. There is the same trend in more resistant species such as *Polygonum aviculare*.

Fig. 1*a* also demonstrates the wide variation between species in their reaction to methyl-chloro-phenoxyacetic acid even in the most susceptible phase of development. The same extensive variation is exhibited at the time of germination; Fig. 2*a* contains data from experiments where the growth substance was applied to the soil and counts were subsequently made of the emerging seedlings. Perhaps the most interesting result is the very marked toxic effect in the germination phase on *Dactylis glomerata*, which, it will be recalled, is completely resistant once the plants have become established seedlings (see p. 290).

On the basis of the very large number of results it is possible to classify annuals into a number of groups with characteristic reactions at different stages of growth to methyl-chloro-phenoxyacetic acid. The first class

Table 8. *The interrelationship between compound, concentration and stage of development on the growth of* Linum usitatissimum. *Exp. I*, 1945

Compound and % concentration		Stage of development							
		Total plant weight (tons/acre)				Fibre production (cwt./acre)			
		Height 2·5–5·0 cm. early	Height 10–15 cm. medium	Height 60 cm. late	Mean	Height 2·5–5·0 cm. early	Height 10–15 cm. medium	Height 60 cm. late	Mean
Control		3·17	3·17	3·17	3·17	7·63	7·63	7·63	7·63
Cupric chloride	1·0	3·10	2·86	2·70	2·89	6·04	5·38	5·94	5·78
	2·0	2·36	2·82	2·90	2·69	4·18	6·71	5·46	5·45
	3·0	2·33	2·50	2·78	2·54	3·59	4·88	5·04	4·50
	Mean	2·60	2·73	2·79	2·71	4·60	5·66	5·48	5·25
Sodium methyl-chloro-phenoxyacetate	0·025	3·19	3·24	2·91	3·11	7·62	8·25	6·43	7·43
	0·05	2·89	3·07	3·33	3·10	5·84	7·64	8·08	7·18
	0·10	3·25	3·32	3·25	3·27	7·14	8·05	6·84	7·35
	Mean	3·11	3·21	3·16	3·16	6·86	7·97	7·11	7·32
	Mean	2·86	2·97	2·98	—	5·73	6·81	6·30	—
Sig. diff. ($P=0·05$):									
Between treatments		0·35				1·32			
Between any treatment and control mean		0·26				1·00			
Between compounds		0·12				0·44			
Between concentration or occasion means		0·14				0·54			
Sig. interactions:									
Compound and concentration		0·29				1·08			
Concentration and occasion		0·35				1·32			
Concentration, occasion and compound		—				5·31			

comprises plants which are equally susceptible at all stages from germination to flowering. At the other end of the scale are species which

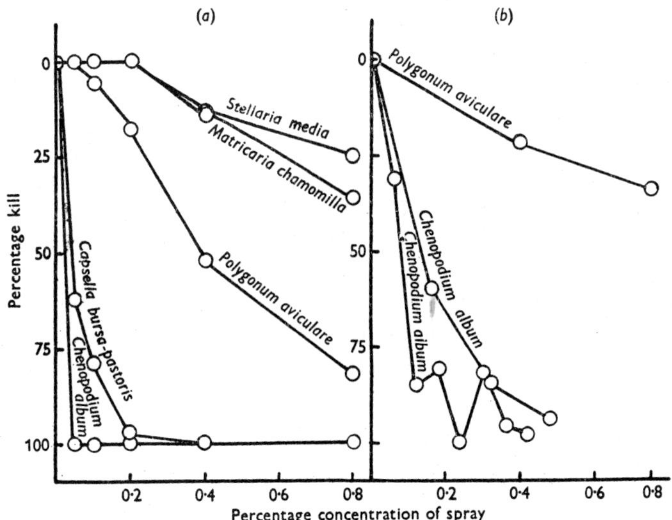

Fig. 1. The toxic effects of sodium methyl-chloro-phenoxyacetate on different species when sprayed (*a*) in the cotyledon stage, (*b*) in the pre-flowering stage.

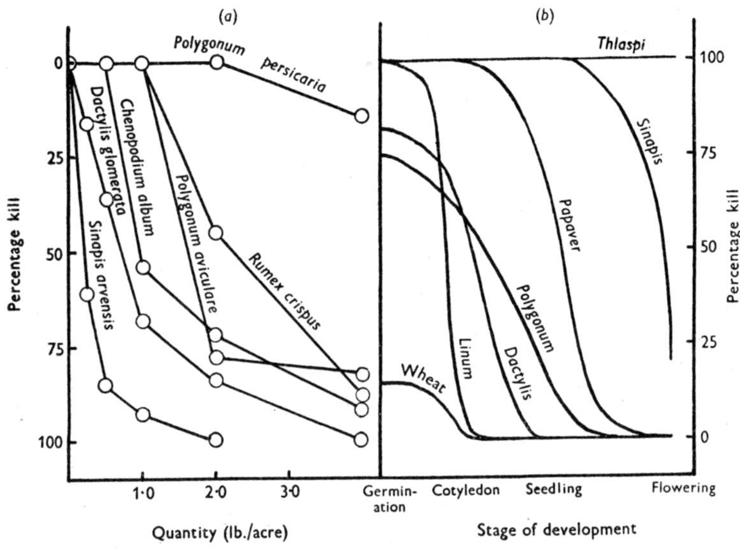

Fig. 2. The toxic effects of sodium methyl-chloro-phenoxyacetate on different species (*a*) in the germination phase, (*b*) at different stages of development. In Fig. 2*b* curves based on degree of kill expected after spraying with a 0·2 % solution.

are susceptible or partially susceptible in the germination phase but have become resistant once the cotyledon stage is reached. There are at least four intermediate grades in which resistance is reached in (*a*) the seedling stage, (*b*) the pre-flowering stage, (*c*) the flowering stage and (*d*) a partial resistance in the flowering stage. It is possible to construct curves for typical plants within each group on the basis of the kill expected at various stages of development when given a standard treatment. Such a series of curves based on the toxic effects of a concentration of 0·2% sodium methyl-chloro-phenoxyacetate is given in Fig. 2*b*.

V. THE INTERRELATIONSHIP BETWEEN SPECIES AND RELATIVE TOXICITY WITHIN A GROUP OF COMPOUNDS

The relative toxicity of chlorinated phenoxyacetic acids

In the initial phase of this aspect of the investigations a detailed comparison was restricted to two compounds, namely, the 2-methyl-4-chloro and 2:4-dichloro derivatives of phenoxyacetic acid. This restriction was based on a twofold consideration: first, the compounds selected are two of the most active at present known; secondly, it was also felt that until the differences between two closely related growth substances had been fully explored no real basis for wider comparisons could be properly planned.

It has already been seen in Table 5 that concentrations of up to 0·1% of sodium methyl-chloro-phenoxyacetate have not affected the seed production of *Linum usitatissimum*. Table 9 shows the effects of increasing the concentration when applications are made at two stages of development,

Table 9. *The interrelationship between stage of development, compound and concentration on the relative effects of chlorinated phenoxyacetic acids on seed production by* Linum usitatissimum (*linseed*). *Exp. I*, 1947

Compound	Stage of development: height in cm.	Production of seed (cwt./acre)				
		Percentage concentration				
		0·1	0·2	0·4	Means	
Sodium methyl-chloro-phenoxyacetate	8–10	15·27	15·06	12·35	14·23	14·17
	25–35	14·56	15·56	12·23	14·12	
	Mean	14·92	15·31	12·29		
Sodium dichloro-phenoxyacetate	8–10	13·10	13·11	7·92	11·38	11·71
	25–35	14·61	13·12	8·37	12·00	
	Mean	13·86	13·11	8·15		
Dichloro-phenoxyacetic acid	8–10	13·31	11·14	6·23	10·23	10·95
	25–35	13·66	12·45	8·93	11·68	
	Mean	13·49	11·80	7·58		
	Mean	14·09	13·41	9·34		

Control = 13·60

Sig. diff. (P = 0·05):

Between treatments	3·80
Between treatments and controls	2·92
Between compounds or concentrations	1·55

i.e. when the plants are 8–10 and 25–35 cm. high. Superimposed on these treatments there is a comparison of the sodium salt of the methyl-chloro derivative with the sodium salt and a suspension of the dichloro derivative.

It is clear from the results that the substitution of a methyl group for one of the chlorines has reduced the depressant effect on seed yield. With the dichloro compound there is no significant difference between the effects of the sodium salt and the acid itself. There is a marked influence of concentration, especially above 0·2%, but the effects are similar at both stages of development.

It has already been seen in Tables 5 and 8 that injurious effects of growth substances may only be manifested in the fibre production, and it is therefore of some interest to note the effects not only on seed production but also on the oil content of the seed. Data of this type are given in Table 10 and reveal that there is no longer a significant difference between compounds but there still remains a significant effect for concentration.

Table 10. *The interrelationship between stage of development, compound and concentration on the oil content of the seed of* Linum usitatissimum. *Exp. I*, 1947

Compound	Stage of development: height in cm.	Oil content of seed (ether extract as % of dry matter) % concentration				
		0·1	0·2	0·4	Means	
Sodium methyl-chloro-phenoxyacetate	8–10	42·7	43·0	43·4	43·0	43·3
	25–35	44·8	43·3	42·6	43·6	
Sodium dichloro-phenoxyacetate	8–10	44·3	42·8	41·9	43·0	43·0
	25–35	43·8	43·3	41·6	42·9	
Dichloro-phenoxy-acetic acid	8–10	44·2	42·4	42·2	42·9	42·7
	25–35	44·3	43·2	39·6	42·4	
	Mean	44·0	43·0	41·9		

Control = 44·2

Sig. diff. (P = 0·05):

Between treatments	1·7
Between concentrations and compounds	0·7
Between compounds or concentrations and control	0·8

It has been shown previously that at the time of germination methyl-chloro-phenoxyacetic acid is toxic to at least one species of grass (*Dactylis glomerata*), and further experiments have been carried out to determine whether this susceptibility is shared by other grasses. The results set out in Table 11 are from one experiment on winter-wheat sown on land containing *Alopecurus agrestis* and *Scandix pecten-veneris* and includes a comparison of sodium methyl-chloro-phenoxyacetate with dichloro-phenoxyacetic acid (the sodium salt was not available in quantity at the time) applied

Table 11. *The interrelationship between compound, concentration and species on the toxic effects in the germination and pre-emergence phases of development*

Compound and % concentration of spray		*Alopecurus agrestis*		Wheat		*Scandix pecten-veneris*	
		% kill	√density	% kill	√density	% kill	√density
Early spraying (germination):							
Methyl-chloro-phenoxy-acetic acid (Na salt)	0·05	0·5	13·5	0·0	16·3	12·0	14·1
	0·10	0·0	14·1	6·8	15·7	51·6	10·6
	0·20	50·9	9·4	21·5	14·3	63·6	9·3
	0·40	47·5	9·7	41·2	12·6	82·6	6·3
Dichloro-phenoxyacetic acid	0·05	0·0	13·9	10·3	15·4	16·0	14·0
	0·10	0·0	14·1	1·9	16·0	13·4	14·1
	0·20	46·4	10·0	8·8	15·5	46·6	11·2
	0·40	57·5	8·8	21·3	14·4	72·6	7·6
Late spraying (emergence):							
Methyl-chloro-phenoxyacetic acid (Na salt)	0·05	23·2	11·9	2·8	16·0	22·4	13·5
	0·10	0·0	13·7	21·1	14·4	60·4	19·4
	0·20	11·3	12·6	12·6	15·2	74·3	7·7
	0·40	6·3	12·8	22·8	14·1	92·9	4·0
Dichloro-phenoxyacetic acid	0·05	0·0	13·9	16·4	14·9	18·2	13·9
	0·10	0·0	13·4	25·0	14·0	53·4	10·5
	0·20	0·0	14·1	13·2	15·1	42·6	11·6
	0·40	9·9	12·9	39·7	12·6	86·1	5·4
Sig. diff. for transformed data (P=0·05):							
Between treatments		3·3		2·1		3·4	
Between concentration means		1·6		1·1		1·7	
Compound effect:							
M.C.P.A.		12·2		14·8		9·4	
v.							
D.C.P.A.		12·6		14·7		11·1	
Sig. diff. (P=0·05)		n.s.		n.s.		1·2	
Occasion effect:							
Early		11·7		15·0		10·9	
v.							
Late		13·2		13·7		9·5	
Sig. diff. (P=0·05)		1·2		n.s.		1·2	

to the soil when (*a*) the wheat had been sown a few days and (*b*) when the coleoptiles were first emerging above the soil surface.

The results are exceedingly complex. The only agreement between the reactions of the three species is a common effect of concentration. For *Alopecurus agrestis* and wheat there is no significant difference between compounds; the methyl-chloro compound is more toxic than the dichloro compounds to *Scandix pecten-veneris*. The difference between early and late spraying is dependent upon the species: for *Alopecurus* early spraying gives a higher kill than late spraying, while for *Scandix* the reverse is true;

but for wheat there is no difference. It is clear that though the grasses are far less susceptible than *Scandix pecten-veneris*, yet both compounds have caused a considerable mortality, especially at the higher concentrations. Finally, for *Alopecurus agrestis* there is a significant interaction between concentration and time of spraying, since while there is little difference for occasion with the two higher concentrations there is an advantage for the late application at the two lower concentrations.

So far the comparisons of the two growth substances have related to their effects on annuals. It was, however, first demonstrated in 1943 that the growth substances were also toxic to perennial plants (Blackman, 1945*a*), and since then data on their comparative behaviour have been accumulated for numerous perennial plants. Perhaps the most complex set of conditions governing relative toxicity is revealed by the investigations on *Lepidium draba*. After preliminary observations had been made in 1945, detailed experiments planned to extend over a number of years were laid down in 1946 to investigate the effects of concentration and time of spraying. The degree of control obtained in the year subsequent to spraying are given in Table 12 for one of the experiments. In order to simplify the issues the data have been split up to show (1) the relative effects of the two compounds at the three concentrations and (2) the interaction between compound and spraying occasion.

It is evident from the top half of the table that a rise in concentration causes a small increase in the degree of kill. The lower half of the table emphasizes that the relative toxicity of the two compounds is dependent upon occasion. Methyl-chloro-phenoxyacetic acid is more toxic when the plants are 1–4 cm. high, but this greater toxicity no longer holds either at the later spraying date or when the plants are sprayed twice. It is also seen that delaying spraying has brought about a higher kill and that double spraying is less effective than a single spraying at the optimum time.

Confirmatory evidence of this relationship between compound and occasion is given in Table 13. In this trial the plants were either sprayed once when (*a*) the flower buds could be seen, (*b*) the flowers were fully open and (*c*) shoots had regenerated in the autumn after the wheat crop had been harvested or the plants were sprayed on two occasions, i.e. *a* and *b*, *a* and *c*, *b* and *c*. Combined with these six occasions were three concentrations for the two growth substances, but since the concentration effects were small these have been omitted from the table.

It is clear that at the earliest spraying methyl-chloro-phenoxyacetic acid is more toxic than dichloro-phenoxyacetic acid, and this difference is accentuated in the autumn spraying. Yet if the plants are sprayed in full flower this position is reversed. Again with double spraying, the methyl-chloro compound is significantly more toxic when the plants are sprayed early in the spring and again in the autumn, but with the other two spraying combinations this compound difference vanishes.

So far the interdependence of the relative toxicity of the two growth

Table 12. *The interrelationship between stage of growth, compound and concentration on the toxic effects of chlorinated phenoxyacetic acids to* Lepidium draba

Percentage concentration of spray solution	Percentage kill of *Lepidium draba*					
	Methyl-chloro-phenoxyacetic acid (Na salt)		Dichloro-phenoxyacetic acid (acid)		Mean	
	% kill	Trans. data	% kill	Trans. data	% kill	Trans. data
Concentration effects:						
0·1	71·3	57·60	71·3	57·61	71·3	57·61
0·2	80·1	63·50	77·8	61·90	79·0	62·10
0·4	84·8	67·04	74·1	59·41	79·7	63·23
Mean	79·0	62·72	74·5	59·64	—	—
Occasion effects:						
Plants 1–4 cm. high	74·0	59·36	59·2	50·32	—	—
Plants 30–35 cm. high	84·6	66·87	85·9	67·95	—	—
Double spraying	77·8	61·92	76·0	60·65	—	—
Sig. diff. for transformed data (arc sin $\sqrt{\%}$ kill) (P=0·05)						
Concentration effects:						
Between concentration means			3·68			
Between compound means			5·21			
Occasion effects:						
Between treatments			5·21			

Table 13. *The relationship between stage of growth and the toxicity of chlorinated phenoxyacetic acids to* Lepidium draba

Stage of development	Percentage kill of *Lepidium draba*			
	Methyl-chloro-phenoxyacetic acid (Na salt)		Dichloro-phenoxy-acetic acid (acid)	
	% kill	Trans. data	% kill	Trans. data
Single sprayings:				
Pre-flowering	77·2	61·45	47·3	43·47
Flowering	58·4	49·84	69·2	56·30
Autumn regeneration	42·5	40·66	14·4	22·27
Double sprayings:				
Pre-flowering and flowering	74·0	59·33	69·8	56·64
Pre-flowering and regeneration	83·4	65·93	66·6	54·67
Flowering and regeneration	75·1	60·04	80·4	63·71
Sig. diff. transformed data (arc sin $\sqrt{\%}$ kill) (P=0·05):				
Between treatments		8·41		

substances and the species leads to the conclusion that methyl-chloro-phenoxyacetic acid is on the whole more toxic to *Lepidium draba* and *Scandix pecten-veneris* and the dichloro acid more injurious to *Linum usitatissimum*, while for *Alopecurus agrestis* and *Triticum* there is no difference. Other experiments have shown that this varying susceptibility holds for other species. The trend for methyl-chloro-phenoxyacetic acid to produce a higher degree of kill applies to *Sinapis arvensis*, *Raphanus raphanistrum*, *Ranunculus repens*, *R. arvensis*, and *Cirsium arvense*. On the other hand, the effects of the two compounds do not appear to be differentiated by *Polygonum convolvulus*, *Galium aparine*, *Convolvulus arvensis*,

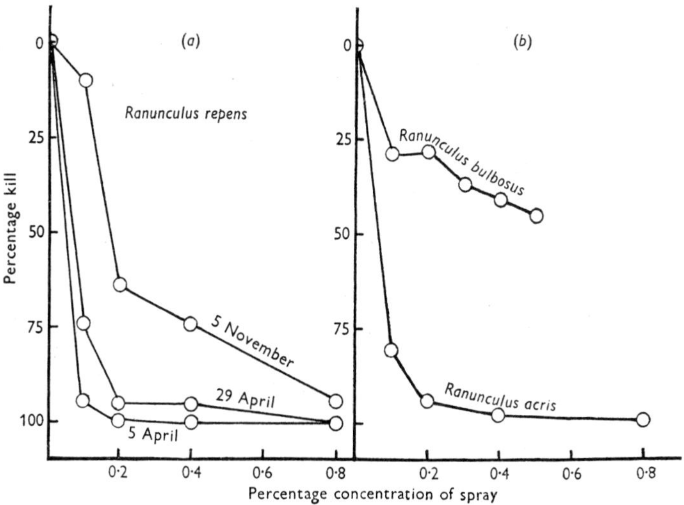

Fig. 3. The toxic effects of sodium methyl-chloro-phenoxyacetate on (*a*) *Ranunculus repens* when sprayed at different times of the year, (*b*) *R. bulbosus* and *R. acris*.

Ranunculus bulbosus or *Taraxacum officinale*. Finally, the results for *Senecio jacobaea* are the most conflicting, as in some experiments the dichloro grouping has been significantly more toxic, while in other trials the reverse has been true.

These further investigations on perennial plants have brought out other characteristics. The interaction between stage of development and toxic effect observed in *Lepidium draba* have been noted to a lesser extent in *Convolvulus arvensis* and *Cirsium arvense*. If the shoots are sprayed soon after they have emerged in the spring then the degree of kill obtained is small, but with increasing leaf area the toxic effect becomes progressively greater until a maximum is reached when flower buds are completely differentiated. Subsequently there is a tendency for the toxic action to decline when the plants are in full flower.

Table 14. *The interrelationship between concentration and the number of applications on the toxic effects of methyl-chloro-phenoxyacetic acid (sodium salt)*

No. of sprayings at 6-weekly intervals	% concentration of spray							
	0·1		0·2		0·3		Mean	
	% kill	√density	% kill	√density	% kill	√density	% kill	√density
Bellis perennis:								
Sprayed once	57·1	6·96	70·1	5·69	62·1	6·31	63·1	6·32
Sprayed twice	78·7	4·77	70·4	5·61	90·2	3·36	79·8	4·58
Sprayed thrice	82·8	4·11	98·8	1·18	98·5	1·26	93·4	2·18
Mean	72·9	5·28	79·8	4·16	83·6	3·64	—	—
Plantago media:								
Sprayed once	69·3	2·72	80·8	1·79	79·6	2·41	76·6	2·31
Sprayed twice	98·7	0·88	98·7	0·88	100·0	0·71	99·1	0·82
Sprayed thrice	98·7	0·88	100·0	0·71	97·4	1·00	98·7	0·86
Mean	88·9	1·49	93·2	1·13	92·3	1·37	—	—

Sig. diff. transformed data ($\sqrt{(x+\frac{1}{2})}$) ($P=0·05$):

Bellis perennis:

Between treatments	2·23
Between concentration or occasion means	1·29

Plantago media:

Between treatments	1·40
Between concentration or occasion means	0·81

In the perennial plants so far considered there is a well-defined growth cycle culminating in cessation of growth at flowering, but in other types of perennial, such as *Ranunculus repens*, active growth may continue throughout the season. The effects of spraying at different periods are shown in Fig. 3*a*, where it is clear that the susceptibility is greater in the spring.

A comparison of Fig. 3*a* with Fig. 3*b* also brings out the fact that even within the *Ranunculus* genus there may be very wide differences in the phytocidal effects of a growth substance. Other experiments have shown that the partial resistance of *R. bulbosus* is due to its power of recovery and regeneration since there is a marked effect of double spraying. Similar results have been obtained for other species. For example, in Table 14 it is evident that a single application of a high concentration is less toxic to both *Bellis perennis* and *Plantago media* than repeated applications of a low concentration at 6-weekly intervals.

The relative toxicity of aryl-nitro compounds

The results obtained by Harris & Hyslop (1942) that the phytocidal action of sodium dinitro-*o*-cresylate is increased by small additions of ammonium sulphate to the spray solution have been amply confirmed in the present investigations. It has, however, been found that the degree of activation is dependent upon the species. For example, it is seen in Table 15 that while the degree of kill has been greatly increased in the case of *Atriplex patula* and *Fumaria officinalis*, there is no significant effect for *Polygonum aviculare* and *P. convolvulus*.

Table 15. *The interaction between species and the activation of sodium dinitro-o-cresylate with ammonium sulphate. Exp. XIII*, 1943

Compound and % concentration		*Atriplex patula*	*Fumaria officinalis*	*Polygonum convolvulus*	*Polygonum aviculare*
Sodium dinitro-*o*-cresylate	0·3	21·0	45·5	93·8	0
	0·5	12·6	38·4	93·8	8·9
	0·7	34·3	88·4	98·4	44·7
	Mean	22·6	57·4	95·3	14·0
Sodium dinitro-*o*-cresylate plus 1·0% ammonium sulphate	0·3	65·5	98·2	98·4	31·7
	0·5	77·4	100·0	90·6	19·1
	0·7	95·5	99·1	96·9	29·7
	Mean	79·5	99·1	95·3*	26·8*

* No significant effect for synergistic action.

Since Harris & Hyslop found that activation could also be produced by sodium hydrogen sulphate, Robbins, Crafts & Raynor (1942) put forward the view that activation was due to an increase in the concentration of undissociated dinitro-*o*-cresol in the solution following on the change in pH, and involved the postulate that that dinitro-*o*-cresol penetrates more rapidly than the sodium salt through the cuticle into the plant.

On the other hand, experiments carried out since 1943 on the comparative toxicity of a suspension of dinitro-*o*-cresol together with its sodium and ammonium salts, do not suggest that the addition of ammonium sulphate acts solely through the change in the hydrogen-ion concentration of the spray solution. It was pointed out by Blackman (1945*b*) that the spray droplets of either the ammonium salt or the sodium salt, activated with ammonium sulphate, might lose ammonia to the air, and that as a result

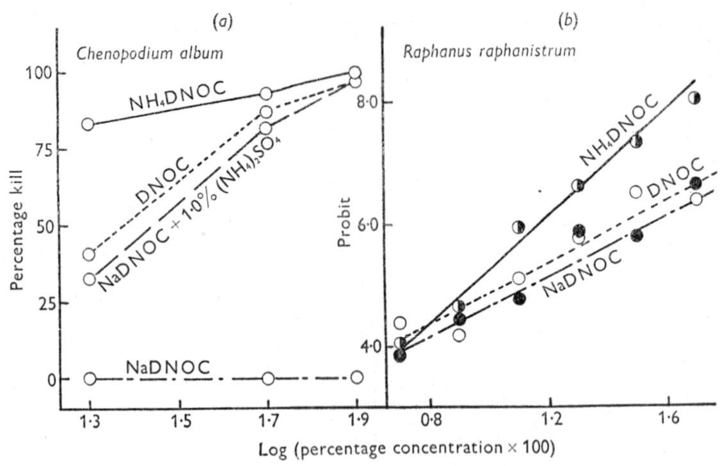

Fig. 4. (*a*) The comparative toxicity of dinitro-*o*-cresol, ammonium dinitro-*o*-cresylate and the sodium salt, with and without the addition of ammonium sulphate, to *Chenopodium album*. (*b*) Probit regression lines of the toxicity of dinitro-*o*-cresol and its ammonium and sodium salts to *Raphanus raphanistrum*.

of this loss dinitro-*o*-cresol itself would be deposited on the leaf surface. If this is so then the activated spray or the ammonium salt should equal but not exceed in toxicity a suspension of dinitro-*o*-cresol. It is, however, clear from Fig. 4*a* that while both the activated sodium salt and dinitro-*o*-cresol are greatly superior to the sodium salt, they in turn are less toxic than the ammonium salt.

This greater activity of the ammonium salt is also evident in Fig. 4*b*, where the experiment was designed to allow of the use of probit analysis (this type of experiment is discussed later). Taking a level of 50% kill the ammonium salt is $1{\cdot}55 \pm 0{\cdot}16$ times as toxic as the sodium salt to *Raphanus raphanistrum*. On the other hand, it is evident from Fig. 5 that the relative toxicity may be dependent upon the species. In the case of *Chrysanthemum segetum*, the suspension of dinitro-*o*-cresol is significantly more active than either the sodium or ammonium salt.

Further experiments have been concerned with a comparison of other salts of dinitro-*o*-cresol, and some of the results are shown in Table 16. Once more it is apparent that the reaction of individual species is an important factor. In the case of *Capsella bursa-pastoris* the potassium and copper salts are significantly less toxic than the other compounds, while for *Urtica urens* the calcium salt is also significantly less toxic. With these two species there are no significant differences between dinitro-*o*-cresol and the ammonium and sodium salts. But only dinitro-*o*-cresol has caused any injury to *Linum usitatissimum*.

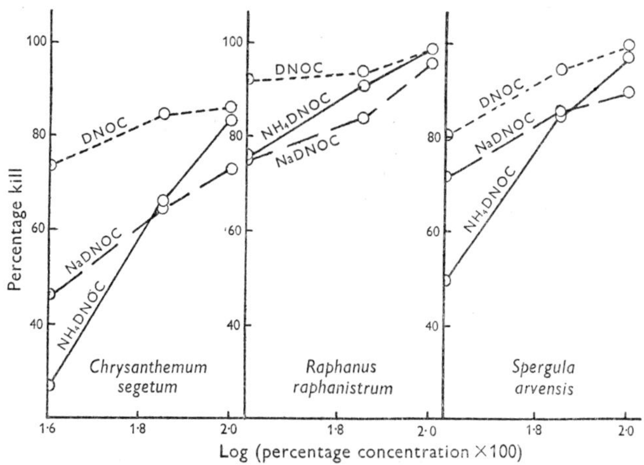

Fig. 5. The interrelationship between species and the relative toxicity of dinitro-*o*-cresol and its ammonium and sodium salts. Only in the case of *Chrysanthemum segetum* is dinitro-*o*-cresol significantly more toxic than either salt.

As the ammonium salt is on occasion more toxic than dinitro-*o*-cresol, it was thought that there might be some linkage between the ammonia and the proteins in the surface cells of the leaf and that this could lead to a greater absorption. Since such a linkage is greater for amines, amine salts should be more active, so a number of experiments have been carried out with ethanolamine salts. In one experiment dinitro-cresol was compared with the ammonium and ethanolamine salts, and the mean percentage kills for the three compounds were 57, 39 and 5 respectively for *Atriplex patula* and 68, 65 and 78 for *Matricaria inodora*. Again, in another trial with *Chenopodium album* where dinitro-*o*-cresol gave a 99% control, the figures for the sodium, mono-ethanolamine and tri-ethanolamine salts were 48, 32 and 55. Lastly, a comparison of the sodium salt against the mono-, di- and tri-ethanol-amine salts resulted in very similar kills of *Papaver rhoeas*. Thus the evidence does not suggest that linkage with the leaf proteins is an important factor.

Table 16. *The interrelationship between species and the toxic effects of dinitro-o-cresol and its salts*

Compound and % concentration of spray		*Urtica urens*		*Capsella bursa-pastoris*		*Linum usitatissimum* plant weight (cwt./acre)
		% kill	√density	% kill	√density	
Dinitro-*o*-cresol	0·35	80·6	5·4	84·9	3·3	52·4
	0·70	92·5	3·3	97·6	1·4	30·1
	Mean	—	4·4	—	2·4	41·2
Ammonium* dinitro-*o*-cresylate	0·35	18·4	10·7	84·9	3·3	56·5
	0·70	95·0	2·6	93·9	2·0	54·1
	Mean	—	6·6	—	2·6	55·3
Sodium dinitro-*o*-cresylate	0·35	77·2	5·8	88·2	2·9	56·8
	0·70	80·4	4·5	97·6	1·4	53·5
	Mean	—	5·1	—	2·2	55·2
Calcium dinitro-*o*-cresylate	0·35	49·8	8·8	83·1	3·3	55·3
	0·70	55·0	7·8	88·2	2·7	58·3
	Mean	—	8·3	—	3·0	56·8
Potassium dinitro-*o*-cresylate	0·35	56·9	8·0	65·2	4·9	50·3
	0·70	53·5	8·1	88·2	2·6	45·5
	Mean	—	8·1	—	3·8	47·9
Copper dinitro-*o*-cresylate	0·35	0·0	12·5	67·1	4·7	57·7
	0·70	46·4	8·3	81·6	3·6	59·2
	Mean	—	10·4	—	4·2	58·4
Sig. diff. (P=0·05):						
Between treatments			3·4		1·8	10·4
Between compounds			2·4		1·3	7·4

* Equivalent dinitro-*o*-cresol concentrations for all salts.

Although the initial field trials involving comparisons of different compounds, each at two or three concentrations, were sufficiently accurate to establish broad differences in relative toxicity, a more accurate technique is called for when a precise analysis of smaller differences is required. Such a need has arisen in the more recent investigations on the relationship between the chemical structure of aryl-nitro compounds and their toxicity.

Much attention has been given by statisticians to the problem of determining the relative toxicity of insecticides, fungicides and bactericides. In investigations of this type it has been demonstrated that the curve relating percentage kill to the logarithm of the concentration of toxicant is of sigmoid form. The technique of probit analysis has been devised to facilitate the treatment of such data, and a monograph on this subject has recently appeared (Finney, 1947).

It has been established that such methods are applicable to the data from suitably designed field experiments with herbicides, and they are of particular value where an accurate estimate of the relative toxicity of chemically related compounds is required. Briefly the method involves that the percentage kill data for the two compounds are transformed to probits, weighted linear regression lines fitted, and from the calculated values of the concentrations required to produce a 50% kill, an estimate of the relative potency is obtained.

By these methods a comparison has been made of the relative potency of 3:5 dinitro-*o*-cresol and 2:4-dinitro-6-secondary-butyl-phenol. According to Crafts (1945) the secondary butyl compound is 1·56 times more toxic than the methyl compound, but it is evident from Fig. 6 that the ratio may be dependent upon the species. Whereas for *Scandix pecten-veneris*,

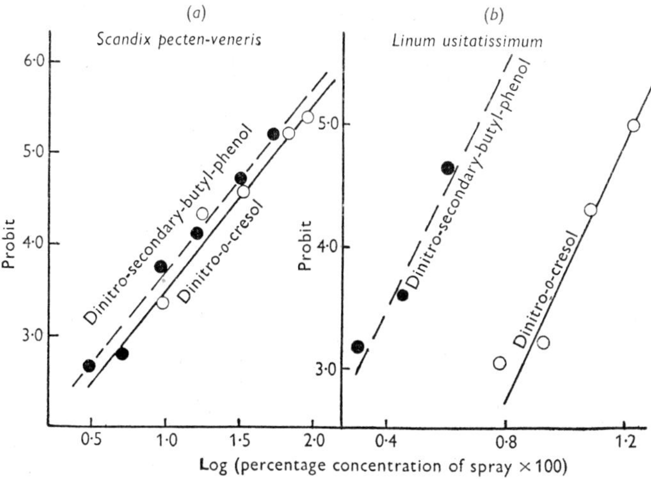

Fig. 6. Probit regression lines of the toxicity of 3:5-dinitro-*o*-cresol and 2:4-dinitro-6-secondary-butyl-phenol to (*a*) *Scandix pecten-veneris*, (*b*) *Linum usitatissimum*.

sprayed in the 5–8 leaf stage, the substitution of a secondary butyl group has increased the toxicity by 30% (1·30 ± 0·21), with *Linum usitatissimum* (sprayed in the greenhouse in the two-leaf stage) the factor is 3·42 ± 0·26. For other species in which the ammonium salts have been compared a value of 2·62 ± 0·57 has been obtained for *Sinapis arvensis* and an approximate figure of 4·6 for *Raphanus raphanistrum*.

From these results, it is therefore evident that an accurate assessment of relative toxicity between compounds cannot be restricted to observations on a single species. In fact, although the results require confirmation, the indications are that in the case of the field pea, the ammonium salt of dinitro-*o*-cresol is more toxic than ammonium dinitro-*o*-butyl-phenate; this is in agreement with American claims (Grigsby & Barrons, 1945; Barrons & Grigsby, 1945).

Finally, in an experiment on barley, the secondary butyl compound at 0·7% has given the same degree of injury (as measured by the depression in the yield of grain) as 1·0% dinitro-*o*-cresol.

The relative toxicity of copper salts

Apart from the data on the effects of cupric chloride and nitrate on flax in Table 5, no evidence has been presented on the variation in toxicity between different copper salts. The results set out in Table 17 give the general trend which has been found in numerous experiments. It is seen that the greatest injury to *Linum usitatissimum* is brought about by cupric chloride and cupric nitrate, while no significant reduction has followed spraying with the cuprammonium chloride.

Table 17. *Comparative effects of copper salts on* Linum usitatissimum. *Exp. XV*, 1944

Copper salt	Total weight of plants (cwt./acre) % concentration of spray*		
	2·0	5·0	Mean
Control			32·1
Cupric chloride	27·4	14·4	20·9
Cupric nitrate	26·7	18·0	22·3
Cupric sulphate	29·5	22·3	25·9
Cupric acetate	32·4	20·9	26·7
Cupric ammonium chloride	28·8	22·3	25·6
Cuprammonium chloride	33·9	32·4	33·2
Significant difference ($P=0·05$):			
Between treatment means		4·3	
Between control and compound means		3·1	

* Salt concentrations on equivalent copper basis to 2·0 and 5·0% cupric chloride.

Table 18. *The comparative effects of cupric chloride and cupric sulphate with and without additions of sodium chloride. Exps. V and VI*, 1943

	Percentage kill of *Sinapis arvensis*			
	Experiment V		Experiment VI	
	% concentration	% kill	% concentration	% kill
Cupric chloride	1·0	37	1·0	95
	2·0	64	3·0	100
Cupric sulphate	2·0	25	1·0	38
	5·0	15	3·0	45
Cupric sulphate + sodium chloride	1·0+0·7	17	1·0+0·7	44
	2·0+1·4	21	3·0+2·1	56

The disparity between the effects of the cupric chloride, the double salt and the cuprammonium complex suggests that the toxicity is not solely a matter of the presence of copper and chloride ions in the spray solution. This assumption is borne out by the divergent behaviour of cupric chloride and mixtures of cupric sulphate and sodium chloride. The results given in Table 18 show that the kills produced by the mixture and cupric sulphate are similar, while cupric chloride is very much more toxic.

VI. DISCUSSION

Though this paper has been restricted to an examination of only four groups of phytocidal substances, yet the results bring out the extreme complexity of the problems involved in a comparison of relative toxicity. It is, however, possible to distinguish some of the factors that must be taken into account in studies of phytocidal action. First, there is clear evidence that even for broad comparisons between groups of compounds no true assessment can be expected unless a wide range of species is examined. For example, erroneous conclusions would have been drawn as to the relative toxic effects of sulphuric acid, cupric chloride, dinitro-*o*-cresol and methyl-chloro-phenoxyacetic acid on dicotyledonous annuals in the seedling phase if the comparison had been restricted to the data in any one of Tables 1, 2 or 3.

Secondly, it has been established that precise comparisons cannot be achieved unless the stage of development is taken into account. It is not sufficient to ensure that the range of test plants is sprayed at a single defined stage of development, as the relationship between the stage of development and the toxic effect is dependent upon both the species and the compound. Thus while for some, but not all, dicotyledonous annuals the susceptibility to sulphuric acid decreases sharply once the early seedling stage is passed, with dinitro-*o*-cresol this decrease is less pronounced while it may be wholly absent with copper chloride. Moreover, with sodium methyl-chloro-phenoxyacetic acid, the change from susceptibility to resistance may be indefinite or it may show a critical change at varying stages of development (see Fig. 2*b*). In consequence, as Tables 6, 7 and 8 show, relative toxicity between groups may be reorientated by a change in spraying occasion.

In species which are largely tolerant to the four compounds, such as the annual grasses, comparisons between compounds will still be vitiated unless the stage of development is taken into account. It is true that there is a common trend for the different compounds since least tolerance occurs either very early in the seedling stage or again when the plants are approaching maturity, but there is nevertheless an interaction between species and compound. Methyl-chloro-phenoxyacetic acid is particularly toxic at the time of germination, but the toxicity is dependent upon the species. Thus while in the germination phase methyl-chloro-phenoxyacetic acid is far more toxic to *Dactylis* than wheat, there is no appreciable difference once the plants have become established and resistant.

The same complexities of species and stage of development arise even when closely allied compounds are being examined. It has been demonstrated that the relative toxicity of dinitro-*o*-cresol and its salts are dependent on the species (Figs. 4 and 5 and Table 16). It has been shown that the ratio of the potency of ammonium dinitro-secondary-butyl-phenate to ammonium dinitro-*o*-cresylate ranges from 1·3 to 4·6 according to the species. The experiments with *Linum* have revealed that the interdependence of compound, concentration and stage of development may affect seed production differentially from fibre production. For example, dinitro-*o*-cresol and its sodium salt may both have no effect on seed formation, yet the recovery of fibre is far less in the case of the acid and the greatest depression is at the higher concentrations (Table 5). Moreover, it is evident from Table 8 and other investigations (not yet reported) that the internal injury as measured by fibre production is a more sensitive measure of the phytocidal effects than plant or seed weight. In consequence the difference in toxicity between compounds may have wholly different values when judged on the criterion of fibre production.

Finally, the phytocidal effects of the phenoxyacetic acid derivatives on perennial plants reveal once more the importance of the species factor and its interaction with stage of development. In the case of *Lepidium draba* the relative toxicity of methyl-chloro- and dichloro-phenoxyacetic acid can have a whole range of values dependent upon the timing of the spray applications (Tables 12 and 13). In other species, with a less well-defined seasonal period of growth, though there may be differences between compounds the interaction with occasion may be lacking.

It is therefore evident that determinations of relative toxicity are liable to serious error if (*a*) they are restricted to a single species of test plant, (*b*) they are carried out at a single stage of development, (*c*) only a single concentration of each compound is employed. Many of the 'sorting tests' described by other workers are not free from these objections, and in consequence they are only capable of assessing large differences in toxicity and eliminating relatively inert substances. The present experiments have demonstrated the essential advantages of using a range of concentrations so that methods of probit analysis may be used to determine relative toxicity. Moreover, other investigations in progress have shown that no real proof of synergistic action can be forthcoming without a proper design of the experiment which will allow of probit analysis of the data.

Apart from a consideration of the factors involved in the assessment of comparative toxicity, there still remain the fundamental problems of the differential toxicity of individual compounds. In the case of sulphuric acid, it has already been pointed out that the reaction of different species can be related to habit of growth, the position of the meristematic tissues in the seedling stage, and the nature of the epidermis and cuticle. The question arises as to how far such factors are of importance for the remaining groups of compounds.

With the copper compounds one of the features observed is that while the most susceptible plants are those which have a thin cuticle and a relatively undifferentiated epidermis, resistant plants have a waxy cuticle (*Chrysanthemum segetum*) or epidermal hairs (*Chenopodium album*, *Papaver rhoeas*). From this there are grounds for considering that differential toxicity may be a question of the ease with which copper penetrates the leaf tissues. Support for this view is forthcoming from the relative toxicity of cupric chloride and nitrate as against cupric sulphate. The first two salts are strongly hygroscopic, and under average conditions the spray droplets do not completely dry out on the leaf surfaces whereas crystals of cupric sulphate can generally be seen. Thus the area of surface contact of the deliquescent copper salts is greater and should allow of a more rapid penetration. Such an assumption is in line with Åslander's (1927) observations that the phytocidal action of ferrous sulphate is increased at high humidities.

If the spray solution is washed off the shoots after varying periods of contact, then it has been found that while 2·0% cupric nitrate gives a complete kill of *Sinapis arvensis* at the end of 3 hr., yet cupric sulphate at 5·0% only causes a 95% mortality at the end of 80 hr. These differences could be explained on the suggested basis of a greater rate of penetration of the copper from the cupric nitrate due to the larger area of surface contact. On the other hand, such an explanation leaves out of account the possible influence of the acid radicle. Other evidence, admittedly of a preliminary nature, suggests that the acid radicle is also involved. For example, the concentration of copper in solution required to bring about a 50% suppression of germination in sand culture is dependent upon the associated anion, i.e. cupric chloride and nitrate are more toxic than cupric sulphate. Again, with different species widely different concentrations of these copper salts are needed to produce the same degree of suppression. Thus the relation between the varying susceptibility of species to different copper salts may not only involve morphological differences in the cuticle and epidermis but also the physical and physiological factors governing rate of penetration and the degree of tolerance within the plant.

It has already been pointed out that the comparative effects of dinitro-*o*-cresol and its ammonium and sodium salts do not fall into line with the postulate put forward by Robbins *et al.* (1942) that the only role of ammonium sulphate in 'activating' the sodium salt is to lower the pH and thereby increase the concentration of undissociated dinitro-*o*-cresol. Moreover, it has been found by us, and is clear from the data of Crafts & Reiber (1945), that additions of ammonium sulphate which cause only a small change in pH, and in consequence only a very slight increase in the concentration of undissociated dinitro-*o*-cresol, may bring about large increases in toxicity. Such large differences can only be accounted for by the assumption that the acid is very many times as toxic as the sodium salt, but the data of Table 16 and Fig. 5 indicate that for several species the relative toxicity of a suspension of dinitro-*o*-cresol and the sodium salt is of a much lower order.

At present no explanation can be put forward to account for the ammonium salt exceeding in toxicity the acid itself (see Fig. 4). On the other hand, the greater toxicity of dinitro-secondary-butyl-phenol over dinitro-*o*-cresol (Fig. 6) is in line with Badger's (1946) conclusions that in a wide range of biological systems the activity of homologous compounds rises as the number of carbon atoms increases up to a variable maximum.

In the discussion of the factors governing the varying susceptibility of species to sulphuric acid and cupric chloride it has been pointed out that the presence of a well-defined waxy cuticle or epidermal hairs appears to be associated with resistance. With dinitro-*o*-cresol, variation in epidermal structure does not seem to be a differentiating factor. Both plants like *Chrysanthemum segetum* and *Fumaria officinalis* with waxy leaves, and species like *Chenopodium album* and *Papaver rhoeas* with an abundance of epidermal hairs, are all readily killed.

In the case of the phenoxyacetic acids the evidence also suggests that differences in morphology are of less importance than metabolic and physiological factors in determining the individual reactions of different species. Associated with the phytocidal effects, species may exhibit accelerated cell division and differentiation. Species such as *Galeopsis tetrahit* may sprout adventitious roots the whole length of the stem and yet recover, while with *Ranunculus arvensis*, for example, cell multiplication in the basal ends of the petioles may be so great that the epidermis splits. Epinasty is another common feature of the action of these compounds. However, in some species these additional effects, particularly those on cell division, are absent though the plants may die.

Other European workers (Lhoste & Ravault, 1946; Edelberg & Thorup, 1947; Aberg & Denward, 1947) have reported that in cereals the phenoxyacetic acids may cause an inrolling of the leaf sheaths, distortion of the rachis, and malformation of the ears. In the present experiments such physiological disturbances have been singularly absent. To account for this discrepancy, the explanations that can be advanced are, first, that varietal differences may be important, and secondly, that these effects on the inflorescence are correlated with higher temperatures than those usual in England. In the rare instances where depressions in yield have been found, the concentrations of the spray have exceeded 0·2%, and the effect has been primarily to reduce the number of fertile shoots; some results for barley are seen in Table 19.

Table 19. *The effects of phenoxyacetic acids on the development of barley. Exp. XIX*, 1943

Compound and % spray concentration		Tillers per foot	Fertile shoots per foot	Grains per ear	1000 seed (wt.-g.)
Control		23·3	23·0	16·6	39·88
Sodium methyl-chloro-phenoxy-	0·27	25·2	24·7	15·8	42·66
acetate	0·54	19·7	18·6	15·2	40·44
Sodium dichloro-phenoxy-	0·38	21·2	20·7	15·8	41·05
acetate	0·81	19·0	18·4	15·8	40·97
Sig. diff. (P=0·05):					
Between treatments		3·0	3·1	1·5	1·93
Between treatments and control		2·5	2·5	1·2	1·58

In these investigations it has been found that the chlorinated phenoxyacetic acids are most toxic to the grasses in the germination phase, and the question arises: What are the changes which alter the level of tolerance between germination and the seedling stage? It is apparent from Fig. 2*b* that this transition from susceptibility to resistance is shared by many dicotyledonous annuals. Although there is no critical period in the growth cycle when this change-over takes place, nevertheless toxicity tends to be maximal in the earliest stages when cell differentiation is active and the metabolic rates are high.

No comprehensive theory of the mode of toxic action of synthetic growth substances has yet been postulated. Smith, Hamner & Carlson (1947) demonstrated in *Convolvulus arvensis* that after treatment with 2:4-dichloro-phenoxyacetic acid the sugar content of the leaves rises rapidly to a peak and then slowly decreases. Simultaneously, there is in the rhizomes a slow fall in the starch content and a slow rise in the sugar content. Associated with these changes the respiration rate increases. These authors, however, point out that the reduction in carbohydrates is not large enough, nor does it take place with sufficient rapidity to account for the death of the plant.

It is clear that whatever physiological factors are involved the position is highly complex, since in species which become resistant with age, such as *Dactylis*, there must be metabolic and physiological changes between germination and the seedling stage which render the mechanism of toxic action inoperative. Again, if respiratory complexes or other types of enzyme systems are involved, such systems must be specific to families and even to species within one genus (e.g. *Ranunculus*, Fig. 3). Thus until far more is known of the metabolic and physiological differences between species, only limited progress can be made.

The studies on perennial plants have also demonstrated that especially in species with well-developed rhizome systems (*Lepidium*, *Cirsium*, *Convolvulus*) the phytocidal effects are maximal when the shoots are approaching the flowering stage and minimal when the shoots are emerging through the ground. A number of factors may operate to bring about this optimum. In the first place the leaf area will reach a peak at the pre-flowering period, and the greatest amount of material may therefore enter the shoot by spraying at this time. Clearly to kill the rhizome there must be translocation downwards, and the efficiency of the destruction of the roots will be dependent upon the amount of growth substance translocated and the extent to which it penetrates in toxic concentration. Thus as a result of spraying late there will be a larger amount of growth substance in the shoot which could be translocated if the mechanism was active. To what extent the stage of growth controls translocation of phenoxyacetic acids is not known. However, if the growth substances tend to concentrate in the meristems of the shoots, a delay in spraying the shoot until differentiation is more or less complete may reduce the amount 'fixed' by the shoot.

In conclusion, while this paper has been restricted to a discussion of the interrelationship between species and the comparative toxicity of four groups of compounds, investigations on other groups, such as the chlorinated phenols or the alkyl-phenyl carbamates, have shown an equal diversity of results according to the species. In fact, until a great deal more has been learnt the reactions of a species cannot be foretold, since it is not at present possible to distinguish the exception from the rule.

Finally, the authors wish to acknowledge grants from the Agricultural Research Council in aid of these investigations. They are also indebted to the major contributions that have been made to these researches for varying periods since 1942 by A. J. Rutter, H. Wild, J. Carpenter, J. B. Goodey, C. P. Whittingham, E. G. B. Gooding, J. L. Crosbie, P. A. Tallantire, P. Greig-Smith, C. Gimingham, A. C. Crundwell, G. W. Ivens, E. W. Simon and E. S. Bunting.

REFERENCES

ABERG, E. & DENWARD, T. (1947). *Ann. Roy. Agric. Coll. Sweden*, **14**, 366.
ÅSLANDER, A. (1927). *J. Agric. Res.* **34**, 1065.
BADGER, G. M. (1946). *Nature, Lond.*, **158**, 585.
BARRONS, K. C. & GRIGSBY, B. H. (1945). *Quart. Bull. Mich. Agric. Exp. Sta.* **28**, 145.
BLACKMAN, G. E. (1934). *Emp. J. Exp. Agric.* **2**, 213.
BLACKMAN, G. E. (1942). *Ann. Appl. Biol.* **29**, 204.
BLACKMAN, G. E. (1945*a*). *Nature, Lond.*, **155**, 500.
BLACKMAN, G. E. (1945*b*). *J. R. Agric. Soc.* **106**, 137.
BLACKMAN, G. E. (1946). *J. Minist. Agric.* **53**, 16.
BLACKMAN, G. E. & TEMPLEMAN, W. G. (1936). *J. Agric. Sci.* **26**, 368.
BLISS, C. I. (1937). *Bull. Pl. Prot. Leningrad*, **12**, 67.
CHABROLIN, C. (1940). *C.R. Acad. Sci., Paris*, **210**, 262.
CHABROLIN, C. (1942). *C.R. Acad. Agric. Fr.* **28**, 625.
CRAFTS, A. S. (1945). *Science*, **101**, 417.
CRAFTS, A. S. & REIBER, H. G. (1945). *Hilgardia*, **16**, 487.
EDELBERG, L. & THORUP, S. (1947). *Ugeskr. Landm.* **92**, 67.
FINNEY, D. J. (1947). *Probit Analysis.* Cambridge University Press.
GILBERT, F. A. (1946). *Chem. Rev.* **39**, 199.
GRIGSBY, B. H. & BARRONS, K. C. (1945). *Quart. Bull. Mich. Agric. Exp. Sta.* **27**, 301.
HAMNER, C. L. & TUKEY, H. B. (1944). *Science*, **100**, 154.
HARRIS, L. E. & HYSLOP, G. R. (1942). *Bull. Ore. Agric. Exp. Sta.* no. 403.
LHOSTE, J. & RAVAULT, L. (1946). *C.R. Acad. Agric. Fr.* **32**, 572.
MITCHELL, J. W. & HAMNER, C. L. (1944). *Bot. Gaz.* **105**, 474.
NORMAN, A. G. (1946). *Bot. Gaz.* **107**, 475.
NUTMAN, P. S., THORNTON, H. G. & QUASTEL, J. H. (1945). *Nature, Lond.*, **155**, 498.
ROBBINS, W. W., CRAFTS, A. S. & RAYNOR, R. N. (1942). *Weed Control.* New York: McGraw-Hill.
SLADE, R. E., TEMPLEMAN, W. G. & SEXTON, W. A. (1945). *Nature, Lond.*, **155**, 497.
SMITH, F. G., HAMNER, C. L. & CARLSON, R. F. (1947). *Plant Physiol.* **22**, 58.
TEMPLEMAN, W. G. & SEXTON, W. A. (1945). *Nature, Lond.*, **156**, 630.
TEMPLEMAN, W. G. & SEXTON, W. A. (1946). *Proc. Roy. Soc.* B, **133**, 300.
TRUFFAUT, G. & PASTAC, I. (1932). French Patent 425,295 (British Patent 15446/33).
TRUFFAUT, G. & PASTAC, I. (1944). *Chim. et Industr.* **51**, 79.
WESTGATE, W. A. & RAYNOR, R. N. (1940). *Bull. Calif. Agric. Exp. Sta.* no. 634.
ZIMMERMAN, P. W. & HITCHCOCK, A. E. (1942). *Contr. Boyce Thompson Inst.* **12**, 321.

THERAPEUTIC INTERFERENCE: SOME INTERPRETATIONS

By ADRIEN ALBERT
Wellcome Research Laboratories, London

Many examples of selective toxicity have been traced to interference phenomena in which the toxic agent competes against a natural metabolite for a vitally important receptor. However, the phenomenon known as *therapeutic interference* is in a different category because it concerns the action of a foreign substance in preventing the normal curative action of a drug. It has become customary to confine the use of the term to cases in which there is no obvious chemical reaction between the two substances. For example, any loss in efficacy noted after giving mercury plus a soluble sulphide, or an organic arsenical plus BAL, would not be described as therapeutic interference. On the contrary, the commonest and most puzzling examples of therapeutic interference are those in which the antagonists have such great chemical similarity to one another that it has been thought impossible that they could react together.

The name *therapeutic interference* was coined by Browning & Gulbransen (1922) to describe their discovery that mice suffering from trypanosomiasis could no longer be cured by the usual injected dose of neutral acriflavine (I) if previously fed on parafuchsin (II*a*). Actually there were some complicating factors in this experiment, because parafuchsin is itself faintly trypanocidal and hence a parafuchsin-fast strain of trypanosomes had been used. However, it was soon found that parafuchsin interfered so powerfully that it was possible to demonstrate its effect even on a parafuchsin-susceptible strain by using a sufficiently small dose (Schnitzer, 1926). The following experimental details illustrate this point.

Mice were infected with the normal strain of *Trypanosomiasis brucei*. Those which were untreated, or merely injected with 0·05–0·25 mg. (per 20 g. mouse) of parafuchsin, all died by the 6th or 7th day. Those which were injected with 0·5 mg. of neutral acriflavine lost all parasites on the 3rd day and remained well. Those which were injected with 0·5 mg. of acriflavine *plus* 0·05–0·25 mg. of parafuchsin all died on the 6th or 7th day. These dramatic results were obtained when the parafuchsin was injected 4 hr. before the acriflavine, but even when both substances were injected together the effect was substantially the same.

Hence 1 part of parafuchsin can annul the action of 10 parts of acriflavine. Other triphenylmethane dyes (e.g. methyl violet (II*b*)) have the property of interfering with acriflavine in this way, although brilliant green

(II*c*) lacks it (Schnitzer & Silberstein, 1926; Browning & Gulbransen, 1927). Although the triphenylmethanes are trypanocidal in higher doses, they do not interfere with the action of one another.

Is the interfering action of parafuchsin exerted on the parasite or on the host? Evidently it is on the parasite, for the interference phenomenon is demonstrable *in vitro*. When trypanosomes were taken from mice, previously injected with parafuchsin, and mixed with that quantity of acriflavine which normally was sufficient to kill them rapidly *in vitro*, they remained alive (v. Jancsó, 1931).

It has also been shown that parafuchsin, methyl violet and ethyl violet, all of which cause interference, actually prevent the parasite from taking up acriflavine, whereas brilliant green, which does not cause interference, does not prevent this uptake (Hasskó, 1935).

Trypanosome-infected animals were injected with a triphenylmethane dye, and then, 1 hr. later, with neutral acriflavine. Trypanosomes were then withdrawn, and it was found that they had taken up only a fraction as much of the acriflavine as had trypanosomes treated with acriflavine only. For example, trypanosomes which had taken up 0·06 % of methyl violet were able to take up subsequently only 0·21 % acriflavine instead of the normal 0·8 % (estimated colorimetrically and calculated on the dry weight).

It has also been shown that the interference phenomenon is reflected in the metabolism of the parasite (Scheff & Hasskó, 1936). Rats, infected with trypanosomes, were injected with parafuchsin and/or neutral acriflavine. Parasites were withdrawn and observed in a respirometer. It was found that acriflavine considerably depressed the sugar consumption and carbon dioxide output of the parasites, and that parafuchsin had a slight adverse effect. However, when the two drugs were administered to the same rat, the metabolism of the trypanosomes withdrawn showed no inactivation whatsoever (see Table 1).

In all the experiments described so far, the issue is clouded by the fact that one or both substances were injected into the host before the parasites were withdrawn. It would be most valuable to demonstrate that interference occurs when *both* substances are applied in turn, or together, to the parasite *in vitro*. This does not seem to have been tried with trypanosomes, although it has been demonstrated on bacteria and yeast cells (*vide infra*).

Let us now see what explanations are available for this interference phenomenon. Browning & Gulbransen (1922) originally suggested that parafuchsin blocked the receptors in trypanosomes that normally take up acriflavine, but later (1927) decided to reserve any explanation. Schnitzer (1926) suggested that parafuchsin and acriflavine are sufficiently similar, chemically, to be bound by common receptors on the parasite. When these receptors are blocked by parafuchsin they are no longer free to combine

Table 1. *Mutual interference by parafuchsin and acriflavine with their normal toxic action on the metabolism of trypanosomes*

(From Scheff & Hasskó, 1936.)

Rat			Previous injections	No. of trypanosomes* per 2·5 ml. portion of test fluid (serum + Locke)	O_2 consumption per hour per 10^6 trypanosomes		Glucose consumption after $5\frac{1}{2}$ hr. (total) (μg.)
No.	Weight (g.)	Ratio of trypanosomes to red blood cells			After 1 hr.	After $5\frac{1}{2}$ hr. (total)	
I	186	1:5·2	None (control)	29,700	13·09	29·95	20·00
II	166	1:20	Parafuchsin (4·9 mg.)	33,700	10·74	22·99	11·59
III	219	1:11	Neutral acriflavine (6·5 mg.)	34,500	10·52	20·48	7·99
IV	219	1:14	Parafuchsin (5·6 mg.) *plus*, after 1 hr., neutral acriflavine (6·6 mg.)	29,500	11·88	26·42	20·42

* Fractionally centrifuged free of blood corpuscles.

with acriflavine. Hasskó (1936) suggested that the triphenylmethane dyes which cause interference do so by saturating the protoplasm of the cell so that its capability for saturation by acriflavine is diminished.

These explanations are substantially the same; let us examine their implications before putting new hypotheses forward. If the triphenylmethanes and acriflavine are both taken up by the same receptor (doubtless an anionic group) on the parasite, these substances will be in free competition with one another, obeying the law of mass action. In so far as the molecular weights of methyl violet and acriflavine are similar and, as Hasskó says, trypanosomes have only a small affinity for triphenylmethane dyes compared with their affinity for acriflavine, it is difficult to see how the uptake of a little methyl violet by the parasites could exclude ten times its weight of acriflavine from being bound by them (*vide supra*). These considerations would seem to dispose of the hypothesis that both drugs combine with the same receptor.

However, before dismissing the hypothesis, it is important to point out what highly unsatisfactory substances acriflavine and the triphenylmethane dyes are for biological experiments. Acriflavine has not been commercially available as a pure substance but always contains about one-third of its weight of the corresponding unmethylated amine (proflavine). None of the authors has stated that they purified the commercial product before using it. Proflavine has little trypanocidal action and may be considered as a harmless diluent, but its presence may complicate colorimetric analysis as it absorbs at the same wave-lengths as acriflavine. The triphenylmethane dyes present a bigger problem because, in solution, they are in equilibrium with their feebly coloured carbinols (e.g. (III) for parafuchsin). Hence it must be considered whether the competition for a receptor is between (I) and (III) rather than (I) and (II). This is not very likely because (III), lacking the high degree of ionic resonance possessed by (II), must be a tetrahedral molecule. Thus it differs radically in shape from flat molecules such as (I) and (II) and also lacks their positive charge. Another peculiar property of the triphenylmethanes (namely, that the carbinol readily forms colourless ethers in cold alcohol) can complicate analytical procedures. For instance, in Hasskó's experiment, the dye was stripped from the parasite with absolute alcohol and the depth of red or violet colour in the extract was determined colorimetrically after making weakly acid with acetic acid. It is quite possible that, under these circumstances, too low an estimate of the amount of dye present would be made. It is therefore possible that the amounts of triphenylmethane dyes bound by the parasites were larger than Hasskó realized. If that proves to be so, the common-receptor hypothesis will have to be reconsidered.

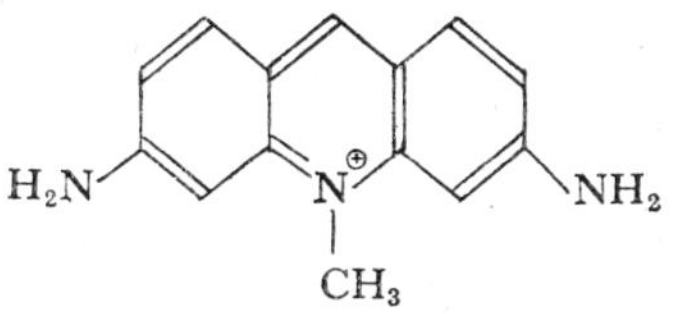

H_2N $N^{\oplus}$ NH_2
CH_3

(I)

Neutral acriflavine (synonyms: trypaflavine, euflavine)

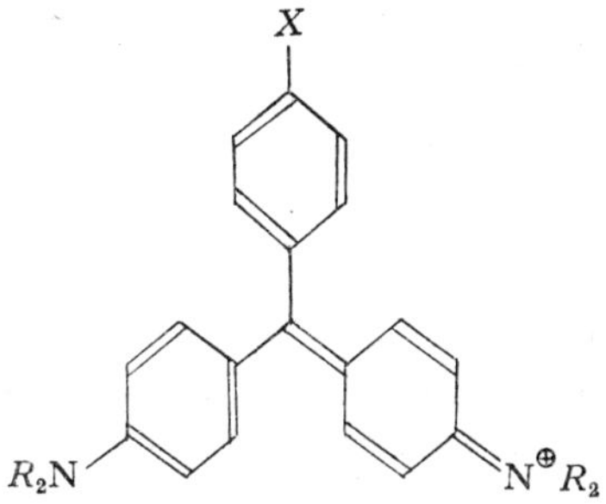

X

R_2N $N^{\oplus}R_2$

(II)

Triphenylmethane dyes. (*a*) Parafuchsin, R=H, X=NH_2. (*b*) Methyl violet, R=CH_3; X=NH.CH_3. (*c*) Brilliant green, R=C_2H_5; X=H.

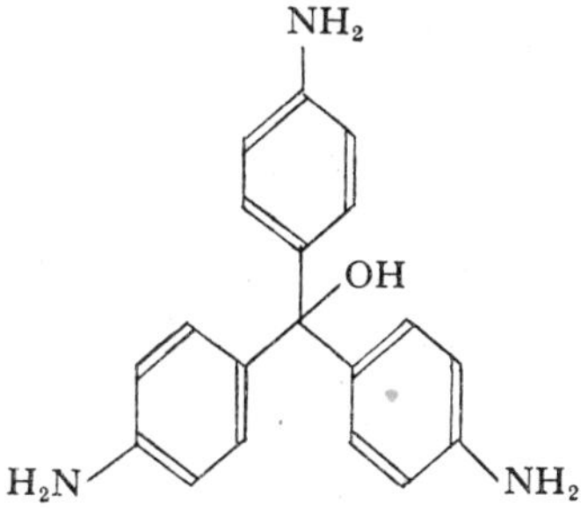

NH_2

OH

H_2N NH_2

(III)

Feebly coloured carbinol of parafuchsin

An alternative hypothesis will now be considered, namely, that parafuchsin and acriflavine are each taken up by a different kind of receptor and that both kinds of receptor occur close together. As parafuchsin is comparatively inert, it would appear that it is taken up by a receptor which has no vital part to play in the parasite's metabolism. The inference is that parafuchsin interferes with the action of acriflavine by sterically obstructing it from combining with a receptor which it is vitally important for the parasite to have free. Given that the two receptors in question are neighbours, the parafuchsin ion (or carbinol) could, by its very bulk, prevent the approach of an acriflavine ion to the vital receptor. On the other hand, the approach of a smaller substrate would not be barred, and hence normal metabolism could go on.

This hypothesis is a new one, except in so far as Hasskó's vague generalization (*vide supra*) will cover any form of blocking on or in the cell. There is little evidence available with which to test this conception, but it must be said on the basis of Hasskó's results with methyl violet that it is not easy to conceive in chemical terms of an arrangement of two kinds of receptors on a surface such that the blocking of one kind by parafuchsin ions could prevent the union of ten times as many acriflavine ions with receptors of the other kind. It is just conceivable that a small number of molecules of parafuchsin could combine with a protein molecule in such a way as to prevent it from unfolding sufficiently to allow acriflavine ions to enter its cavities in which the drug receptors may well be sited.

Instead of pursuing this highly speculative line of thought, a more promising hypothesis will be put forward, one which should lend itself well to experimental verification. This hypothesis is one which opens up an entirely new field of thought in the interpretation of interference phenomena. It seeks to prove that interaction between two interfering substances can occur in the complete absence of the parasite. Because interfering pairs of substances so often resemble one another chemically, it has not been suspected that they may react with one another in a rather subtle way, namely, that they may form, in solution, a mixed micelle devoid of toxic properties.

This idea occurred to the author after reading some experimental work by Valko & DuBois (1944) in which dodecyl dimethyl benzyl ammonium chloride (a common kationic disinfectant which kills *B. coli* in 10 min. at a dilution of 1:40,000) had its effective concentration doubled when it was previously mixed with one-quarter of its weight of a higher homologue, hexadecyl dimethyl benzyl ammonium chloride. The last-named substance has a slight antibacterial action of its own, and it obviously cannot destroy the more active homologue by an orthodox chemical reaction. Nevertheless, it is becoming increasingly clear that there is a tendency for such long-chain paraffin salts to form mixed micelles, that is to say, aggregates of ions (Klevens, 1946; Klevens, 1948; Corrin & Harkins, 1946).

It is necessary to make the logical assumption that usually the monomer is the active form of a drug and that there is a competition for monomer between the parasites' receptors and any micelles which should happen to be built up at the expense of free monomer. An example has been discovered by Alexander & Trim (1946) in the interaction of hexyl-resorcinol, soaps and the worm, *Ascaris lumbricoides*. When the concentration of the soap was sufficiently high for soap micelles to be formed in quantity, the drug entered the micelles in preference to reacting with the worms, and hence its activity dropped to zero. In these experiments, at least one molecule of soap is needed for each molecule of phenol inactivated. In the examples of therapeutic interference which we are considering, the two kinds of molecules are sufficiently similar in structure to form much more stable interlocking micelles in which a quite small amount of interfering substance could tie up a large amount of the active drug.

The important physico-chemical constant for studies of this kind is the *critical micelle concentration* (C.M.C.) which is the concentration above which micelles begin to form. An example of the general principle that substances lower the C.M.C. of lower homologues is that potassium myristate (C_{14}) develops micelles at a concentration of one-quarter of that at which potassium laurate (C_{12}) does, and that 15% of potassium myristate halves the C.M.C. of potassium laurate and reduces the C.M.C. of potassium caprate (C_{10}) to one-quarter of its former value (Klevens, 1948, 1946).

With these facts in mind, it is not difficult to see how an addition of 25% of the C_{16} homologue to the C_{12} disinfectant, in Valko & DuBois's experiment, could lower the effective concentration of the drug by tying it up in micelles.

Taking this conception back to the trypanosome problem, the hypothesis becomes this, that parafuchsin is preventing acriflavine from acting, by lowering the latter's critical micelle concentration, so that it is no longer available to any extent as an active monomer but is largely tied up in biologically inert micelles.

Two pieces of evidence offer some support. First, Goldacre (unpublished) has measured the C.M.C. of a number of triphenylmethane dyes, including parafuchsin, and found them to be of the order of 10^{-5} M, whereas those of a selection of acridine dyes were found to be 10^{-3} M. Now, the usual curative dose of acriflavine in a 200 g. rat is 5 mg. and, taking blood volume as 20 ml., the concentration in the blood for the first few minutes after injection would be 1×10^{-3} M. The C.M.C. which is thus momentarily reached, may prove to be greatly prolonged by the parafuchsin. It will be interesting to see whether measurements on mixtures will establish the reality of such mixed micelles. The second piece of evidence is that the deleterious action of parafuchsin on the metabolic processes of trypanosomes was annulled by acriflavine and the deleterious action of the latter was annulled by parafuchsin (see Table 1). This phenomenon, for which no explanation has previously been advanced, is found again in some *in vitro* studies of the mutual interference by pairs of dyes with the toxicity of either dye towards yeast (Wright & Hirschfelder, 1930).

These authors found that acriflavine, parafuchsin, methyl violet and brilliant green, when tested singly, all inhibit the production of carbon dioxide by yeast. But when acriflavine and methyl violet were used together, their toxic effects were cancelled out. Whereas 1:8000 of acriflavine inhibited the respiration by 41 %, all inhibition disappeared if 1:120,000 of methyl violet were also present. Reciprocally, whereas 1:40,000 of methyl violet caused a 30 % inhibition, this fell to 16 % in the presence of 1:40,000 of acriflavine.

Even if the experimental data establishes the likelihood of this explanation of parafuchsin-acriflavine interference, several related problems will remain to be solved. (1) Why does parafuchsin not lower the toxic dose of acriflavine for mice (Schnitzer, 1926)? This is likely to be answered along the lines of relative affinity constants, the vital tissue in mice apparently forming stronger bonds with monomeric acriflavine than parafuchsin does, this in turn binding the drug more tightly than do trypanosomes. (2) Why do all three triphenylmethane dyes (II) interfere also with the trypanocidal action of organic arsenicals such as arsacetin and salvarsan (Schnitzer & Rosenberg, 1926; Browning & Gulbransen, 1927)? There is some evidence that organic arsenicals tend towards the colloidal state in aqueous solution. All organic arsenicals are transformed to the corresponding arsenoxides before they act, but there is no actual evidence as to whether these tend to associate in water, although it is well established that they are associated in other solvents (Blicke & Smith, 1930). However, the structural similarity between arsenicals and (II) is slight, unless the presence of a benzene ring with hydrophilic groups attached is the sole criterion. If so, the phenomenon of therapeutic interference is likely to be found to be extraordinarily widespread, a menace lurking in almost every bottle of medicine.

The surprising thing is that brilliant green interferes with the action of arsenicals but not with that of acriflavine (Schnitzer & Rosenberg, 1926). Brilliant green certainly does not prevent the uptake of acriflavine by trypanosomes (Hasskó, 1935), but the literature does not record its influence on the uptake of arsenic. Brilliant green also interferes with the toxic action of acriflavine on yeast, but parafuchsin does not do so (Wright & Hirschfelder, 1930). The resemblances between these interference phenomena on trypanosomes, on the one hand, and yeast, on the other, must not be taken too far, because the triphenylmethane dyes are more toxic than acriflavine for yeast, whereas the reverse is the case for trypanosomes.

The discussion so far has been confined to a few examples of therapeutic interference. Other cases are known which have been explained along different lines, and most satisfactorily so. For instance, there is the set of twenty-two oxidation-reduction indicators, of which all the members with potentials between +0·12 and −0·06 V. at pH 7 interfere with the therapeutic action of arsenicals and antimonials on trypanosome-infected rats (v. Jancsó & v. Jancsó, 1936). The effect reaches a maximum at about +0·01 V., i.e. in the region of methylene blue and toluidine blue. The authors' interpretation was that this interference by the oxidation-reduction active substance was caused by its carrying out the work of the enzyme inhibited by the arsenic. This is highly likely, for methylene blue has long been used by experimental biologists as a substitute for certain respiratory enzymes when normal respiration has been arrested by cyanides. It should be noted that ascorbic acid (+0·07 V.) also interferes with the action of arsenicals, antimonials and acriflavine (Fischl & Fischl, 1934) and probably for the same reason. This incompatibility may be significant in clinical medicine.

A well-defined competition exists between hydrogen ions and the kations of antibacterial acridines for a negatively charged receptor (possibly a phenolic group) on bacteria (Albert, Rubbo, Goldacre, Davey & Stone, 1945). This can produce a therapeutic interference of clinical significance because acid-producing organisms occurring in wounds can increase the ratio of hydrogen ions to drug kations. Clinically, the situation is dealt with by perfusing the wound with sodium bicarbonate.

Therapeutic interference is also known outside the realms of chemotherapy. For example, *N*-allyl-morphine can prevent the analgesic action of morphine in mammals (Unna, 1943). Again, the higher quaternary amines $Me_3NR^{\oplus}Cl^{\ominus}$, where R covers the range C_7–C_{16}, can antagonize the acetylcholine-like action of the lower homologues ($R = C_1$–C_5) (Raventos, 1938). These phenomena may prove to be further examples of the formation of mixed micelles as discussed above.

In conclusion, it may be said that therapeutic interference obviously comprises phenomena which are complex in nature, but not so complex that it is beyond the wit of man to solve their mysteries. The study of this subject will become increasingly rewarding in proportion as fresh examples of interference are sought and reported. Not only are there obvious clinical implications, but implications of interest for the experimental biologist are also evident. At the very least, the study of therapeutic interference seems likely to teach us something new about the mode of action of drugs and other selective toxic agents, whether given singly or in mixtures.

REFERENCES

ALBERT, A., RUBBO, S., GOLDACRE, R., DAVEY, M. & STONE, J. (1945). *Brit. J. Exp. Path.* **26**, 160.

ALEXANDER, A. & TRIM, A. (1946). *Proc. Roy. Soc.* B, **133**, 220.

BLICKE, F. & SMITH, F. (1930). *J. Amer. Chem. Soc.* **52**, 2946.

BROWNING, C. & GULBRANSEN, R. (1922). *J. Path. Bact.* **25**, 395.

BROWNING, C. & GULBRANSEN, R. (1927). *J. Path. Bact.* **30**, 513.

CORRIN, M. & HARKINS, W. (1946). *J. Coll. Sci.* **1**, 469.

FISCHL, V. & FISCHL, L. (1934). *Z. ImmunForsch.* **83**, 324.

GOLDACRE, R. (unpublished).

HASSKÓ, A. (1935). *Z. Hyg. InfektKr.* **116**, 669.

v. JANCSÓ, N. (1931). *Zbl. Bakt.*, Abt. I. Orig. **122**, 393.

v. JANCSÓ, N. & v. JANCSÓ, H. (1936). *Z. ImmunForsch.* **88**, 275.

KLEVENS, H. (1946). *J. Chem. Phys.* **14**, 742.

KLEVENS, H. (1948). *J. Phys. Coll. Chem.* **52**, 130.

RAVENTOS, J. (1938). *Quart. J. exper. Physiol.* **27**, 99.

SCHEFF, G. & HASSKÓ, A. (1936). *Zbl. Bakt.*, Abt. I. Orig. **136**, 420.

SCHNITZER, R. (1926). *Z. ImmunForsch.* **47**, 116.

SCHNITZER, R. & ROSENBERG, E. (1926). *Z. ImmunForsch.* **48**, 23.

SCHNITZER, R. & SILBERSTEIN, W. (1926). *Z. ImmunForsch.* **49**, 551.

UNNA, K. (1943). *J. Pharmacol.* **79**, 27.

VALKO, E. & DUBOIS, A. (1944). *J. Bact.* **47**, 15.

WRIGHT, H. & HIRSCHFELDER, A. (1930). *J. Pharmacol.* **39**, 39.

SOME EFFECTS OF DIET ON TOXICITY—THE INFLUENCE OF DIET ON THE INDUCTION AND INHIBITION OF TUMOURS

By L. A. ELSON

The Chester Beatty Research Institute, The Royal Cancer Hospital, London, S.W. 3

The influence of the diet of the animal on the toxicity of substances administered for therapeutic or other purposes is not usually considered very seriously, and, except in a few special cases, no very marked effect on the action of most drugs has been observed.

In the case of a prolonged toxic action such as the production of tumours by carcinogenic substances, however, the diet of the animal may have a very considerable effect. The time of appearance of the tumour may be hastened or delayed, or tumour induction even prevented altogether by dietary factors. Recently, it has been found that the inhibitory action of a number of carcinogenic substances on body growth and tumour growth is greatly influenced by the nature of the diet, being, in fact, inversely proportional to the protein content. This dietary influence may also be apparent in the response of tumours to treatment by X-radiation. Some of the factors involved in these effects will be considered, and their bearing on the nature of the process of carcinogenesis and on growth inhibition briefly discussed.

Intake of calories. The effects of varying caloric intake upon tumour incidence and growth has been extensively investigated by Tannenbaum (1947*a*, *b*). He finds that restriction of calories inhibits the genesis or incidence of all types of mouse tumours he has studied. These include chemically induced skin and connective tissue tumours and spontaneous mammary, liver and lung tumours. On the restricted diets fewer mice develop tumours, and these appear, on the average, at a later time. The growth of established tumours is inhibited to some extent by restriction of food intake, but the host also loses weight, and Tannenbaum concludes that 'present evidence does not suggest that caloric restriction may effect the growth of tumours in a practical, useful way'.

Induction of liver tumours by azo compounds, etc. The discovery that the ability of some azo compounds, particularly *p*-dimethylaminoazobenzene, to induce liver tumours in rats is greatly influenced by dietary conditions, has stimulated a large number of investigations into the nature of the factors concerned. From observations of the protective effect against tumour formation afforded by addition to the diet of substances such as liver and yeast the range has now been narrowed to more clearly defined entities such as vitamins (particularly of the B group), proteins and fats. As the work progresses, however, it is becoming apparent that the influence of diet is by no means simple. Some factors have been found which increase resistance to tumour production, some which enhance tumour formation, and some which act either as accelerators, or as retarders, depending on the composition of the basal diet used. The various aspects have been reviewed by Burk & Winzler (1944), Miller (1947), Kensler (1947), White, White & Mider (1947), and Opie (1947).

Effect of diet on the body growth and tumour growth inhibiting action of carcinogenic substances. A number of carcinogenic substances, including aromatic hydrocarbons such as 1:2:5:6-dibenzanthracene, stilbene derivatives such as 4-dimethylaminostilbene, etc., have been found to inhibit the growth of young rats and also to have a marked inhibitory effect on the growth of malignant tumours (Haddow, Scott & Scott, 1937; Badger, Elson, Haddow, Hewett & Robinson, 1942; Haddow, Harris, Kon & Roe, 1948).

In an investigation of the nature of this growth-inhibitory action (Elson & Warren, 1947) the first experiments failed to show any immediate inhibitory effect on the growth of young rats after intraperitoneal injection of 50 mg. 1:2:5:6-dibenzanthracene. The animals continued to grow at approximately the same rate as before the injection for at least 10 days, after which time some fluctuations occurred and then the majority of the animals began to lose weight after varying intervals of time, and the loss of weight continued until the death of the animal which occurred within approximately 20–70 days after injection (Fig. 1). These results did not support the conclusions of Haddow *et al.* (1937) that carcinogenic hydrocarbons produce immediate and long-continued inhibition of growth, but it was noticed that the average growth rate of their rats was 1·5–2 g./day, whereas the rats used in our investigation were maintained on a 20% protein diet and had a growth rate of 3–4 g./day. When the experiment was repeated with rats maintained on a 10% protein diet on which their average growth rate was about 2 g./day, immediate inhibition of growth was observed. This inhibition was prolonged until the death of the animal; in some cases almost complete cessation of growth occurred, and the animals maintained a practically constant weight for periods of up to 160 days after the injection of 1:2:5:6-dibenzanthracene (Fig. 2). Although the growth-inhibitory action of the compound is greater when the animals are maintained on the 10% protein diet than when they are fed a 20% protein diet, the toxicity

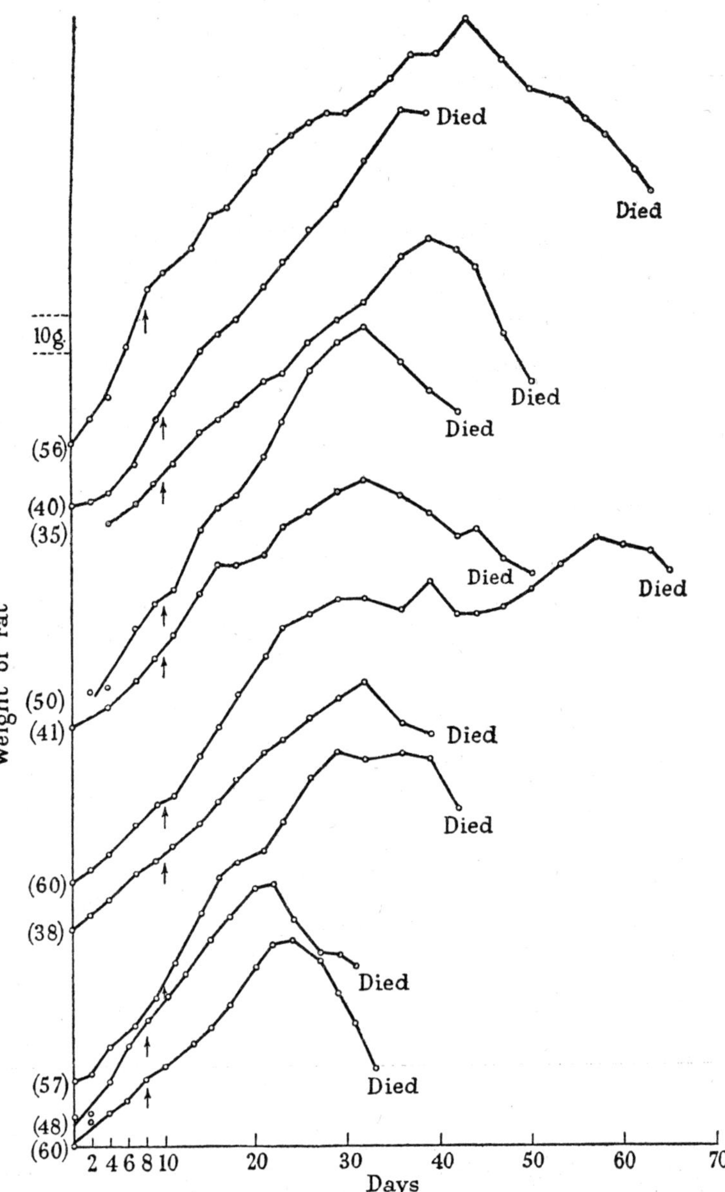

Fig. 1. Growth rates of individual rats maintained on a high (20%) protein diet before and after treatment (shown by arrow) with 1:2:5:6-dibenzanthracene (50 mg. in 1 c.c. arachis oil per rat i.p.). Figures in parenthesis give the initial weight (g.) of the rat.

as judged from the survival time of the animals is less. The average survival time of seventeen animals maintained on the 10% protein diet was 67 days,

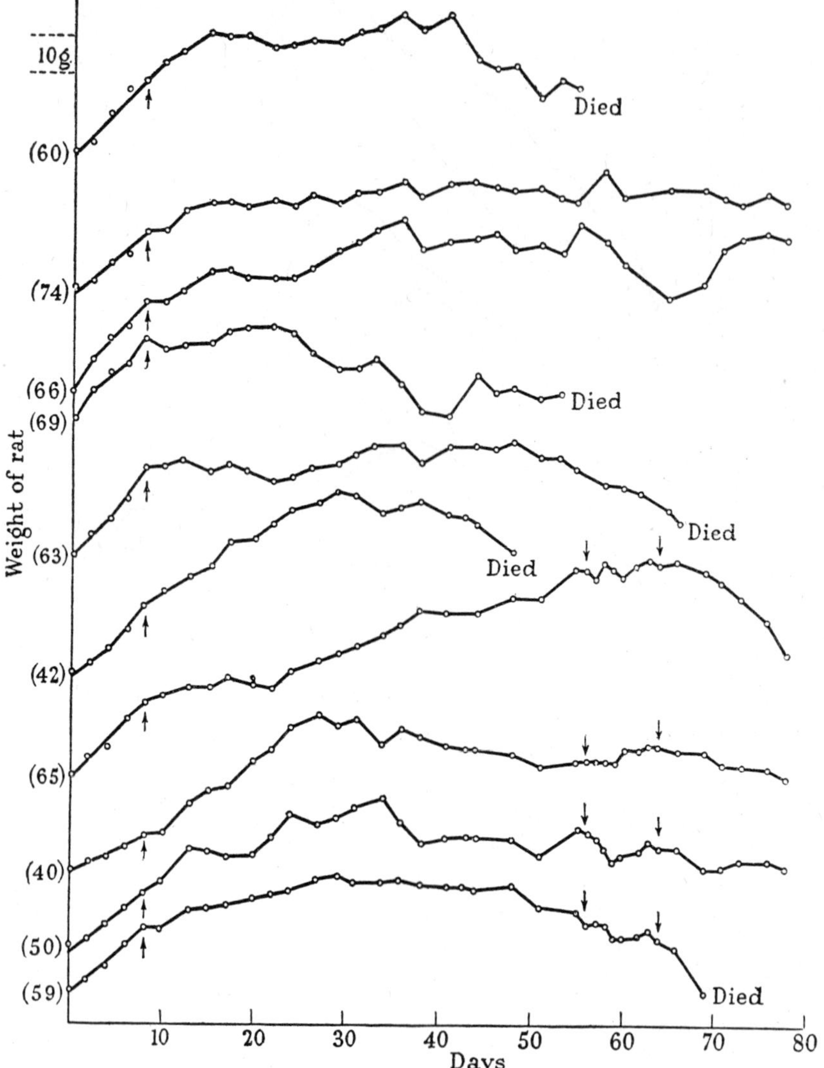

Fig. 2. Growth rates of individual rats maintained on a low (10%) protein diet before and after treatment (shown by arrow) with 1:2:5:6-dibenzanthracene (50 mg. in 1 c.c. arachis oil per rat i.p.). 1 c.c. of anterior pituitary growth hormone was administered (s.c.) daily during the period shown by the arrows. Figures in parentheses give the initial weight (g.) of the rat.

but was only 44 days for the same number of animals maintained on the 20% protein diet.

The difficulty of assessing the toxicity of substances of this type is thus apparent. If, for instance, the test animals were maintained on a high-protein diet and observed for a period of say 14 days after injection, a dose of 500 mg./kg. 1:2:5:6-dibenzanthracene would be found completely non-toxic, whereas if the experiment were prolonged for a further 40 days or so this dose would be seen to be considerably greater than the M.L.D. A probable explanation of this effect of diet on the survival time of the animals is that the actual toxic agent is a metabolic product of 1:2:5:6-dibenzanthracene, and that in animals the low-protein diet on the rate of absorption and metabolism of the carcinogen is slower than in those maintained on the high-protein diet, so that a dynamic equilibrium is attained between the rate of production of the toxic metabolite and the total metabolism of the animal.

The question of whether the tumour-inhibiting action of these carcinogenic substances is also dependent on the protein content of the diet was then investigated. In an experiment using Walker Rat Carcinoma 256 implanted into young rats the inhibitory effect of 1:2:5:6-dibenzanthracene (500 mg./kg. given by intraperitoneal injection the day after implantation) was observed in three groups of animals, one group being maintained on 20% protein diet, one on 10% protein diet and one on 5% protein diet. The ratio of the average weight of the tumours of the control animals to the average weight of the tumours of the dibenzanthracene-treated animals (C/T), when the animals were killed 13 days after implantation of tumours, was 1·6 for those maintained on the 20% protein diet, 2·5 for those on the 10% protein diet and 4·0 for those on the 5% protein diet. Thus very little tumour inhibition was obtained with the animals fed 20% protein diet, definite inhibition with those on the 10% protein diet and marked tumour inhibition with the animals on the 5% protein diet (Elson & Haddow, 1947).

Still more striking effects of the influence of diet on tumour inhibition have been observed with the amino-stilbene derivatives. Details of an experiment on the inhibition of Walker Rat Carcinoma 256 by 2′-chloro-4-dimethylaminostilbene are given in Table 1. Practically no inhibition of tumour growth was obtained in the treated animals fed on the high-protein diet (C/T = 1·1), whilst a very remarkable hold up of tumour growth was observed in those, similarly treated, but maintained on the 5% protein diet (C/T = 19·3). This same effect of diet is also shown in the inhibition by these stilbene derivatives of the growth of established tumours.

In considering the growth-inhibiting action of these carcinogenic substances, the fact that the nature of this inhibition depends entirely on the protein content of the diet of the animals suggests that it is brought about by an interference with the availability, or with the actual synthesis of protein necessary for growth.

Table 1. *Action of 2′-chloro-4-dimethylaminostilbene on Walker rat carcinoma in animals maintained on 20 and 5% protein diets*

Dose: 150 mg./kg. Weight (g.) of tumours after 11 days.

20% protein diet		5% protein diet	
Control (C)	Treated (T)	Control (C)	Treated (T)
23·0	8·3	9·2	1·4
9·0	4·0	16·0	1·3
16·5	19·6	16·3	0·1
10·4	18·0	7·0	0·2
4·5	9·1	13·1	0·1
10·3	17·8	7·2	0·9
9·3	0·8	14·5	1·2
16·5	1·1	17·1	0·4
9·7	14·8	5·7	0·1
—	20·2	10·2	—
Mean *12·2*	*11·4*	*11·6*	*0·6*
C/T = 1·1		C/T = 19·3	

Such an interference could conceivably take place by antagonism of an essential amino-acid. This possibility is being investigated by testing whether the protective action against the growth-inhibitory effect of the carcinogen afforded by the 20% protein diet can also be obtained by incorporating in the 5% protein diet any one amino-acid in sufficient quantity to ensure that its total amount is at least as great as that present in the 20% protein diet. The Walker rat carcinoma under conditions similar to those described (Table 1) is being used. So far neither cystine nor methionine have shown any protective effect.

Another possible method of interference with protein metabolism, which may have a more direct bearing on the carcinogenic action of these substances, is concerned with the relation of nucleic acids to protein synthesis. According to the work of Caspersson and others, protein synthesis is associated with high concentrations of nucleic acids in the cell (Caspersson & Santesson, 1942; Caspersson, 1947; Brachet, 1947*a*). Nucleic acids are intimately connected with cell division. It seems probable, therefore, that an interference with protein synthesis by carcinogens may occur through disturbances in the nucleoprotein metabolism, and that these disturbances may be directly connected with the process of carcinogenesis.

With this in view, the influence of 1:2:5:6-dibenzanthracene and some aminostilbene derivatives on the nucleic acid content of the liver of rats maintained on high- and low-protein diets has been investigated (Elson & Harris, 1947).

Davidson & Waymouth (1944) have shown that the concentration of pentose nucleic acid in the livers of rats falls if the animals are kept without food for 24 hr. The total nucleic acid, however, rises, owing to increase in the deoxypentose nucleic acid fraction. The overall effect is thus to produce a decrease in the ratio, pentose nucleic acid:deoxypentose nucleic acid (P.N.A.:D.N.A.). Similar results for rats maintained on a low-protein diet were found by Kosterlitz & Cramb (1943).

In our experiments (Table 2) the control series A confirms this decrease in PNA:DNA ratio, although no depletion of the pentose nucleic acid has occurred here, and the decrease is a result of an increase in the deoxypentose nucleic acid.

Table 2. *Distribution of lipid and nucleic acid phosphorus (N.A.P.) in the livers of rats maintained on high- and low-protein diets and treated with carcinogenic and non-carcinogenic hydrocarbons*

% protein in diet	No. of animals	Liver weight as % body weight	Lipid P (mg./100 g. liver)	Total N.A.P. (mg./100 g. liver)	P.N.A.P. (mg./100 g. liver)	D.N.A.P. (mg./100 g. liver)	P.N.A.P./D.N.A.P.
A. Control series							
20	10	4·8±0·08	128·6±5·2	125·2±4·6	97·2±3·6	28·0±1·6	3·54±0·16
5	9	4·9±0·26	100·6±4·9	132·2±4·9	99·5±4·4	32·8±0·75	3·04±0·16
B. Treated with 1:2:5:6-dibenzanthracene							
20	8	6·3±0·40	144·45±6·4	115·1±5·6	94·2±4·4	20·95±0·42	4·53±0·1
5*	9	6·5±0·09	122·9±5·3	106·95±5·0	87·0±2·9	19·95±1·7	4·40±0·2
C. Treated with pyrene							
5*	5	4·65±0·63	104·5±5·9	124·3±4·5	94·3±1·6	30·0±3·1	3·20±0·26

* Two animals in series B and one in series C had livers with high neutral fat content.

In contrast with this diet effect, administration of 1:2:5:6-dibenzanthracene produced a marked increase in the ratio (B). This increase (which occurred in animals on either diet) is the result of a decrease in the concentration of deoxypentose nucleic acid. In rats on the lower protein diet the total nucleic acid concentration is also decreased with a concomitant decrease in the pentose nucleic acid.

Administration of the non-carcinogenic, non-growth-inhibiting hydrocarbon pyrene produced no significant change in the nucleic acid balance.

The mechanism of the growth-inhibiting and carcinogenic action of 1:2:5:6-dibenzanthracene may therefore be directly connected with this property of increasing the ratio of pentose to deoxypentose nucleic acid. Deoxypentose nucleic acid is accumulated on the chromosomes prior to cell division, so that prevention of the formation of this nucleic acid by the carcinogenic hydrocarbon may directly suppress cell division.

Most of the information on the relation between nucleic acid and protein synthesis has been deduced by Caspersson and others from the prevalence of pentose nucleic acid in cells which are actively synthesizing protein. Increase in protein synthesis has been correlated with increased amounts of cytoplasmic pentose nucleic acid. If the idea that carcinogenic substances prevent protein synthesis is correct, we now have, as a result of our investigations, direct information on the nucleic acid distribution in the liver under conditions of inhibition of growth by suppression of protein synthesis. The results show that in the treated animals there is a rise in the ratio of pentose to deoxypentose nucleic acid caused mainly by a decrease in the proportion of the deoxypentose fraction; the proportion of pentose nucleic acid remains practically unchanged. If protein synthesis is related to cytoplasmic pentose nucleic acid a decrease in this fraction instead of the deoxypentose fraction might have been anticipated, under conditions of suppression of protein synthesis.

It seems possible, however, that it is not the absolute amount of nucleic acid, but the rate at which biochemical transformation of the nucleic acid is taking place that should be directly related to protein synthesis. Although it cannot be considered that there is as yet any very definite evidence for conversion of pentose to deoxypentose nucleic acid, it has been assumed that such a transformation may occur (Mitchell, 1942; Brachet, 1947*b*). This would have to take place through a series of intermediate steps involving various enzymatic reactions, and it is conceivable that the energy required for protein synthesis could be made available through these transformations.

On this view, depriving the animals of food for 24 hr. in the experiments of Davidson & Waymouth would not interfere with the conversion of pentose to deoxypentose nucleic acid but might prevent the renewal of cytoplasmic pentose nucleic acid, thus resulting in the fall in this fraction and the rise in deoxypentose nucleic acid observed. In our animals the amount of protein in the diet (5%) is apparently just sufficient to maintain the concentration of pentose nucleic acid.

The effect of treatment with the carcinogen, however, might be to prevent the formation of deoxypentose from pentose nucleic acid and thus also to suppress protein synthesis.

Mitchell (1942) has suggested that one of the effects of X-radiation is to inhibit conversion of pentose nucleic to deoxypentose nucleic acid. If this is the case it is possible that the control of growth of tumours by radiation may also be influenced to some extent by the protein content of the diet. Preliminary experiments on the X-ray treatment of established Walker rat carcinoma have shown in general a greater inhibiting action of radiation on the growth of the tumour in animals maintained on the 5% protein diet than in those fed on the 20% protein diet and irradiated with the same dosage (Elson & Lamerton, 1948). The protective effect of the high-protein diet on the inhibition of tumour growth by radiation is, however, not so great as that observed in treatment with the aminostilbene derivatives.

The response of the tumours to radiation is not constant, and, with increased dosage, it is possible to cause complete tumour regression in some animals. It is found that, although maintenance of the animals on a 5% protein diet favours the growth-inhibitory action of radiation, the number of regressions is often greater in the animals fed on the 20% protein diet. This suggests that two effects are concerned in the response of tumours to radiation: (1) inhibition of tumour growth and (2) ability of the animal to rid itself of the inhibited tumour. Process (1) is apparently favoured by a low-protein diet and process (2) by a high-protein diet. This opens up a field for further investigations which may well be of importance for both the practical and theoretical aspects of radiotherapy.

REFERENCES

BADGER, G. M., ELSON, L. A., HADDOW, A., HEWETT, C. L. & ROBINSON, A. M. (1942). *Proc. Roy. Soc.* B, **130**, 255.

BRACHET, J. (1947*a*). *Symp. Soc. Exp. Biol.* **1**, 207.

BRACHET, J. (1947*b*). *Cold Spr. Harb. Symp. Quant. Biol.* **12**, 18.

BURK, D. & WINZLER, R. J. (1944). *Vitamins and Hormones*, **2**, 305. New York: Acad. Press.

CASPERSSON, T. (1947). *Symp. Soc. Exp. Biol.* **1**, 127.

CASPERSSON, T. & SANTESSON, L. (1942). *Acta radiol., Stockh.*, Suppl. XLVI.

DAVIDSON, J. N. & WAYMOUTH, C. (1944). *Biochem. J.* **38**, 379.

ELSON, L. A. & HADDOW, A. (1947). *Brit. J. Cancer*, **1**, 97.

ELSON, L. A. & HARRIS, R. J. C. (1947). *Brit. J. Cancer*, **1**, 327.

ELSON, L. A. & LAMERTON, L. F. (1948). In the Press.

ELSON, L. A. & WARREN, F. L. (1947). *Brit. J. Cancer*, **1**, 86.

HADDOW, A., HARRIS, R. J. C., KON, G. A. R. & ROE, E. M. F. (1948). *Philos. Trans.* A, **241**, 147.

HADDOW, A., SCOTT, C. M. & SCOTT, J. D. (1937). *Proc. Roy. Soc.* B, **122**, 477.

KENSLER, C. J. (1947). *Ann. N.Y. Acad. Sci.* **49**, art. 1.29.

KOSTERLITZ, H. W. & CRAMB, I. (1943). *J. Physiol.* **102**, 18 *P*.

MILLER, J. A. (1947). *Ann. N.Y. Acad. Sci.* **49**, art. 1.19.

MITCHELL, J. S. (1942). *Brit. J. Exp. Path.* **23**, 296, 309.

OPIE, E. L. (1947). *Approaches to Tumor Chemotherapy*, p. 128. Washington D.C.: A.A.A.S.

TANNENBAUM, A. (1947*a*). *Ann. N.Y. Acad. Sci.* **49**, art. 1.5.

TANNENBAUM, A. (1947*b*). *Approaches to Tumor Chemotherapy*, p. 96. Washington D.C.: A.A.A.S.

WHITE, J., WHITE, F. R. & MIDER, G. B. (1947). *Ann. N.Y. Acad. Sci.* **49**, art. 1.14.

EXTERNAL METABOLITES AND ECOLOGICAL ADAPTATION

By C. E. LUCAS

Department of Oceanography, University College, Hull

W. C. Allee (1934) was of the opinion that 'once formed, aggregations of aquatic organisms condition the medium surrounding them by the addition of secretions and excretions the nature and biological effect of which form one of the important problems of mass physiology'. There have been earlier statements by other workers which have had some such general ideas in mind, but this one appears to be the first definite formulation of an important biological principle. I have suggested (1938, 1944, 1947) that this principle has an even wider significance in that it also has extensive terrestrial application, and in that it suggests a new approach to ecology, particularly to the problems of community formation and integration. I was led to an interest in it when reconsidering some of the interrelationships between organisms in planktonic communities, but planktonic evidence is not really plentiful, and it soon appeared that more evidence could be obtained from other fields; the great mass of information now available from the study of 'antibiotics' is a most important part of that evidence. It is in no sense intended to belittle its importance, either from the strictly biological or from the medical point of view, if it should be suggested here that the phenomenon which has been called 'antibiosis' may be usefully associated with a wide variety of processes which themselves have seldom been considered as interrelated. To link these together, however, enables us to see each within a wider biological field, and to draw attention to a fundamental aspect of communal life. An attempt will be made here to consider such processes and to draw together some of the characteristics which are common to them. Most of the examples will be very familiar, and the number of references to the literature will be reduced to the minimum; to try to compile an exhaustive bibliography would be a very formidable task.

That living organisms do secrete and excrete products of their metabolism into their environment has long been obvious. Certain of these products, the oxygen of plants and the carbon dioxide of animals, for example, have provided classical material for physiological research. Perhaps particularly in the water, but undoubtedly also on land, there has been no doubt as to their general significance, not only for the organisms producing them but also for their neighbours in the community. Other examples, particularly the more typical animal excreta, have equally been recognized as playing important biological roles. They have, in fact, provided key features of the familiar cycles of life. It would probably be true, however, to say that each instance had been viewed in a peculiar, somewhat anthropomorphic way. From these earlier points of view, at least, excreta were simply excreta. They were 'waste products' which were ultimately subject to bacterial action, leading to the production of nitrates and phosphates which provided important materials for the subsequent growth of plants. These came to be eaten by animals and so the cycle was complete. Carbon dioxide, too, might so be produced, and it occurred regularly as a metabolic by-product of respiration. To this extent it was seen as an excretion, although oxygen, so similarly produced by plants, was seldom seen in such a light. Whilst the associated production of other elements in one combination or another, particularly sulphur compounds, was recognized quite early, their significance was rarely considered, whilst it was scarcely recognized that plants might contribute anything other than oxygen and 'natural manure' to their environment.

Despite this, biologists have for many years recognized the continual external secretion of other materials which we should suitably term metabolites. Examples are the nectar of flowers and the skin secretions of animals; others will be mentioned later. Their significance for other organisms in the community has also been recognized, but they have seldom been thought of as having anything in common, in their manner of arising or in their subsequent influences, either among themselves or with metabolites of the first type. Some have been the subject of intensive interest and research, e.g. the dependence of ants and bees on one form of secretion or another, but the interest has usually been specialized and isolated. Perhaps it would not be too biased to suggest that this has also been true of that remarkable example of team work which has been primarily concerned with the external secretion of 'antibiotics'.

In this contribution I should like briefly to recall to mind some of the more familiar and simpler instances of the addition by organisms to their environment of metabolites previously elaborated near the surface or deep within the tissues. With these are coupled some of the less obvious ones, as well as some of the elaborate ones which have only recently been recognized for the parts which they play. These last are important in several ways, particularly in the roles they play within the bodies of their producers. It was not until such clear demonstrations were available, and not until the important work of the inter-war period had been done on the nature and actions of metabolites within the body and during its growth (chemical morphogenesis and endocrinology), that a clear appreciation could be obtained of the widespread external production of potent metabolites and of their ecological significance in this state. It is for such ecologically potent metabolites that the term 'ectocrines' has been suggested (Lucas, 1947).

The Protozoa provide numerous examples, particularly among flagellates and ciliates, which have been worked out in varying detail. One can now have little doubt not only that it is typical of all Protozoa to secrete and excrete into their environment—almost always an obviously liquid one—but that these metabolites are ecologically potent. Early work will come to mind in which the numbers of organisms per unit volume of fluid appear to have a significance for growth and reproduction which can no longer be attributed to the usual prey-predator relationships. We can now interpret these processes in terms of essential enzymes diffusing from the cells which play fundamental parts during the 'lag' phase and during logarithmic growth. With them may be associated familiar examples of the harmful effects resulting from increased quantities of 'excreta' in dense cultures—a form of 'conditioning' which is found in the most divergent forms of life. In Protozoa, too, as in protophyta and bacteria, there is widespread evidence that, whilst such 'conditioning' may be harmful for the producer and perhaps some other forms, yet others, its successors in time, literally 'succeed' where it had failed. At least they can tolerate if not actually benefit from this condition, although they in their turn become susceptible to appropriate concentrations of their own metabolites. This, indeed, is fairly accepted as the basis of ecological succession in the bacteria at least, and most probably in the protista—the successive production of metabolites toxic to some earlier members, but apparently beneficial to some later ones. Except in some of the bacteria, however, little is known of the natures of the successive metabolites in such series, and even in the bacteria far too little is known.

Another type of interaction at this biological level is provided by those organisms which produce a certain metabolic by-product when growing in pure culture, and a different one when grown in the presence of yet another form which itself does not produce that by-product. Whilst chemical interaction of the products may sometimes be concerned, this is not the only type of interaction, and in some examples at least the metabolites are ecologically potent for one or both of the producers. We know, for example, of organisms which appear to be incapable of growth 'in isolation', which can yet grow together, since each produces a vitamin (e.g. pyrimidine or thiazole) which the other cannot synthesize. The two are, in fact, complementary, although either could be grown artificially if the appropriate vitamin were supplied, and either can doubtless satisfy its need in the neighbourhood of *any* organisms producing and eliminating an excess of it. Here we have at once the shedding of biologically potent substances and examples of the communal integration which may thereby arise, since the released substances in this instance are far from specific. A hint of a moral, too, may be derived, since we know full well that it is not natural for organisms to grow 'in isolation' and by no means always can we make them do so in the laboratory.

Yet other instances from micro-organisms are mentioned later as illustrating one or another type of external metabolite of special interest, but, in general, the existence of such non-predatory relationships has been demonstrated clearly in bacteria, yeasts, actinomycetes, and in various algae and Protozoa (Waksman, 1945). 'Micro-organisms, during their growth in common media, are in particularly intimate contact and are found to exhibit mutual interaction to a high degree, both in the sense of symbiosis (for example, through the production by one organism of substances which must be obtained preformed by the other) and antibiosis (for example, in the production of gramicidin, penicillin, or iodinin, which inhibit the growth of many other organisms)' (McIlwain, 1944, p. 303).

In the absence of adequate investigation examples are not too readily found among the less advanced cellular animals, but in the plants, particularly in the fungi, they are certainly common enough. They have provided the subjects of various papers in this symposium, and it is almost presumptuous for a marine biologist to review them. Briefly, along with some bacteria, yeasts and a wide range of plant tissues, numerous fungi provide examples of metabolites having so evident a toxic potency for certain other organisms (often, but not only bacteria) that they have been termed 'antibiotics'. Whilst some are effective at moderate concentrations, others are still effective at very great dilutions—so much so that the natural co-existence of the fungus, say, and the relevant susceptible organism is almost inconceivable. Some of the metabolites, again, are highly specific, whilst others have a more general toxic effect. Undoubtedly the most intensive investigations of these have been in the medical field, with the ultimate aim of alleviating the susceptibility of man (and domesticated animals) to their bacterial parasites—a point of some ecological significance —but such substances are produced and exert their influence in a variety of ways. Dr Brian, for example, will show us that a wide field is now opening in the subject of soil ecology, whilst I have tried to indicate some of the possibilities in the aquatic environment and particularly the sea (1947). Dr Templeman and his colleagues have already shown us that certain substances which are usually regarded as beneficial, e.g. plant-growth regulators, may yet be antagonistic to certain organisms under certain conditions.

This is not the first time (see, for example, 'Espinasse, 1944) that attention has been drawn to the misleading nature of the term 'antibiotic'. It may seem pedantic, but it is not really so, to point out that a metabolite so toxic to the *Pneumococcus* as to save the life of many human beings cannot with propriety be termed 'antibiotic'. To revert immediately to the natural community, the localized production of penicillin may save not only *P. notatum* from the competition of Gram-negative bacteria and others, but any other local organism which is in the slightest degree intolerant either of the secretions of these bacteria or of their direct competition for nutrients, etc. This seems *a priori* certain, but the literature already provides numerous instances of it, and not a few in which the metabolite itself, whilst toxic to one section of the community, is directly beneficial to another either at the same or at another concentration (in nature, concentration may be a direct function of both time and space). It has been suggested (Lucas, 1944) that even penicillin may benefit the associated or subsequent growth of some organism or organisms quite apart from man. Whilst a specific instance of this cannot yet be quoted, we do know that at extreme dilution it can have a stimulating effect on bacteria which it otherwise would harm, whilst one of the penicillins, apparently in the pure state, has already been shown to stimulate cambial growth and root formation (de Ropp, 1946). The point should not be laboured, but it seems essential to maintain, wherever possible, the wider ecological approach which alone will permit an adequate development of the subject, in preference to the necessarily limited 'antibiotic' or even 'antibacterial' view needed for the speedy solution of an urgent human problem. A relevant instance is provided by the excretion or secretion from an actinomycete which appears to be toxic to *Fusarium oxysporum cubense* (Thaysen & Butlin, 1945), this being the 'Panama' disease of the banana. The metabolite as well deserves the name 'antibiotic' as any other to which it has been given, but the ecologist studying a community containing these three organisms would hesitate to use it. The new field opened by Dr Brian and others seems likely to prove that many normal soil saprophytes produce metabolites which are antagonistic to pathogenic fungi, but which may consequently be beneficial to their angiosperm hosts.

The fungi, too, provide material for studying another type of relationship mediated by external metabolites. This, in one form or another, is the relationship of 'mycorrhiza'. We know now that the orchid seed, which cannot develop in the absence of the appropriate fungus, is unable to synthesize a vitamin of the aneurin complex which the fungus does synthesize and which it also excretes. Doubtless a similar relationship applies to the mycorrhizae of conifer trees and others, a relationship which may be disturbed, as it appears, by the free metabolites of any soil inhabitants which are antagonistic to the fungi. There are instances, too, and not only among the fungi, in which two organisms may live together in that, each being unable to synthesize some different essential for proper growth whilst the other normally excretes it, the two have become complementary.

It is now being realized, in fact, that mycology alone is providing a fresh insight into the nature of ecological relationships. In addition to those relationships involving the inorganic elements of the environment, and those of prey and predator or parasite and host, external metabolites, ectocrines, regularly occur which may be beneficial in some circumstances (e.g. particular concentrations) or to some organisms (those which do not produce them, or enough of them), and toxic in other circumstances (e.g. higher concentrations) or to other organisms (those living by metabolic processes which are blocked or diverted by the substance in question).

Knowledge already available suggests that such relationships are widespread in plants and animals. McIlwain (1944) considered that the 'known properties of enzyme systems, and the observed variations in the occurrence and structures of metabolically functioning substances of living organisms, provide a basis for processes in one organism being affected by products from another'. Not only do orchid seeds, for example, depend on the metabolites of fungi, but there are other seeds, spores, and even pollen grains, which cannot synthesize all their requirements (Brown, 1946). Whilst these requirements are specific, however, the agents may often be obtained from one of a variety of sources which liberate them into the environment. These agents are frequently as essential for their producers as for the germs which need them; they are distinctive only by their overproduction and elimination. Here is a new basis for symbiosis and one which must be widespread, whatever other processes may also be concerned. Between the components of lichens, and between certain animals and their associated algal cells, for example, there may well be simpler agents concerned (e.g. the carbonic acid and nitrogenous by-products of the animals, which are themselves external metabolites), but more subtle ones of the type just mentioned are inherently likely. It has, for example, been suggested that the plants may, in particular, provide carotenoid by-products which may be required by the animal for effective reproduction or sexuality.

It is appropriate to consider, too, all those instances in which internal parasites liberate metabolites essential for the processes of their hosts. We do not yet know just how common such relationships are, but they include hosts ranging from insects to mammals, and there is reason to believe they may be very widespread indeed. It may not be unreasonable to include here the relationships between leguminous plants and their nodule bacteria, although the most obvious agent supplied by the bacteria, 'nitrogen', is of an elementary nature compared with many others. In so far as the parasites themselves are so often living within the excretions and secretions of their hosts, it is unlikely that they are dependent merely on the 'food' so provided (and in this they should not fundamentally differ from their fellows parasitic within the body tissues). They may well have certain synthetic deficiencies to make good, particularly when we consider their unlit environment. On the other hand, in some instances at least, we know that substances antagonistic to them are liberated by their hosts. It is further relevant to the present thesis that their harmful effects upon their hosts, if any, are by no means limited to their need for food; they are often mediated by external metabolites. But it is particularly relevant that many, perhaps most, parasitic relationships are regularly in a more or less balanced equilibrium—it is normal for the one organism to be adapted to the processes of the other and the relationship may even be obligatory.

Another type of association is provided by the attractions of some organisms for others which are mediated by external metabolites. Mosquitoes and other pests of animals are susceptible in varying degrees to the 'smell' of their hosts. Whilst most members of the host species are attractive, in distinction from the many other organisms which are neutral, some of them, as in humans, may be repellent, apparently as the result of an air-borne odour. Many vegetable pests, too, are attracted by the smells of their hosts (which, beyond a very limited range, must be at an extreme dilution even when assessed in terms of molecules), whilst the many external parasites of aquatic organisms seem likely to be attracted by water-borne metabolites. Similar relationships must have arisen in all those internal parasitic associations which are first mediated through the active location of a host by a temporary or permanently external parasite. It would be interesting to know how far these 'attractive' metabolites are simply attractive, and presumably the agents of a conditioned response, and how far they are themselves beneficial—how far they directly fulfil a potential or absolute need.

It has long been known that natural odours have mediated other types of relationship. It is only a short step from the attraction of, say, the carrot 'smell' for the carrot fly to that of many flowers for so many insects (and humans). At the same time one may note the relationships built upon the desirability of the external secretions of the nectaries for many of these insects, as well as for some birds. There are also the apparently obnoxious odours of a few flowers for which only a small number of insects have a tolerance, as well as the more diffuse odours of many aromatic plants which appear to repel a majority of insects. With our present knowledge, the attraction exerted by certain female moths for the males of the same species, over great distances, is also likely to be due to a potent and volatile metabolite, and this is matched, probably at long and short range in varying degree and kind, by the scents of male and female mammals (e.g. musk) and probably other animals. In both of these an external metabolite, a by-product, has during the course of evolution come to mediate an important biological process. It must be by some comparable process that, in aquatic organisms, the undifferentiated form of *Bonellia* is attracted towards the differentiated female and comes to develop as a male, and similarly in the parasitic males of deep-sea angler fish and the other examples of complementary sexuality (*Diplozoon*, *Crepidula*, etc.).

At a quite different biological level such relationships are matched by the familiar attractions (usually by water-borne metabolites) exerted by eggs upon sperms, in both plants and animals of all types which have been investigated, and both within and without the parent body. In some plants and animals these seem to concern substances of a carotenoid nature, although other substances may well be concerned. At an apparently lower level of sexuality, an extremely subtle process has been revealed by the brilliant work of Moevus, Kuhn and others (e.g. Moevus, 1933, reviewed by Cook, 1945) in their analysis of the sexuality of *Chlamydomonas eugametos*, where the potency of the gametes is determined by the liberation into a naturally illuminated environment of the *cis*- and *trans*-dimethyl esters of crocetin. Not only within this species, but as between this and related species, the proportions of male gametes appear to depend directly upon the relative proportions of the *cis*-ester in the illuminated medium. It has appropriately been said that this and related work on the carotenoid metabolites is 'only the beginning of what promises to be the most fascinating story of chemistry' (Mayer, 1943, p. 77).

Reference has already been made to the attractive and repulsive potency of bodily odours. In so far as these are products of external secretions, it is also particularly relevant to recall the specialization of some of these secretions, in which the modified skin glands of mammals produce a food essential for their young. Here again is clear evidence of an external metabolic by-product which has come during evolution to mediate communal relationships, in this case, perhaps, first one and then another type of relationship. This, in turn, recalls the special crop secretions with which a number of birds feed their young. Whilst such a secretion might conventionally be regarded as internal, it undoubtedly has come to mediate an external relationship—its potency is external, at least in part. In both of these, too, there is good reason to suppose that vitamins, as well as food, may be important factors. Then there are the well-known ant-aphid associations, whilst other members of the Hymenoptera have come, via their own secretions, to exert a fundamental influence on the whole organization and life of their offspring. Of bees, we now know that, by varying the period during which the larvae are fed with the secretion 'royal jelly', the development of a neutral worker or a sexual insect may be determined. We further know that the vitamin pantothenic acid, among others, is extremely abundant in this secretion, and that its influence on growth and longevity is great (Gardener, 1946). In its apparent control of sex development and length of life this secretion, and the external relationship which it mediates, provides an example of biological potency which fairly parallels those of sex determination in *Chlamydomonas*.

With the exception of 'antibiosis' and parasitism, these examples have mainly been concerned with beneficial relationships in the widest sense, or relationships which in general have furthered rather than hindered the life of another member of the community or family. There are, however, plenty of examples of relationships additional to 'antibiosis' which are harmful at first hand, although, as with beneficial ones, the complexity of the ecological network is such that one man's food may be literally another man's poison, whilst the nexus of indirect harm or benefit is almost impossible to gauge. Those harmful parasitic relationships which are not primarily based on food requirements are recognized as being mediated by the release of toxic by-products into the hosts' tissues. The host is directly poisoned, and by some natural metabolite which blocks its own metabolic processes. This is also true of 'antibiosis', but it has been familiar to us for much longer in the secretion and direct external application of poisons by both animals and plants. There is the widest variety of the mechanisms whereby contact is effected, ranging from the minute cnidoblasts of coelenterates to the much larger poison glands of arthropods and snakes, and including the relatively simpler structures of plants (e.g. the nettle). With these may be coupled these examples in which a secretion has come to facilitate external parasitism, ranging from the salivary secretion of mosquitoes (which not only facilitates the drawing of blood as in many other insects, but also bears the deadly malaria parasites) to that of the ichneumon which paralyses but does not kill the host of its eggs and larvae.

Poisons of a more general nature, undoubtedly resembling 'antibiotics' in their action whilst as yet unidentified for certain, must be looked for in the presumed external secretions of diatoms and other marine phytoplanktonts which appear to mediate the 'animal exclusion' postulated by Hardy (1935). In this exclusion it has been found that many, but not all, planktonic animals tend to be scarcer in the presence of dense growths of phytoplankton than when the numbers are only moderate. Whilst this must in part be simply due to the fact that the plants can multiply more successfully in the absence of their consumers, the relationship appears to apply to carnivores as well as herbivores, and even to fish such as the herring, which are regularly scarce in the midst of great diatom patches. There is, too, a small amount of experimental evidence that zooplanktonts may avoid waters in which dense phytoplankton is actively photosynthetic, and that they may even succumb in such conditions (Lucas, 1936), whilst instances occur from time to time of the wholesale death of associated fish and other aquatic animals when unusual growths of one planktonic alga or another have occurred. The natures of the secretions are as yet unknown, but they are known to exist and to be responsible for the mortality (Otterstrøm & Nielsen, 1940), and they may be accompanied by a distinctive odour which is said to affect human beings (Smith, Gunter & Williams, 1947). In another marine field it has been observed that barnacles become inactive in the presence of dense numbers of dinoflagellates, possibly owing to the by-products of the flagellates (Allison & Cole, 1935). In contrast, it has been found that the eggs and very young stages of some at least of the zooplanktonts and fish may be unusually abundant in the midst of some phytoplankton patches, and it has been suggested that the abundant oxygen supplies of such waters may be advantageous to the young (Wimpenny, 1936). One can also imagine other by-products of such growths to which adaptations might have been developed.

Although there is as yet far too little information available, evidence is being produced of the locally harmful secretions released from the roots of flowering plants (e.g. Pickering, 1917, among early workers); in contrast we have the apparently beneficial examples of legume-root extracts influencing the growth of nodule bacteria in the free state, as shown by Chen, Nicol & Thornton (1940). Salisbury (1944) has already commented upon the great importance of antagonistic secretions in the ecological competition of higher plants. There seems, in fact, good reason to suppose that detailed investigation of soil biochemistry might reveal as complicated a set of interrelationships mediated by angiosperm and other secretions as by bacteria and fungi, relationships contributing not only to competition but probably also to various degrees of symbiosis and perhaps partly mediating succession (Lucas, 1944). Developments may reasonably be anticipated in this field as more is discovered about the production by plants of such growth substances, etc., as might be expected to be more easily, though not solely, eliminated through the root system rather than above ground.

Whilst they are not conventionally regarded as poisons we may be reminded that the digestive juices of most animals (usually and reasonably regarded as internal) are logically and historically derived from secretions which, in the Protozoa and the more primitive cellular animals, were not so obviously internal; certain organisms (e.g. coelenterates) might fairly be said to pour out a fluid for the external digestion of their prey. This type of prey-predator relationship is also familiar in those carnivorous plants which hasten the digestion of their prey by undoubtedly external secretions. In some forms the secretion may even prove to exert an attraction for the insect prey. In contrast to this a wide variety of parasites have adapted themselves to the digestive secretions of animals and have become the inhabitants of one part or another of the alimentary canals of their hosts. Unlike parasites of the bodily tissues themselves, which have adapted themselves to the internal secretions, these may be considered as of the world outside their hosts—a part of that world which has been almost surrounded by the host but not entirely so, since their entry into it and passing from it need not include any physical destruction of the tissues. Their ecological affinity, or niche, rather more resembles that of the organisms which have come to live in the more obviously external secretions, frequently mucoid, which are released by so many animals. Such secretions may well supply chemicals necessary for the life of the parasites, since so often they cannot live a normal life apart from them, but at least it is clear that they have overcome any harmful effects the secretions may have, although in some cases they harm the host by their own secretions.

There remain, after these fairly specific instances, a number of biological processes which are relevant in one aspect or another, and which may be dealt with briefly.

Allen (1914) observed that the growth of phytoplankton in aquaria was normally poor, but that it was much improved if the aquarium water was treated with hydrogen peroxide (or chlorine), thus implying that some substance was thereby oxidized which in its natural state tended to inhibit the growth of the plants. On the other hand, however, both in the sea and in fresh water, evidence is rapidly accumulating of the necessity, for the growth of plants, of substances (other than the familiar nutrients) which may not always be present and which may be replaced in laboratory cultures by chemicals ranging from cystine to thiamin (Harvey, 1939). Of more natural products, the growth-stimulating effect of 'soil solutions' have been known for some time (Schreiber, 1927), and also that of algal extracts (e.g. Allen, 1914; Harvey, 1939). More recently, it has been shown that the waters in which certain coastal algae have grown exert a stimulating effect on the laboratory growth of other algae (De Valera, 1940; Kylin, 1943), whilst some phytoplankton extracts may be stimulating in some conditions and inhibiting in others (Levring, 1945). Of particular interest has been the observation that carotenoid pigments occur, freely and widely, in the sea, both in sediments and as solutions or suspensions (Fox, 1944; Kalle, 1938). Such free carotenoids may well have considerable potency as part of the environment of marine animals, in view of the well-established potency of some, at least, of the group. Again, it has been suggested that, as also on land, biologically potent sterols may well be excreted by animals or plants into the water.

On the land, there is evidence that one or another of the plant hormones, which largely determine modes of growth in their producers, may be released into their environment. As administered experimentally, these may extensively influence the growth and success of other organisms. Thus, Templeman & Sexton (1946) have shown how certain of these have a differential influence on the growth of monocotyledons and dicotyledons, scarcely affecting the former but proving lethal at appropriate concentrations to some of the latter. There is, too, the newly discovered influence of such plant hormones, *at appropriate concentrations*, on the root growth of shoot cuttings and seeds. This brings to mind an old practice in certain parts of the country of inserting a wheat grain in the tissues of a shoot which it is desired shall root. The method appears to be successful, although no detailed tests seem to have been made. Certainly it might now be expected to give rise to certain growth substances for root stimulation. Again, some of these growth substances, or related compounds, have been found to be remarkably stimulating to the growth of fruits such as some tomatoes and apples, etc., causing the soft parts of the fruits to swell and ripen even in the absence of fertilization. Whilst this is an unlikely example of a natural relationship being mediated by an external metabolite, it may well have close parallels in some communities where the free-growth hormones of one can directly affect the 'fruits' of another, and it remains as a good example of the manner in which the tissues of an advanced type of plant may be influenced when subject to the external application of a potent metabolite.

There are also the familiar and important instances of fruits (apples, pears, etc.) releasing gases in their ripening (e.g. carbon dioxide and ethylene) which at once tend to accelerate the ripening of neighbouring fruits, yet in general have a toxic effect on associated young plants. Probably there have as yet been insufficient tests of such effects, since there are also instances of favourable stimulation of other tissues (e.g. root formation), as well as induction of epinasty, which suggest that both the relative concentrations and the number of the by-products may be important.

Again, the mycorrhiza relationships between higher plants and fungi appear not only to involve the supply of necessary metabolites to the higher plant, but probably other external metabolites which appreciably influence the neighbouring tissues, for example, the coral-like roots of such infected plants. Such modifications may be compared with those arising in association with more harmful fungal growths and with certain insect pests. Each is most easily explained as being mediated by an external metabolite.

Finally, lest it may appear in relation to older and more familiar views of ecological relationships that the possible significance of external metabolites is exaggerated, this review may reasonably include a reference to one process in which that significance has long been known and recognized. The gases oxygen and carbon dioxide are the best known of the contemporary by-products of metabolic activity, and their significance both within and outside the body is great. In so far as it is correctly believed that at an earlier stage in the earth's history the atmosphere was almost or completely devoid of free oxygen, so it will appear that much of the course of evolution must have been determined by the comparatively early adoption by some forms of processes leading to the production of oxygen as a metabolic by-product. Within so wide a conception of non-predatory interrelationships can also be seen the by-production of carbon dioxide, and these two gases, together, form links by which most, if not all, communities are directly related.

The examples just given form only a small fraction of our knowledge of the ways in which the external metabolites of ectocrines of different organisms have come to mediate ecological relationships with their fellows, and there are surely many more remaining to be discovered. Many of them have long been familiar, so familiar, perhaps, that insufficient attention may have been paid to those features which they possess in common and with a number of less familiar and mainly more recent observations. External secretions are, and must always have been, as commonplace as excretions in the environment. It may be useful, therefore, to consider more generally these processes as features of community life.

Excretion, however defined, is a fundamental feature of animal life, and from certain points of view it is typical too of plants. Its significance was once thought to be obvious, and limited. As we have come to know more of the processes inside the body, however, it is clear that there may be excreted not only quite complex chemicals but those which are among the most potent we know. The recent advances in our knowledge of the internal organizers, hormones, vitamins, enzymes, etc., have shown, among other things, how frequently they may be excreted, and sometimes in large quantities. This is true of some of the sex hormones, for example. In so far as they are known to be basic to one or other vital process, and as relatively few of them are in any sense specific, we can expect that when released, in water or on land, they will retain some potency (perhaps after slight modification), this time to other organisms. But we must include here many more than those metabolites which have so far been shown to have significance for internal processes. 'Many, perhaps all, living cells are able to produce substances which can be shown to affect the vital processes of other neighbouring, or even distant cells' (Heilbrunn, 1943).

The present view of the evolutionary history of such metabolites *within* the body sees them as by-products of metabolism, sometimes the result of some loss of cellular function, which have come to mediate relationships between one cell and others, or between one tissue and others. Such by-products of *internal* metabolism are well understood as parts of the *internal* environment. During evolution and ontogeny, tissues have been subject to the action of such metabolites. Those which could just tolerate them were presumably little affected, whilst harmful metabolites would reduce the chances of certain tissues in the internal competition of the developing body. Those tissues, however, which could not only tolerate a metabolite but could adapt themselves to it, and come to take advantage of it, would also have at least this advantage over their fellows in the body, and it is typical that in so many instances we see one tissue changing and growing relative to the rest when subjected to the stimulus of metabolites from another. It is typical, too, that during ontogeny there is a succession of such processes. There is first the mutual influence of egg and sperm. Here are two separate cells with one or both stimulating the other by their external secretions until they unite. The resulting cellular growth quickly leads to a stage when the metabolites of one part of the cell group begin to organize the growth of the rest, and, as differentiation proceeds, first one and then another potent metabolite (one to which a tissue has become responsive) is released and influences the next stage in the growth of some organ. We believe that bodily evolution in large part has followed the successive production of internal metabolites to which some tissues have responded and so come to change their development, their places in the body, and their metabolism. Later ontogenetic growth, leading to bodily maturity, is a succession of cellular proliferations each conditioned by the metabolites of those prior to it in history, and each, as far as we know, producing some by-products of its own metabolism. In maturity we see a cyclical series of such processes. Among such metabolites, in varying degree, a wide variety of internal parasites have come to live. Like the body tissues, they have been influenced by them and, whilst some might for a time be merely tolerant, others must either have been stimulated to further development by them or have found them a further disadvantage in their struggle for existence.

It might have seemed reasonable, therefore, simply to deduce that, if such biologically potent chemicals are released regularly into the communal environment, they would immediately influence some members of the community. Whilst some chemicals might be directly changed by chemical action, they would be unlikely to become completely neutral. Some of the metabolites by their very fundamental nature might, at appropriate concentrations, be expected to influence almost any forms of life, whilst others are, in the body, much more tissue-specific. Here, on theoretical grounds alone, can be seen a whole new field of ecology and a completely new insight into the course which communal evolution must have taken.

The examples just reviewed include ectocrine metabolites which range from fundamental enzymes, through vitamins and hormones, to a wide variety of other metabolic products, as simple as oxygen and as complex as carotenoids, sterols, and penicillin. Some are specific in their actions and others have widespread effects. Most of these effects may, according to our knowledge, be described as beneficial (perhaps essential for existence) or harmful (even to death) to the organisms investigated. A metabolite may be harmful or beneficial to one and the same organism according to its concentration or the age of the organism. Yet others may not readily be so described, although it is certain that they have ecological and communal significance. Many would appear to have little direct effect on the producer or others of its species, whilst others may affect at least the latter, and there are few metabolites which have been at all adequately investigated which have no clear ecological influence. The indirect effects, via the ecological nexus, are as yet incalculable in most cases.

In its earlier development ecology was first concerned with the influence on organisms of the simple physical and chemical factors of the environment. Ecology then had a rapid growth when the work of Elton and others demonstrated how these external relationships were vastly complicated by the network of feeding interrelationships. Recently, more precise analysis has shown, as in the sea, how the activities of organisms may appreciably modify the inanimate environment, concentrating within themselves many of the elements a thousand-fold and more in relation to their relative proportions outside, whilst they have long been known as being capable of appreciably modifying the bulk structure of the earth's surface. It now seems that we must further complicate the ecological picture by a range of ectocrine relationships which we are only just beginning to understand, but which have already been shown to have great importance for man.

The tendency to eliminate a variety of metabolites from the body seems likely to have been typical of the earliest form of life we can conceive, and it must have had important consequences. Whether or not the substances were immediately potent to associated forms of life, they were, in any event, part of their environment. In so far as the products were harmful they might tend to inhibit associated growth. In fact, only these forms which were unharmed by them could continue to live and multiply in association. Whilst some might immediately be able to benefit from the association, the others must, if they were not at least tolerant, either have migrated, or succumbed, or adapted themselves. It appears that the last must have been important, as in other aspects of evolution. Continued tolerance is biologically improbable, since merely tolerant forms must quickly be overtaken in evolution by any others which acquire even the slightest adaptation to the external conditions. Ultimately, and most important in evolution, only those forms could have 'succeeded' which possessed or developed an escape mechanism or which were capable of or developed an associated life; in some cases the latter appears to have become to a varying extent obligatory.

In this connexion, reference should be made to the work of Lwoff (1943), on bacteria, protista and fungi chiefly, in which he puts forward the theory that primitive organisms of the past (and some present-day representatives) were capable of synthesizing all their metabolic needs, including the 'essential metabolites', those substances which seem to be common to all living processes. During the course of evolution many forms lost the ability to synthesize one or more of these, which they then (as most do now) needed to acquire exogenously. 'Le metabolite devient alors un facteur de croissance ou une vitamine', and is characteristically eliminated by those forms which can synthesize it. From this point of view, the need for growth factors is merely a consequence of a loss of function. This is not the occasion to discuss fully the theories of Lwoff. Much of his evidence is in support of the present suggestions as regards the biological significance of some external metabolites. From the point of view of evolution, however, it is necessary to stress that the present suggestions visualize a continuous process whereby new metabolites are produced from time to time by organisms which themselves are living in the presence of the metabolites of their fellows and are reacting to them. There has been specialization and loss or exchange of function, but, just as the organisms have evolved, so have their metabolites, even though some of them, perhaps the most essential, must have been in existence and played their vital roles throughout much of the course of evolution.

This is not to say that all substances excreted or secreted have come to mediate an ecological relationship; even were this so it would be as impossible to demonstrate as it is to demonstrate whether organisms are responsive to or influenced by all the other factors of which we know. Yet the available evidence regarding environmental relationships and evolutionary change might make most of us chary of believing in the absolute neutrality of any known factor. Just as in the organism itself the slightest changes may have ultimate survival value, negative or positive, so it is in the environment. The organisms producing any metabolite in question are themselves subject to it and various others, but, except in individual old age or perhaps community ageing, it is uncertain how far they are directly influenced by their own metabolites. In so far as these become harmful or beneficial to their fellows, however, they might indirectly be influenced, although this is not necessarily self-evident.

Certainly within the body we have reason to believe that cellular metabolites have had such a history and, at the risk of repetition, it is important to emphasize this. To do so is not to suggest that there are direct analogies between the community and the organism, except in so far as certain fundamental biological properties and processes are common to the two. The important and helpful thing is that we know more of such processes in the body than in the community. We know that many of the internal processes are either typical of a wide variety of life, or that they are relatively slight evolutionary modifications of a general metabolic theme, and we know of the general tendency to eliminate the products of such processes. Even when modified or degenerated by external chemical action, the resulting compounds will tend, as they do in the laboratory, to retain a biological relevance and potency in so far as they fit, or nearly fit, or perhaps block, some biological reaction.

If, then, we can accept as probable the very early origin of such interrelationships, we have a basis for community integration which is at least as important as that based on the prey-predator relationship. In particular, it is one which may well account much more suitably for certain communal peculiarities than any other. One example is in the familiar process of succession. Both in water and on land, its chief characteristic is a succession of different species, not always in precisely the same order but always having a certain regularity, which has never been shown to relate directly to feeding relationships among animals, or to general nutrition in plants, or yet to changes in the inanimate environment. There is already considerable evidence that external metabolites may be potent for the typical succession of bacteria and probably some protista, and it may well have further application in these and in cellular plants, particularly in the colonization of virgin land. The small-scale temporal succession of ecology fairly parallels that large-scale succession which is evolution itself, and it is important that both may also be seen as having a horizontal component which in ecology is colonization and in evolution is a reflexion of geographical distribution. The knowledge we are now acquiring makes it seem clear that in both processes the spread, or succession of one form after another (literally its success in survival), could have been possible only after the ground had been made ready for it by predecessors, or in the absence of some antagonistic form. This is most evident when thinking of some more primitive species, but there are numbers of higher ones to which it may also apply in ecology and in evolution.

Another aspect of communal life in which ectocrines may play a part concerns the familiar 'edges' which characterize communities. However communities may be defined, it is usual for them, perhaps particularly in plants but also in many animal groupings, to have surprisingly sharp boundaries. Seldom can these be correlated with inanimate environmental factors and, whilst there are well-established examples which are due to predation, relatively few of them are in any conclusive sense due to this process. On the other hand, many could well be imagined as due to the balance of control via external metabolites, or as due to the threshold value which some of these would appear to exert—the more so when we consider the fact that some metabolites may be neutral or even stimulating at low concentrations, but inhibitory at higher ones. Here reference may also be made to many of the observations of Allee and his school (1931, 1934) demonstrating the apparently beneficial effects of grouping on animals, and probably plants. Some at least of these appear to be due to their own excretions in the wider sense, and others to the organisms' ability when massed together to overcome some toxic or harmful influence in the environment. However these may be, the net result is to render variably sharp the boundary of the group or community.

Lest it should seem that these very general considerations have too little to do with 'selective toxicity and antibiotics', it may be well to summarize the points put forward:

(1) It is characteristic of cells to liberate certain metabolites, and these are known in a variety of instances to have great influence as endocrines within the organism. Such influences may be beneficial or furthering to some tissues, and harmful or hindering in others; probably one and the same substance may have either effect in suitable concentrations and tissues. They are biologically potent.

(2) It is now well known that a number of these potent metabolites are eliminated as secretions or excretions by the organisms themselves, and many other chemicals are eliminated which are not yet known to have any specific effects within the body.

(3) Particularly in so far as any of these metabolites are of the stuff of vital processes, and have become parts of the environment of other organisms, they may be expected to have immediate potency for many of them, perhaps most specially those metabolites which are known to be

internally potent. The term 'ectocrines' has been suggested for such metabolites.

(4) More generally, however, in so far as any such chemicals have been environmental to organisms, the organisms may be expected to have experienced them as furthering or hindering their own processes. Mere toleration is biologically and statistically improbable, whilst the capacity for adaptation of most organisms suggests that further differentiation between beneficial and antagonistic relationships would be likely to have developed between the producers and those affected. In the extreme instances escape, exclusion, or death must be expected on the one hand, and obligatory association (parasitism, symbiosis) on the other.

(5) Such processes are believed to be important in evolution, and they are considered to mediate communal relationships in ecology, which is the contemporary aspect of evolution. They should be seen as part of the nexus which also includes physical and chemical relationships as well as those of prey and predator.

'Antibiotic' relationships, in the sense in which the term has come to be used, are among the best examples of the processes in mind. They are, in fact, special instances of a widespread type of relationship. We now know something of the manner in which these metabolites may be specific or generalized in their action, and we shall come to know much more. It is certainly reasonable to consider them in terms of selective toxicity, and this alone may suggest the need for a reconsideration of the propriety of their familiar name.

We know, too, how the harmful effects of a free metabolite on one organism may indirectly benefit another, and we are coming to know now a little of how one and the same 'antibiotic' metabolite may also be beneficial in some ecological grouping, perhaps even to the same organism at different concentrations (or occasions in its history). Such ambivalence in fact should be expected, in the light of all that we know of the adaptative powers of living forms, and the further detailed investigations which may be expected from teams of biologists and chemists are likely to produce many more examples. In so far as this viewpoint is substantiated by the evidence, the term antibiotic may seem to be increasingly inappropriate, and with it those more specific terms 'antibacterial' and 'antiprotozoal', etc. Perhaps some will say that the exact meaning is accepted as being inappropriate, that the meaning is understood and the use justified by custom. Yet we should not consider such terms suitable for quite comparably differentiated reactions within the body, and much of the present comprehension of these reactions might have been delayed or rendered impossible if we had done so. There is at least a similar danger, in large-scale or small-scale ecology, if a wider ecological perspective is not retained than the exigencies of warfare and human disease may have required. We must be continually watching not only for selective toxicity, but also for true adaptation, if we are not to have a very incomplete understanding of the ecological significance of external metabolites. The practical risks here seem to be no less serious than those of unbalanced ecological principle.

REFERENCES

ALLEE, W. C. (1931). *Animal Aggregations*. Chicago University Press.

ALLEE, W. C. (1934). Recent studies in mass physiology. *Biol. Rev.* **9**, 1.

ALLEN, E. J. (1914). On the culture of the plankton diatom *Thalassiosira gravida* in artificial sea water. *J. Mar. Biol. Ass. U.K.* **10**, 417.

ALLISON, J. B. & COLE, W. H. (1935). Behaviour of the barnacle *Balanus balanoides* as correlated with the planktonic content of the sea water. *Bull. Mt. Desert Is. Biol. Lab.* p. 24.

BROWN, R. (1946). Biological stimulation in germination. *Nature, Lond.*, **157**, 65.

CHEN, H. K., NICOL, H. & THORNTON, H. G. (1940). The growth of nodule bacteria in the expressed juices from legume roots bearing effective and ineffective nodules. *Proc. Roy. Soc.* B, **129**, 475.

COOK, A. H. (1945). Algal pigments and their significance. *Biol. Rev.* **20**, 115.

DE VALERA, M. (1940). Note on the difference in growth of *Enteromorpha* species in various culture media. *K. fysiogr. Sällsk. Lund. Förh.* **10**, 1.

'ESPINASSE, P. G. (1944). Effects of secretions. *Nature, Lond.*, **154**, 610.

FOX, D. L. (1944). Fossil pigments. *Sci. Mon., Lond.*, **59**, 394.

GARDENER, T. S. (1946). Nucleic acid and mice. *J. Gerontology*, **1**, 445 (quoted in *Discovery*, **9**, 1, 2, 1948).

HARDY, A. C. (1935). In HARDY, A. C. & GUNTHER, E. R. (1935). The plankton of South Georgia whaling grounds and adjacent waters. *Discovery Rep.* **2**, 1.

HARVEY, H. W. (1939). Substances controlling the growth of a diatom. *J. Mar. Biol. Ass. U.K.* **23**, 499.

HEILBRUNN, L. V. (1943). *An Outline of General Physiology*. Philadelphia.

KALLE, K. (1938). Zum Problem der Meereswasserfarte. *Ann. Hydrog., Berl.*, **66**, 1.

KYLIN, H. (1943). Über die Ernährung von *Ulva lactuca*. *K. fysiogr. Sällsk. Lund. Förh.* **13**, 1.

LEVRING, T. (1945). Some culture experiments with marine plankton diatoms. *Medd. fran. Ocean. Inst. Goteborg*, **3**, no. 12.

LUCAS, C. E. (1936). On certain interrelations between phytoplankton and zooplankton under experimental conditions. *J. Cons. int. Explor. Mer*, **2**, 343.

LUCAS, C. E. (1938). Some aspects of integration in plankton communities. *J. Cons. int. Explor. Mer*, **8**, 309.

LUCAS, C. E. (1944). Excretions, ecology and evolution. *Nature, Lond.*, **153**, 378.

LUCAS, C. E. (1947). The ecological effects of external metabolites. *Biol. Rev.* **22**, 270.

LWOFF, ANDRÉ (1943). *L'Évolution Physiologique*. Paris.

MCILWAIN, H. (1944). Origin and action of drugs. *Nature, Lond.*, **153**, 300.

MAYER, F. (1943). *The Chemistry of Natural Colouring Matters* (translated and revised by A. H. Cook). New York.

MOEVUS, F. (1933). *Arch. Protistenk.* **80**, 469.

OTTERSTRØM, C. B. & NIELSEN, E. STEEMAN (1940). Two cases of extensive mortality in fishes caused by the flagellate *Prymesium parvum* Cater. *Rep. Danish Biol. Sta.* **44**, 5.

Pickering, S. (1917). The effect of one plant on another. *Ann. Bot., Lond.*, **31**, 183.

Ropp, R. S. de (1946). Penicillin as a plant hormone. *Nature, Lond.*, **158**, 555.

Salisbury, S. J. (1944). Antibiotics and competition. *Nature, Lond.*, **126**, 472.

Schreiber, E. (1927). Die Reinkultur von marinem Phytoplankton und deren Bedeutung für die Erforschung der Produktionsfähigkeit des Meerwassers. *Wiss. Meeresuntersuch.*, N.F., Abt. Helgoland, **10**, no. 10.

Smith, F. G. Walton, Gunter, G. & Williams, R. H. (1947). Mass mortality of marine animals on the lower west coast of Florida, November 1946–January 1947. *Science*, **105**, no. 2723.

Templeman, W. G. & Sexton, W. R. (1946). The differential effect of synthetic plant growth substances. *Proc. Roy. Soc.* B, **133**, 300.

Thaysen, A. C. & Butlin, K. R. (1945). Inhibition of the development of *Fusarium oxysporum cubense* by a growth substance produced by Meredith's actinomycetes. *Nature, Lond.*, **156**, 781.

Waksman, S. A. (1945). *Microbial Antagonisms and Antibiotic Substances.* New York.

Wimpenny, R. S. (1936). The distribution breeding and feeding of some plankton organisms in the North Sea. *Fish. Invest.* (Minist. Agric. Fish.), ser. II, **15**, no. 3.

THE PRODUCTION OF ANTIBIOTICS BY MICRO-ORGANISMS IN RELATION TO BIOLOGICAL EQUILIBRIA IN SOIL

By P. W. BRIAN

Imperial Chemical Industries Ltd., Butterwick Research Laboratories, Welwyn, Herts

I. INTRODUCTION

The production of antibiotic substances by micro-organisms has been known for fifty years or more. The recent discovery of the unique chemotherapeutic properties of penicillin led to a greatly increased interest in antibiotics. It is now known that substances toxic to other micro-organisms, which usually diffuse into the surrounding environment, are produced under certain circumstances by a wide range of micro-organisms, particularly by bacteria, fungi and actinomycetes. At least one case is known of production of an antibiotic substance by a unicellular green alga (*Chlorella*) (Pratt, 1944), and the production of the 'killer' substance, paramecin, by certain strains of *Paramecium aurelia* (Sonneborn, 1939; van Wagtendonk & Zill, 1947), may be regarded as an example of production of antibiotics by Protozoa.

The motivating idea behind most of the research on antibiotics during recent years has undoubtedly been the quest for drugs complementary or supplementary to penicillin for use in chemotherapy. This search for new antibiotics has been empirical, and large numbers of organisms in the classes mentioned above have been examined by various sorting techniques. Many have been shown to produce antibiotic substances under conditions of artificial culture, and a proportion of these have been isolated in substantially pure form. Though the organisms studied in these surveys have frequently been obtained from culture collections and not selected as belonging to any particular natural habitat, it is now recognized that the capacity to produce antibiotics is particularly characteristic of micro-organisms whose natural habitat is in the soil.

The concept of a dynamic equilibrium between the various constituents of the soil microflora and microfauna has been widely accepted for many years; in the past, competition between members of the soil association has been conceived usually in terms of space competition, competition for nutrients and predatory relationships. The discovery of a frequent capability to produce antibiotics has suggested that it may be in some cases a factor concerned in the maintenance or change of microbiological balance in soil, and thus indirectly concerned with soil fertility. The evidence in favour of this view is not conclusive, a fact forcibly pointed out by Waksman (1945), and regrettably little experimental work bearing directly on this point has been published. It is the purpose of this communication to indicate the evidence in favour of the view that production of antibiotics by micro-organisms is of significance in soil processes, and to indicate the major gaps in our present knowledge.

The subject will be dealt with in the following sequence:

(i) Evidence that soil organisms are particularly concerned in the production of antibiotic substances.

(ii) A discussion of a number of types of microbiological competition or antagonism in soil, with particular reference to cases where there is reason to suppose that chemical agencies may be involved.

(iii) Consideration of the probable relative importance of known antibiotics in soil processes from the point of view of the abundance and distribution of organisms producing them and in relation to their physical and chemical properties.

II. PRODUCTION OF ANTIBIOTIC SUBSTANCES BY SOIL MICRO-ORGANISMS

Soil is an excellent culture medium for micro-organisms, which are usually present in large numbers. While it is true that methods of counting are in some respects inadequate, and while it is true that numbers vary widely both from soil to soil and in the same soil at different times, it is possible to give some idea of the number of viable units as determined by plate counts, usually present per gram of surface soil:

Bacteria	1 to 100 millions
Actinomycetes	1 to 10 millions
Protozoa	100,000 to 1 million
Fungi	50,000 to 1 million
Algae	10,000 to 50,000

In the case of each of these groups of organisms, the commoner species are fairly constantly found in soils from all parts of the world. Among the bacteria, as Waksman & Horning (1943) have pointed out, various heterotrophic sporeformers (*Bacillus subtilis*, *B. mycoides*, etc.) and non-sporeformers (*Pseudomonas fluorescens*, *Radiobacter* spp., etc.), the nitrogen-fixing forms *Azotobacter* and *Clostridium pastorianum* and the nitrifying organisms are of universal occurrence. The work of Sandon (1927) has shown that the commoner soil Protozoa are universally distributed. Soil fungi have been studied less, but, from surveys reported from different parts of the world

(Bisby, James & Timonin, 1933; Bisby, Timonin & James, 1935; Dale, 1910, 1914; Gilman & Abbott, 1927; Jensen, 1912, 1931; Swift, 1929) it seems certain that, of the moulds isolated by plating techniques, certain species of the genera *Penicillium*, *Aspergillus*, *Trichoderma*, *Gliocladium*, *Fusarium*, *Mucor* and *Zygorrhyncus* are widely distributed; of the fungi not usually isolated by plating techniques, the Hymenomycetes, characteristic of forest soils, and species of *Pythium* and *Phytophthora*, are found in all parts of the world. The commoner species of saprophytic actinomycetes are found in soils everywhere; these are not found commonly except in soil.

From the constancy in their composition we may conclude that the microflora and microfauna of the soil are characteristic and not a chance collection of organisms, and that it is justifiable to speak of a typical soil association. At the same time, it must be recognized that there is a qualitative variation in the soil microflora and microfauna from soil to soil and that under certain extreme conditions the microflora and microfauna may be specific, e.g. the dominance of hymenomycetous fungi in forest soils, of anaerobic bacteria in peat bogs or of actinomycetes in alkaline sands. Many of the organisms we have considered are not only characteristically found in the soil association but are rarely found elsewhere, i.e. soil is their main natural habitat.

If we now consider those groups of organisms known to produce antibiotics the correlation between the soil habitat and the possession of the capacity to produce antibiotics becomes much more impressive. Among the bacteria, two groups of closely related species are particularly concerned with the production of antibiotics—the aerobic sporeformers (*Bacillus subtilis—mycoides—mesentericus* group) and the *Pseudomonas fluorescens* group. These are characteristic, universally distributed soil organisms. Among the fungi, production of antibiotics is most notable among species of the genera *Penicillium*, *Aspergillus*, *Trichoderma* and *Fusarium*, the species in almost every case being typical soil inhabitants, and among many species of Hymenomycetes, which are characteristically found in forest soil and litter. Production of antibiotics is frequent among the saprophytic actinomycetes; all those species of actinomycetes reported as producing antibiotics have been originally isolated from soil.

It is true that soil organisms have received most attention; particularly is this true of bacteria and actinomycetes. If, however, one studies the surveys of antibiotic activity among fungi of all kinds made by Wilkins (1946) Wilkins & Harris (1942, 1943, 1944*a*, 1944*b*, 1944*c*) and Brian & Hemming (1947) it becomes quite clear that production of antibiotics is much more common among soil-inhabiting fungi than, for instance, among fungi characteristically parasitic on the aerial parts of higher plants.

A number of plant-pathogenic fungi have been reported as producing antibiotics, but most of these are fungi which invade plants from the soil; as examples may be quoted several species of *Fusarium* (Arnstein, Cook & Lacey, 1946; Cook, Cox, Farmer & Lacey, 1947) and *Myrothecium roridum* (Brian, Hemming & Jefferys, 1948).

We may conclude that there is strong evidence that the capacity to produce antibiotics is frequent among, and characteristic of, the micro-organisms of the soil association.

III. ANTAGONISMS BETWEEN SOIL MICRO-ORGANISMS

Factors affecting the relative numbers or relative metabolic activity of constituents of the soil microflora and microfauna have been much studied, and it has frequently been found that the presence of one organism is beneficial to another or that their association is mutually beneficial or, in other cases, that the presence of one organism is detrimental to another, the effect being shown by a decrease in numbers or a lowering of some metabolic activity of the organisms adversely affected. It is this latter type of relation —antagonism—that will now be considered.

The considerable literature on microbiological antagonisms in soil has been reviewed by Weindling (1938), Garrett (1944) and Waksman (1937, 1941, 1945). A few examples only will be considered here under these headings:

(i) Reduction in virulence of soil-borne plant-pathogenic fungi by saprophytic antagonists.

(ii) Soil-sickness and the effect of partial sterilization of soil.

(iii) Biologically induced toxicity in Wareham Heath soil.

(iv) The rhizosphere effect.

(i) *Reduction in virulence of soil-borne plant pathogens by saprophytic antagonists*

One of the main centres of early research was in Canada, in connexion with root- and foot-rots of cereals. Sanford & Broadfoot (1931), working on the wheat parasite *Ophiobolus graminis*, observed that if soil were inoculated with the fungus, the virulence of the inoculum decreased with time much more rapidly in natural soil than in a similar soil previously sterilized by heat. They assumed that the loss of virulence in normal soil was due to antagonism of the natural soil micro-organisms. By simultaneous inoculation of soil with pairs of organisms, one being *Ophiobolus* in each case and the other a soil fungus or bacterium, they were able to show that the virulence of *Ophiobolus* was much reduced or even eliminated by certain saprophytic soil fungi and bacteria. When considering possible mechanisms by which the antagonism was effected, they concluded that competition for nutrients could not play a major part, since of two saprophytes inoculated simultaneously with *Ophiobolus*, both capable of growing vigorously in the soil, one might be markedly antagonistic and the other without effect. They suggested, therefore, that the antagonism might be due, at least in part, to toxic effects on *Ophiobolus* of the 'staling-products', or as we should now say, 'antibiotic substances', produced by the antagonist. They were able to show in certain instances, that, when grown on liquid media, the antagonistic organisms secreted into the medium substances inhibiting to *Ophiobolus*. More recently Slagg & Fellows (1947) found a number of soil fungi antagonistic to *Ophiobolus* in pure culture and, of a selected list of these most were found to be antagonistic in soil, and this activity was usually associated with the power to produce substances toxic to *Ophiobolus* when grown in pure culture on liquid media. In fact, strains of most of the antagonistic fungi used by Slagg & Fellows produce known antibiotics.

The early observations of Sanford & Broadfoot were followed by many similar ones in connexion with other soil-borne plant diseases. A commonly occurring soil fungus, *Trichoderma viride*, was found by several workers to have pronounced antagonistic powers. This fungus was intensively studied by Weindling, who in 1932 found it to be antagonistic to the pathogenic fungus *Rhizoctonia solani*, causing damping-off of *Citrus* seedlings (Weindling, 1932, 1934, 1937, 1941; Weindling & Fawcett, 1936; Weindling & Emerson, 1936). Sterile soil inoculated with *Rhizoctonia* caused severe damping-off of *Citrus*. If the soil were inoculated, at the same time or subsequently, with *Trichoderma*, virulence of attack by *Rhizoctonia* was much reduced. In this case the *Rhizoctonia* hyphae are killed by the *Trichoderma* as the result of a type of parasitic attack, the *Trichoderma* hyphae coiling round the *Rhizoctonia* hyphae and eventually penetrating them. By direct observation Weindling was able to show that no active parasitic penetration of the *Rhizoctonia* hyphae took place until they were already dead and that the *Trichoderma* hyphae appeared able to induce changes resulting in death of the *Rhizoctonia* hyphae from a distance, presumably as a result of production of a diffusible toxic substance. In confirmation, it was found that *Trichoderma* secreted a 'lethal principle' into liquid culture media on which it was grown, which, if applied to *Rhizoctonia* hyphae, induced the same degenerative changes as proximity of *Trichoderma* hyphae. The 'lethal principle' was sensitive to pH, aqueous solutions being relatively stable at pH 3·0 but becoming progressively more unstable as the pH was raised and, confirming the importance of the lethal principle in the parasitism of *Trichoderma*, it was found that parasitic activity was more marked under pH conditions where the lethal principle was relatively stable, the vigour of growth of *Trichoderma* not varying in the pH range studied sufficiently to confuse the issue. Eventually the lethal principle from one of his strains of *Trichoderma* was isolated in pure, crystalline form and named gliotoxin; this antibiotic, first isolated by Weindling & Emerson in 1936, is now known to be produced by a number of soil fungi—*Aspergillus fumigatus*, *Penicillium terlikowskii*, *P. obscurum*, *P. cinerascens* and *Trichoderma viride*—which suggests that it may have special significance in natural processes. Gliotoxin is a generally toxic substance, strongly antibacterial and antifungal. More recently (Brian, Curtis, Hemming & McGowan, 1946) other strains of *T. viride* have been shown to produce another antibiotic, viridin, which is highly antifungal but not antibacterial. It appears likely that both viridin-producing and gliotoxin-producing strains of *Trichoderma* were used in Weindling's work, which offers strong evidence that the antagonistic powers of *Trichoderma viride*, a ubiquitous soil fungus, are closely associated with production of the antibiotic substances gliotoxin and viridin.

The researches of van Luijk (1938) were analogous to those of Weindling. Van Luijk isolated a number of soil fungi antagonistic to *Pythium* spp. causing damping-off of seedlings of lucerne and various grasses. *Penicillium expansum* was particularly effective and, like *Trichoderma*, its activity was found to be associated with the production of a diffusible toxic substance. This was finally isolated in pure form and named expansine (Nauta, Oosterhuis, van der Linden, van Duyn & Dienske, 1946); this antibiotic is identical with substances previously isolated by other workers and named variously patulin, clavatin, clavacin and claviformin. It is also produced by a variety of soil fungi including *Penicillium patulum*, *P. expansum*, *P. claviforme*, *P. urticae*, *Aspergillus clavatus*, *A. giganteus*, *A. terreus* and *Gymnoascus* sp.

Without considering further examples, we may conclude that there is evidence that the well-known antagonism between certain soil saprophytes (fungi and bacteria) and soil-borne plant pathogenic fungi is associated with the production by the saprophytes of antibiotic substances. Further confirmation, from the reverse direction, is provided by the work of Grossbard (1947) who has shown that two fungi (*Aspergillus clavatus* and *Penicillium patulum*) are strongly antagonistic in soil to certain phytopathogens (*Phytophthora* spp., *Bacillus carotovorus*). At the same time there is no doubt that other factors are involved, as indicated by recent work of Slykhuis (1947), who found that, of a number of saprophytes antagonistic to the root parasite *Fusarium culmorum*, the most active forms, species of *Phialophora* and *Acremonium*, showed no tendency to produce antibiotics in artificial culture.

(ii) *Soil-sickness and partial sterilization of soil*

Heavily fertilized soils, such as those in sewage farms (Russell & Golding, 1912) or in glasshouse beds (Russell & Petherbridge, 1912), frequently show a condition described as soil-sickness. This is characterized by high nitrogen content, low bacterial numbers and poor plant growth. If such 'sick' soils are partially sterilized by steaming or by chemical treatment (with formaldehyde, cresylic acid or toluol), there characteristically follows a sustained rise in ammonia nitrogen, a decrease followed by a rapid increase of bacterial numbers and vigorous plant growth. The situation is sometimes complicated by the fact that such treatment also eliminates plant pathogens from the soil, but the benefits to plant growth are usually in excess of what would be expected to result from disease control alone.

The effects of partial sterilization of 'sick' soil have been elucidated in the classical researches of Russell & Hutchinson (1909, 1913), who showed that in the sick soil, multiplication of bacteria responsible for breakdown of organic nitrogen is inhibited by some toxic factor; they also showed conclusively that the toxic factor is of biological origin. They suggested that the biological factor concerned was the soil Protozoa which, in sick soils, were, by predatory activity, reducing the numbers of soil bacteria. While it has been shown that Protozoa do have a depressive action on bacterial numbers it has never been found possible to prove conclusively that the soil Protozoa play a major part in production of the symptoms of soil-sickness. The effect of partial sterilization was attributed to killing the Protozoa selectively, being less resistant to heat or disinfectants than many of the soil bacteria; their predatory activity being thus reduced, multiplication of bacteria and decomposition of organic nitrogen was made possible, the ammonium salts produced being available for plant growth.

The protozoal theory has been frequently criticized (Waksman, 1932), and alternative mechanisms have been suggested. It is probably true that the Protozoa assumed special importance in the earlier discussions of this problem, since their known predatory habit suggested a definite mechanism by which they could reduce bacterial numbers. It is known (Waksman & Starkey, 1923) that partial sterilization not only kills Protozoa but also the soil fungi and most of the actinomycetes. It was not appreciated at the time of Russell's investigation how fungi or actinomycetes could prevent the multiplication of bacteria in conditions of abundant food supply; the newer knowledge of production of antibiotic substances by soil fungi and actinomycetes suggests a mechanism. Most of the observed facts could be explained on this basis, and a reinvestigation of this particular type of microbiological antagonism would be very valuable.

(iii) *Biologically induced toxicity in Wareham Heath soil*

A particularly interesting case of microbiological antagonism is concerned with a growth failure of coniferous trees in parts of an afforestation area on Wareham Heath, Dorset. This has been the subject of a series of investigations by Rayner (1934, 1936, 1939, 1941) and Neilson Jones (1941). On the affected parts of the heath *Pinus* spp. fail to grow normally, both roots and shoots being severely stunted. In addition, the typical mycorrhizae fail to develop. Hymenomycete mycelium, including that of the usual mycorrhizal fungi, is scarce, of reduced activity or even absent from the soil. It has been established beyond doubt that in areas where growth failure is observed a toxic factor is present in the soil which reduces the activity of fungal mycelia; this prevents formation of the normal associations between mycorrhizal fungi and the pine roots, such an association being generally regarded as essential for vigorous growth of the trees. Observations also suggest that this toxic factor may affect root development directly, as well as indirectly through its effect on the mycorrhizal fungi. The soil toxicity may be removed in several ways. If the soil is sterilized by heat or by disinfectants, and a suitable mycorrhizal fungus is introduced, pine seedlings will develop vigorously. If sterilized soil is reinoculated with a little naturally toxic soil the original toxicity rapidly develops throughout the soil, indicating that it is of biological origin. The toxicity may also be removed by certain compost treatments, whose effects cannot be explained in terms of manurial value but presumably by an effect on microbiological balance. Unlike the other forms of microbiological antagonism discussed above, in the case of Wareham Heath soil toxicity clear proof was obtained that a diffusible antifungal substance was present in all soils showing symptoms of toxicity. The technique used was simple but convincing. Over a layer of the soil in a Petri dish was poured a film of agar; where toxic soils were used fungi inoculated on the exposed agar surface failed to grow but with normal soils, or Wareham Heath soil detoxicated by sterilization or compost treatment, mycelial growth was vigorous on the agar. We have then, in this example of Wareham Heath soil, a proved case of microbiological antagonism, having pronounced consequences on soil fertility, associated with production of a diffusible toxic substance in the soil.

The nature and origin of the toxic substance or substances is still in doubt. It was noted by Neilson Jones that bacterial and fungal numbers were low, whether total number or number of species was considered, and that general metabolic activity in the soil was at a low level. An exception was noted in that the numbers of sulphate-reducing bacteria were at a level that might be considered normal in clay soils where their activity is usually most noticeable, and it was suggested that the toxic factor might be sulphide produced by reduction of sulphate. More recently Brian, Hemming & McGowan (1945) confirmed the earlier observation of low bacterial and fungal counts and noted that the mould flora was restricted to a number of species of *Penicillium*, all of which could be regarded as typical soil Penicillia. The dominant species were *P. janczewskii*, *P. terlikowskii* and a form falling into the *P. nigricans-janczewskii* series but not assignable to any described species. Each of these three species was shown to produce antibiotics, and it was suggested that the particular soil conditions at Wareham favoured initial development of these Penicillia, which then upset the soil microbiological balance still further, with the effects described above, by production of these antibiotics, which included both antifungal and antibacterial substances. Subsequent work has defined the antibiotics concerned. *P. terlikowskii* produces the antifungal and antibacterial substance gliotoxin (Brian, 1946); *P. janczewskii* produces griseofulvin (Brian, Curtis & Hemming, 1946; McGowan, 1946; Grove & McGowan, 1947; Brian, 1949), a substance which disorganizes fungal growth and inhibits hyphal extension at very low concentrations and is also highly toxic to roots of higher plants; the form from the *P. nigricans-janczewskii* series produces a fungistatic and bacteriostatic red pigment (Curtis & Grove, 1947). The dominance in the soil flora of these three species, if confirmed by extended surveys, considered together with their pronounced power to produce antibiotics and the direct proof that diffusible antifungal substances are produced in the soil, is extremely suggestive, but much further work is necessary.

(iv) *The rhizosphere effect*

Lochhead and his colleagues, in Canada, have established that the growing plant may exert a marked effect, both quantitative and qualitative, on the micro-organisms in the soil adjacent to the root system (rhizosphere). The total number of bacteria is usually greater in the rhizosphere than in the soil more distant from the plant; this is the quantitative aspect of the rhizosphere effect. Another and more interesting aspect is the preferential stimulation of certain nutritional groups of bacteria.

Lochhead & Thexton (1947) divide soil bacteria into seven nutritional classes, the first three of these being numerically most important:

(*a*) those which will grow on a simple basal medium containing glucose and inorganic salts,

(*b*) those requiring amino-acids in addition to the basal medium,

(*c*) those requiring known vitamins in addition to the basal medium,

(*d*) those requiring the addition of both known vitamins and amino-acids,

(*e*) those requiring unidentified factors present in yeast extract,
(*f*) those requiring unidentified factors present in soil extract,
(*g*) those requiring both soil and yeast extract.

In the argument that follows, classes (*a*), (*b*) and (*c*) only are considered. Of these, though all are more abundant in the rhizosphere than elsewhere in the soil, class (*b*) is preferentially stimulated by rhizosphere conditions. In one case quoted by Lochhead & Thexton, whereas the total number of bacteria was 14 times greater in the rhizosphere than in the soil not immediately adjacent to roots, the number of class (*b*) organisms was 50 times greater, class (*a*) 25 times greater and class (*c*) only 10 times greater. Lochhead & Thexton have shown that this differential increase in amino-acid requiring bacteria may be due to two causes:

(i) secretion of amino-acids by the root, creating conditions specially favourable for class (*b*) organisms;
(ii) secretion of substances by class (*a*), non-exacting, bacteria which have a differential effect on bacteria of classes (*b*) and (*c*).

In the latter case they were able to show that culture filtrates of class (*a*) bacteria have two tendencies, to be toxic to many class (*c*) bacteria but not to class (*b*) bacteria, and to be frequently stimulatory to class (*b*) bacteria but only rarely to class (*c*) bacteria. The net effect of this complex system of antibiosis and mutual stimulation is an increase in numbers of class (*b*) organisms. This example gives an interesting indication of the complex chemical and biological relationships that may exist in soil.

IV. LIMITATIONS IN THE SIGNIFICANCE OF ANTIBIOTICS IN SOIL PROCESSES

Having established a case for supposing that one of the mechanisms responsible for biological antagonisms in soil is the production of antibiotics by certain micro-organisms, a number of criticisms have to be met. Waksman (1945) has made the following points in this connexion:

(i) The fact that an organism produces an antibiotic in artificial culture is no evidence that it is capable of doing so in soil, particularly since relatively small changes in a nutrient medium may fundamentally affect the production of antibiotics in pure culture.

(ii) Many known antibiotics are extremely unstable and could not be expected to remain unchanged in soil for sufficiently long to have any effect.

(iii) There is no evidence that production of antibiotics at all affects the survival of organisms producing them.

These points are discussed in turn below.

(i) *Production of antibiotics in soil*

It is true that no published evidence is available that organisms capable of producing antibiotics in artificial culture can also produce them in soil. Nevertheless, it is not so unlikely as Waksman suggests; it is true that modifications of a culture medium may make great changes in the quantity of antibiotic produced, or more often in the time relations of its production cycle, but where growth takes place it is rare for production of an antibiotic to be completely suppressed. Many of the cases where it is said that antibiotics are not produced are really cases of failure to accumulate, as, for instance, when unfavourable pH relationships result in decomposition of the antibiotic as fast as it is produced. The question of stability of antibiotics in soil is discussed below.

It is true that instances of the production of antibiotics in natural soils as a result of microbiological activity are very few; the case of Wareham Heath soil already mentioned is a clear example, others less satisfying may be found in the work of Newman & Norman (1943) on subsurface soil populations and of Waksman & Woodruff (1942) on the survival in the soil of certain bacteria pathogenic to man. It should not be experimentally difficult to study the production of antibiotics in soil and this merits attention. Grossbard (1948) has described some work of this type and has shown that *Penicillium patulum* produces a diffusible antibiotic in soil supplemented with straw or glucose, but not, under the experimental conditions chosen, in unsupplemented soil.

(ii) *Stability of antibiotics in soil*

Waksman cites the case of penicillin, pointing out that many soil bacteria produce penicillinases, and that it would not be expected that penicillin could exist in soil in sufficient concentration to affect the growth of bacteria. No work on the stability of known antibiotics in soil has yet been published, but a brief resumé of certain results obtained by Jefferys (1948), in the course of work still in progress, may be usefully given here. Seven antibiotics have been studied—gliotoxin, glutinosin, griseofulvin, gladiolic acid, patulin, penicillin II (benzyl penicillin) and viridin—very dilute solutions being exposed to soil in conditions of low and high aeration. All of these are inactivated to some degree in soil, most often more rapidly in fresh soil than in sterile soil, indicating some biological attack. Glutinosin, gladiolic acid, griseofulvin and patulin are only inactivated slowly, appreciable activity remaining even after several weeks of contact with soil. Gliotoxin is less stable, its persistence in soil being mainly dependent on pH; it disappears rapidly in soils near neutrality, but in acid soils (pH 3–4) it is fairly persistent. Viridin shows the same relations as gliotoxin but disappears rather more rapidly under all circumstances. Penicillin also is unstable, in this case being inactivated instantly in acid soils but persisting for an appreciable period in certain neutral soils. Results are far from complete, but sufficient evidence has been obtained to indicate that certain antibiotics are quite stable in soil, and that even so unstable a material as penicillin may last sufficiently long in certain soils to exert a biological effect.

Even if the rate of inactivation of an antibiotic is high, at the surface of the cell producing it a local concentration may be maintained sufficient to have a marked local biological effect. Thus if the antagonistic powers of the fungus *Trichoderma viride* depend on production of gliotoxin or viridin, its habit of producing hyphae curled round or adpressed against the hyphae of other fungi ensures that a local concentration of the antibiotic shall have a maximum biological effect.

(iii) *Antibiotics in biological competition*

It has been suggested that the production of antibiotics is biologically advantageous in that an organism capable of producing antibiotics has a greater chance of survival in a competitive environment than one without that power. It has even been suggested that the function of antibiotics lies in their survival value. Waksman is undoubtedly right in condemning a teleological approach to the problem, and no experimental work has yet been published bearing on the survival value of antibiotic production in soil. There is much evidence to the contrary. Many fungi and bacteria common in soil show no tendency to produce antibiotics, so far as our knowledge stands at present. Similarly, most species of fungi, bacteria or actinomycetes known to produce antibiotics exist in nature in strains varying in their capacity to produce the antibiotic, yet there is no evidence that high-yielding strains are more common than low-yielding strains; in fact, the reverse is more frequently the case. The survival in soil of strains of a micro-organism producing high, moderate and negligible yields of an antibiotic could be experimentally studied with little difficulty, and such experiments should yield results of considerable value.

Throughout this review no attempt has been made to show that production of antibiotics has resulted in a greater population of the organisms producing them in natural conditions; this may well at times be the case, but it has been thought sufficient to produce evidence that, in certain circumstances, the reduction in numbers or metabolic activity of some organisms or groups of organisms in the presence of other antagonistic organisms is due, at least in part, to the production of antibiotic substances by the latter.

Date Due

REFERENCES

ARNSTEIN, H. R. V., COOK, A. H. & LACEY, M. S. (1946). *Brit. J. Exp. Path.* **27**, 349.

BENEDICT, R. G. & LANGLYKE, A. F. (1947). *Ann. Rev. Microbiol.* **1**, 193.

BISBY, G. R., JAMES, N. & TIMONIN, M. I. (1933). *Canad. J. Res.* **8**, 253.

BISBY, G. R., TIMONIN, M. I. & JAMES, N. (1935). *Canad. J. Res.* **13**, 32.

BRIAN, P. W. (1946). *Trans. Brit. Mycol. Soc.* **29**, 211.

BRIAN, P. W. (1949). *Ann. Bot., Lond.* (in the Press).

BRIAN, P. W., CURTIS, P. J. & HEMMING (1946). *Trans. Brit. Mycol. Soc.* **29**, 173.

BRIAN, P. W., CURTIS, P. J., HEMMING, H. G. & MCGOWAN, J. C. (1946). *Ann. Appl. Biol.* **33**, 190.

BRIAN, P. W. & HEMMING, H. G. (1947). *J. Gen. Microbiol.* **1**, 158.

BRIAN, P. W., HEMMING, H. G. & JEFFERYS, E. G. (1948). *Mycologia*, **40**, 363.

BRIAN, P. W., HEMMING, H. G. & MCGOWAN, J. C. (1945). *Nature, Lond.*, **155**, 637.

COOK, A. H., COX, G. F., FARMER, T. H. & LACEY, M. S. (1947). *Nature, Lond.*, **160**, 31.

CURTIS, P. J. & GROVE, J. F. (1947). *Nature, Lond.*, **160**, 574.

DALE, E. (1910). *Ann. mycol., Berl.*, **10**, 452.

GARRETT, S. D. (1944). *Root Disease Fungi.* Waltham, Mass. U.S.A.

GILMAN, J. C. & ABBOTT, E. V. (1927). *Iowa St. Coll. J. Sci.* **1**, 225.

GROSSBARD, E. (1947). *Rep. Exp. Res. Sta. Cheshunt*, 1946, p. 41.

GROSSBARD, E. (1948). *Nature, Lond.*, **161**, 614.

GROVE, J. F. & MCGOWAN, J. C. (1947). *Nature, Lond.*, **160**, 574.

JEFFERYS, E. G. (1948). Personal communication to author.

JENSEN, C. W. (1912). *Bull. Cornell Agric. Exp. Sta.* no. 315, p. 414.

JENSEN, H. L. (1931). *Soil Sci.* **31**, 123.

KAVANAGH, F. (1947). *Advances in Enzymology*, **7**, 461.

LOCHHEAD, A. G. & THEXTON, R. H. (1947). *Canad. J. Res.* **25**, 20.

MCGOWAN, J. C. (1946). *Trans. Brit. Mycol. Soc.* **29**, 188.

NAUTA, W. T., OOSTERHUIS, H. K., VAN DER LINDEN, A. C., VAN DUYN, P. & DIENSKE, J. W. (1946). *Rec. Trav. chim. Pays-Bas*, **65**, 865.

NEILSON JONES, W. (1941). *J. Agric. Sci.* **31**, 379.

NEWMAN, A. G. & NORMAN, A. G. (1943). *Soil Sci.* **55**, 377.

PRATT, C. (1944). *Science*, **99**, 352.

RAYNER, M. C. (1934). *Forestry*, **8**, 96.

RAYNER, M. C. (1936). *Forestry*, **10**, 1.

RAYNER, M. C. (1939). *Forestry*, **13**, 19.

RAYNER, M. C. (1941). *Forestry*, **15**, 1.

RUDERT, F. J. & FOTER, M. J. (1947). *J. Bact.* **54**, 793.

RUSSELL, E. J. & GOLDING, J. (1912). *J. Agric. Sci.* **5**, 27.

RUSSELL, E. J. & HUTCHINSON, H. B. (1909). *J. Agric. Sci.* **3**, 111.

RUSSELL, E. J. & HUTCHINSON, H. B. (1913). *J. Agric. Sci.* **5**, 152.

RUSSELL, E. J. & PETHERBRIDGE (1912). *J. Agric. Sci.* **5**, 86.

SANDON, H. (1927). *The Composition and Distribution of the Protozoan Fauna of the Soil.* Edinburgh.

SANFORD, G. B. & BROADFOOT, W. C. (1931). *Sci. Agric.* **11**, 512.

SLAGG, C. M. & FELLOWS, H. (1947). *J. Agric. Res.* **15**, 279.

SLYKHUIS, J. T. (1947). *Canad. J. Res.* **25**, 155.

SONNEBORN, T. M. (1939). *Amer. Nat.* **73**, 390.

SWIFT, M. E. (1929). *Mycologia*, **21**, 204.

VAN LUIJK (1938). *Meded. Phytopath. Lab. Scholten*, **14**, 42.

van Wagtendonk, W. J. & Zill, L. D. (1947). *J. Biol. Chem.* **171**, 595.
Waksman, S. A. (1932). *Principles of Soil Microbiology*, 2nd ed. Baltimore, U.S.A.
Waksman, S. A. (1937). *Soil Sci.* **43**, 51.
Waksman, S. A. (1941). *Bact. Rev.* **5**, 231
Waksman, S. A. (1945). *Microbial Antagonisms and Antibiotic Substances.* New York.
Waksman, S. A. & Horning, E. S. (1943). *Mycologia*, **35**, 47.
Waksman, S. A. & Starkey, R. L. (1923). *Soil Sci.* **16**, 137.
Waksman, S. A. & Woodruff, H. B. (1942). *Soil Sci.* **53**, 233.
Weindling, R. (1932). *Phytopathology*, **22**, 837.
Weindling, R. (1934). *Phytopathology*, **24**, 1153.
Weindling, R. (1937). *Phytopathology*, **27**, 1175.
Weindling, R. (1938). *Bot. Rev.* **4**, 475.
Weindling, R. (1941). *Phytopathology*, **31**, 991.
Weindling, R. & Emerson, O. H. (1936). *Phytopathology*, **26**, 1068.
Weindling, R. & Fawcett, H. S. (1936). *Hilgardia*, **10**, 1.
Wilkins, W. H. (1946). *Ann. Appl. Biol.* **33**, 188.
Wilkins, W. H. & Harris, G. C. M. (1942). *Brit. J. Exp. Path.* **23**, 166.
Wilkins, W. H. & Harris, G. C. M. (1943). *Brit. J. Exp. Path.* **24**, 141.
Wilkins, W. H. & Harris, G. C. M. (1944*a*). *Trans. Brit. Mycol. Soc.* **27**, 113.
Wilkins, W. H. & Harris, G. C. M. (1944*b*). *Brit. J. Exp. Path.* **25**, 135.
Wilkins, W. H. & Harris, G. C. M. (1944*c*). *Ann. Appl. Biol.* **31**, 261.

factor'), citrinin, aspergillic acid and penicillic acid are common in soil. All of these antibiotics with the exception of penicillin and viridin are relatively stable; even penicillin and viridin would be relatively stable in certain special soil conditions. Most of these antibiotics are toxic to mammals and, since most experimental work on antibiotics has had a pronounced medical bias, have as a consequence been little studied. In view of their potential importance in soil, study of these antibiotics should no longer be neglected.

It is difficult to assess the importance of the antibiotics produced by Hymenomycetes, only a few of which have as yet been isolated in pure form. These fungi are not generally distributed in soil, but in forest litter and the upper layers of forest soil they are locally dominant, and in these circumstances their capacity to produce antibiotics may assume importance.

VI. SUMMARY

1. The commoner species among the Protozoa, bacteria, actinomycetes and fungi of the soil have a wide distribution and constancy of occurrence. It is possible, therefore, to speak of a characteristic soil association of micro-organisms. A considerable proportion of these organisms possess the power of producing antibiotic substances in artificial culture.

2. Several examples of microbiological antagonism in soil are discussed, and evidence is presented that in each case all the observed facts can be explained by assuming the production of antibiotic substances by one organism or group of organisms having a suppressive effect on the multiplication or metabolic activity of another organism or group of organisms. In certain cases there is strong evidence in favour of this explanation.

3. Evidence that organisms capable of producing antibiotics in pure culture can do so in soil is not strong. Nevertheless, there is some evidence that antibiotics are produced in soil in certain instances. Consideration of the stability in soil of certain pure antibiotics indicates that instability cannot be regarded as being a factor precluding them from certain biological effects in soil.

4. Those antibiotic substances produced by organisms common in soil and which are relatively stable chemically are most likely to be of significance in connexion with biological equilibria in soil. On this basis those known antibiotics most worthy of study are indicated.

V. ANTIBIOTIC SUBSTANCES OF PROBABLE SIGNIFICANCE

It remains to consider which of the well-characterized antibiotics are likely to be of significance in soil. Those are likely to be of greatest significance which are produced by organisms common in soil and which are relatively stable chemically. On this basis the following assessment can be made.

Antibiotics produced by bacteria

The two common groups of antibiotic-producing soil bacteria are the aerobic sporeformers and the *Pseudomonas fluorescens* group. The spore-formers are a difficult group taxonomically and there is a considerable confusion among the various antibiotics that have been reported. The following appear to be distinguishable on the basis of their reported properties, but their degree of chemical relationship is obscure: aerosporin, bacillin, *B. polymyxa* antibiotic, *Bacillus simplex* factor, gramicidin, gramicidin-S, licheniformin and subtilin. As a class they are probably of great importance, but further investigation of their individual properties is necessary. Rudert & Foter (1947) in a survey of soils from all parts of the U.S.A., have found that an organism producing the antibiotic bacillin is widespread and abundant. The pseudomonads produce such substances as hemipyocyanine, pyocyanine and pyocyanase, all of which may be of some significance.

Antibiotics produced by actinomycetes

The antibiotics produced by actinomycetes include such substances as streptomycin, streptothricin, actinomycin, chloromycetin, litmocidin, pro-actinomycin, lavendulin, actinorubin, sulfactin, actidione and grisein. Though a high proportion of soil actinomycetes produce antibiotics, we lack sufficient knowledge at present to indicate whether certain species are particularly involved or whether certain antibiotics are of greater significance than others.

Antibiotics produced by fungi

Here one can select potentially important antibiotics with more confidence, since speciation is more definite in fungi, and the antibiotics produced by fungi have, in general, been more accurately characterized, chemically and biologically, than those produced by bacteria or actinomycetes. About thirty well-characterized antibiotics produced by fungi have been reported; the most recently published reviews of this subject are those of Kavanagh (1947) and Benedict & Langlyke (1947).

Organisms producing penicillin, patulin (also known as claviformin, clavacin, clavatin and expansine), gliotoxin, viridin, griseofulvin ('curling-